BIOINSPIRED PHOTONICS

BIOINSPIRED PHOTONICS
Optical Structures and Systems
Inspired by Nature

Viktoria Greanya, PhD

CRC Press
Taylor & Francis Group
Boca Raton London New York

CRC Press is an imprint of the
Taylor & Francis Group, an **informa** business

CRC Press
Taylor & Francis Group
6000 Broken Sound Parkway NW, Suite 300
Boca Raton, FL 33487-2742

First issued in paperback 2021

© 2016 by Taylor & Francis Group, LLC
CRC Press is an imprint of Taylor & Francis Group, an Informa business

No claim to original U.S. Government works

Version Date: 20150417

ISBN 13: 978-0-367-86703-4 (pbk)
ISBN 13: 978-1-4665-0402-8 (hbk)

This book contains information obtained from authentic and highly regarded sources. Reasonable efforts have been made to publish reliable data and information, but the author and publisher cannot assume responsibility for the validity of all materials or the consequences of their use. The authors and publishers have attempted to trace the copyright holders of all material reproduced in this publication and apologize to copyright holders if permission to publish in this form has not been obtained. If any copyright material has not been acknowledged please write and let us know so we may rectify in any future reprint.

Except as permitted under U.S. Copyright Law, no part of this book may be reprinted, reproduced, transmitted, or utilized in any form by any electronic, mechanical, or other means, now known or hereafter invented, including photocopying, microfilming, and recording, or in any information storage or retrieval system, without written permission from the publishers.

For permission to photocopy or use material electronically from this work, please access www.copyright.com (http://www.copyright.com/) or contact the Copyright Clearance Center, Inc. (CCC), 222 Rosewood Drive, Danvers, MA 01923, 978-750-8400. CCC is a not-for-profit organization that provides licenses and registration for a variety of users. For organizations that have been granted a photocopy license by the CCC, a separate system of payment has been arranged.

Trademark Notice: Product or corporate names may be trademarks or registered trademarks, and are used only for identification and explanation without intent to infringe.

Publisher's Note
The publisher has gone to great lengths to ensure the quality of this reprint but points out that some imperfections in the original copies may be apparent.

Visit the Taylor & Francis Web site at
http://www.taylorandfrancis.com

and the CRC Press Web site at
http://www.crcpress.com

To my beloved husband
who told me I was crazy
when I naively said "yes" to writing a book,
but supported me through it anyway.

Viktoria Greanya

Contents

PREFACE		xi
ACKNOWLEDGMENTS		xvii
AUTHOR		xix

CHAPTER 1 INTRODUCTION TO BIOINSPIRED PHOTONIC SYSTEMS   1
   1.1 Biological and Bioinspired Photonics   1
   1.2 Evolution   7
   1.3 Historical Perspective and the Advent of Microscopy   13
   1.4 Tools of the Trade   18
       1.4.1 Microscopy   19
       1.4.2 Spectroscopy and Scatterometry   22
       1.4.3 Challenge of Working with Live Specimens   23
       1.4.4 Fabrication Approaches   24
   1.5 Bioinspired Photonics in the Twenty-First Century and the Challenge of Multidisciplinary Science   25
   References   27

CHAPTER 2 STRUCTURAL COLOR I: LOW-DIMENSIONAL STRUCTURES   31
   2.1 Next Generation Applications Inspired by Ancient Structures   31
   2.2 Sparkly, Vibrant, Bright, and Shiny—Light and Biology in Action   34
   2.3 Describing Biological Photonic Structures   36

	2.4	One-Dimensional Layered Structures	38
		2.4.1 Single Layer Thin Films	38
		2.4.2 Simple Multilayers	41
		2.4.3 Chirped Multilayers	47
		2.4.4 Sculpted Multilayers	51
	2.5	Two-Dimensional Structures	55
		2.5.1 Arrays in Peacock Feathers	56
		2.5.2 Templated Growth and Replication	58
		2.5.3 Quasi-Ordered 2D Structures and Quasicrystals	60
		2.5.4 Improving Light Extraction	62
	References		66
CHAPTER 3	**STRUCTURAL COLOR II: COMPLEX STRUCTURES**		**71**
	3.1	Quasi Two-/Three-Dimensional Structures	71
		3.1.1 Tilted Structures and Narrow Angle Reflectance	71
		3.1.2 Wings of the Morpho Butterfly	76
		3.1.3 Helicoidal Multilayers	79
		3.1.4 Intercalated Structures	82
	3.2	Three-Dimensional Structures	85
		3.2.1 Cubic Structures	89
		3.2.2 Diamond Structures	92
		3.2.3 Gyroid Structures	92
		3.2.4 Inspired Synthetic 3D Structures	100
	3.3	Nanostructures in Black and White	107
		3.3.1 White	107
		3.3.2 Black	111
	References		116
CHAPTER 4	**DYNAMIC, ADAPTIVE COLOR**		**121**
	4.1	Color Changing Organisms as Inspiration	121
	4.2	The Expanding Display Industry	121
	4.3	Nature's "Unconventional" Display Technologies	124
	4.4	Cephalopods	126
	4.5	Architectures of Dynamic Biological Photonics	130
	4.6	Chromatophores	134
	4.7	Chromatophore-Inspired Structures	138
	4.8	Dynamic Structural Color: Iridophores and Leucophores	144
		4.8.1 Iridophores	144
		4.8.2 Leucophores	146
	4.9	Actuating Structural Color	149
		4.9.1 Refractive Index Modulation	151
		4.9.2 Mechanical Deformation	153
		4.9.3 Field-Induced Modulation	156
		4.9.4 Other Approaches	162
	References		163

CONTENTS

CHAPTER 5	**VISION SYSTEMS**		**167**
	5.1	Inspiring Vision	167
	5.2	Biological Eyes: The Front-End Optics	170
		5.2.1 Simple Eyes	170
		5.2.2 Gradient Index Lenses	175
		5.2.3 Compound Eyes	176
		5.2.4 Apposition Eyes	177
		5.2.5 Superposition Compound Eyes	181
		5.2.6 Other Variants on the Compound Eye	182
		5.2.7 Brittle Star: A Strange Compound Eye	183
	5.3	Photoreceptors: The Imager's Back End	184
	5.4	Spectral Sensitivities	188
	5.5	Secondary Structures	191
	5.6	Applications	197
		5.6.1 GRIN Lenses	197
		5.6.2 Artificial Eye Prosthetics	200
		5.6.3 Inspired Compound Eye Lens Arrays	201
		5.6.4 Compound and Simple Eye Imaging Systems	209
		5.6.5 Polarization Sensors	216
		5.6.6 Antireflective Structures	216
	References		219
CHAPTER 6	**BIOMATERIALS FOR PHOTONICS**		**223**
	6.1	Chitin	225
	6.2	Silk	231
	6.3	Biosilica	236
	6.4	Reflectins	241
	6.5	Luciferins and GFP—Bioluminescence and Fluorescence	245
		6.5.1 Luciferase and Luciferin	247
		6.5.2 Green Fluorescent Proteins	248
		6.5.3 Applications for Bioluminescence	250
	6.6	Opsins	255
	References		259
CHAPTER 7	**SENSORS**		**265**
	7.1	Introduction	265
	7.2	Infrared Sensing	267
		7.2.1 Snakes	270
		7.2.2 Bats	274
		7.2.3 Beetles	275
		7.2.4 Thermal Sensors Inspired by the Fire-Beetle	278
		7.2.5 Butterflies	284
		7.2.6 Thermal Expansion and Optical Sensor Structures	285

7.3	Gas and Vapor Sensors	292
	7.3.1 Diatoms	295
	7.3.2 Butterfly Wings as Sensors	300
References		314

Chapter 8 Energy from Light — 319

8.1	Insatiable Appetite for Power and Energy	319
8.2	Harvesting Solar Power	320
8.3	Photosynthesis	321
	8.3.1 Quantum Biology	326
8.4	Photovoltaics	327
8.5	Antireflective Structures	331
8.6	Dye-Sensitized Solar Cells	335
	8.6.1 Biophotonic Crystal Structures	336
	8.6.2 Molecular Antennas	341
8.7	Solar Fuels and Artificial Photosynthesis	345
8.8	Hybrid Systems	349
8.9	Nanoantennas	352
References		357

Chapter 9 The Future of Bioinspired Photonics: Challenges and Opportunities — 361

9.1	Inspiration from Natural Systems for Conventional and Unconventional Applications	361
9.2	Fabrication is Still a Challenge	363
9.3	Biological Fabrication	367
9.4	STEM Education and Outreach	373
9.5	Importance of Multidisciplinary and Basic Research	376
References		379

Index — 383

Preface

This book is intended to be an interdisciplinary introduction to the study of the captivating and diverse photonic systems seen in nature, and how we take inspiration from them to create new photonic materials and devices. The idea for this book came from my time at the Defense Advanced Research Projects Agency, or DARPA, where we were charged with building research and development programs in high-risk, high-potential impact science. When I joined DARPA, I did so with an idea of building a research program that would take advantage of the tremendous progress occurring in the field of biophotonics, in fabrication and characterization capability, and advance the state of the art in dynamic functional photonic devices by looking at biological systems for inspiration: the Bioinspired Photonics program. During this time, I had a conversation with the person who is now my editor at Taylor & Francis Group, about how the field of bioinspired photonics was exciting and growing, and in need of a book that could introduce the topic to a multidisciplinary audience. It was really that conversation, and her foresight, that started this book. *Bioinspired Photonics* is not meant to be an exhaustive review of any single technology space. Rather, it is intended to be a scientific description of the wonders of the natural world for a technical audience. This book is written for a multidisciplinary audience with the goal of providing an introduction to the wide range of topics encompassed by bioinspired photonics and it is impossible to cover everything within a single book. I attempted to illuminate specific concepts in an engaging and readable manner without sacrificing technical depth and detail. This book is geared toward illustrating the diversity of the subject matter with well-chosen examples. I have included pictures and figures wherever possible to illustrate the beauty and diversity of these natural phonic systems and to see and understand the bioinspired structures and devices being developed. Examples are provided to illustrate principles, recent discoveries, cutting-edge science, and application development. It is my hope that this book is broadly appealing to scientists, technologists, and lay people. This book should enable a person just entering into the field to acquire the minimum background necessary to begin exploration of this fascinating subject.

Work on bioinspired photonic systems involves several fields of study, such as nanofabrication, photonics, structural color, neuroscience, etc. The rapid growth in our understanding of the biological systems is leading to rapid growth in the number of devices and applications built from these principles and creates a rich area for discussion. This book will introduce readers to natural structures and outline important examples of how combining biological inspiration with state-of-the-art nanoscience is resulting in the emergence of a new field focused on developing real improvements in materials and devices. We will walk through examples taken from nature, delve into their characterization and performance, and describe the unique and fascinating features of their performance. This is interwoven with discussion of attempts to fabricate synthetic versions of these systems, as well as the specific aspects of the biological examples that researchers are attempting to leverage in their own work.

This field is of potential interest to people in a wide range of technical disciplines. The intended audience for this book includes biologists, zoologists, materials scientists, chemists, physicists, and engineers, and the subject matter touches on all of these areas and more. It is intended that this book provides a distilled description of bioinspired photonics sufficiently engaging to both experts in any of the related fields and novices just beginning their journey. A challenge is that communicating across so many technical boundaries requires speaking multiple scientific languages. Descriptions and discussion are provided with minimal jargon, and sufficient explanation to enable all readers to follow the text. By providing an authoritative text that synthesizes the cutting-edge research being performed in this field, and linking it to engaging examples of biological systems, there is an opportunity to engage a new audience of scientists and technologists. This material will be valuable to anyone interested in incorporating biological inspiration into their own work, working in related fields, or interested in engaging in new study. The technical material is also appropriate for graduate and undergraduate students seeking an introduction into an intriguing area of study, as well as educators teaching undergraduate and graduate courses in varying science departments.

This book is written so that each chapter encapsulates research and development into a specific aspect or application space relevant to bioinspired photonics. Some chapters are focused on a type of biophotonic mechanism, such as structural color and static color displays or lenses and image forming eyes. Other chapters focus on the application space such as sensors, and energy. The chapters are more or less independent of each other, although linked by some common structures, systems, and techniques under discussion. Each chapter describes both basic principles and multidisciplinary approaches to state-of-the-art biomimetic and bioinspired photonic systems. We discuss why these systems are of interest from an application standpoint and walk through the wide variety of fascinating examples of photonic systems we find in nature, our efforts to understand how they work, and the multitude of ways that scientists are attempting to replicate and take inspiration from these systems for fabrication and application. Chapters focus on a class of examples from nature, providing a discussion of our understanding of their functionality to date, a comparison emphasizing the similarities and unique aspects of various systems in different environments, and illustrating examples of the specific inspiration researchers are taking in new materials development.

The goal of the first chapter is to introduce you to the field of biological photonics and bioinspired photonic systems. In this chapter I hope to get you excited about the complex photonic systems we will discuss in the book. We established some common terms and tools to lay the groundwork for further in-depth discussion. I provide a brief history of the science and technology development that led to the emergence of this area of study.

Chapters 2 and 3 focus on structural color and begin with the immediately interesting examples of photonic micro- and nanostructures seen in insects, birds, plants, and more (including fossil evidence), examining structurally dependent color reflection in nature. We examine these passive, complex, and hierarchical color systems including 1D, 2D, and 3D structures, multiwavelength structures, and the way in which multiple coincident structures and the inclusion of pigment layers cooperatively drive the optical response. We discuss ways in which researchers are attempting to achieve the resolution required to fabricate structures for these wavelengths, and replicate the structures found in nature. Finally, we discuss some of the applications being suggested and developed based on these systems, such as displays, paints, and more.

Chapter 4 focuses on describing dynamic, active photonic systems, such as actuated photonic crystals and color-changing chameleons and cephalopods, such as squid and octopuses. The primary interest thus far in these systems is for display technologies, and the chapter illustrates several interesting examples of both commercial application, and others still at the bench-level proof of concept stage. Significant time is also devoted to discussing examples of dynamic color change in animals with specific focus on the astounding ability of cephalopods. Description of our understanding of how their anatomy drives this color change, and some of the neural, muscular, and structural components act in concert, are included.

In Chapter 5, we move to the optical structures found in animal eyes, like the lens systems and antireflection structures. In this chapter we discuss the variety of both simple and compound eyes. The antireflective surfaces found on the eyes of some insects, the fabrication of similar structures, and examples of their incorporation onto devices are also included. Understanding the benefits and weaknesses of animal and insect eye structures allows us to design synthetic analogs with applications in medicine, and technologies for imaging. Many examples of application development for imaging are presented as well as some other applications such as ocular implants.

Biomaterials are the focus of Chapter 6. We discuss the static, structural biological materials such as chitin, keratin, and silk and optically active materials such as reflectin protein and fluorescent proteins. Some time is spent on description of the advantages and disadvantages of biomaterials in applications (such as biocompatibility, biodegradation, refractive index limits, and tolerances). We discuss why the specific properties of these biomaterials are of interest from a photonic perspective, and why we are interested in them for applications. We address how they are being adapted for use as materials for medical optics, inexpensive display screens, photonic crystals, and more. We examine the tremendous impact that bioluminescent materials such as firefly luciferin, green fluorescent protein, and other optical tags are having as biological and medical probes and tags, and are opening up entirely new ways to study cellular behavior in areas such as neuroscience.

Chapter 7 focuses on sensor systems. We discuss not only photonic sensor systems found in nature but also how some researchers are taking photonic structures and using them in ways the animals never did. This chapter discusses the increasing need for novel sensors. It focuses primarily on infrared and thermal sensing, such as those found in beetles and pit vipers, as well as gas and vapor sensors, such as the ones developed using butterfly wing structures.

In Chapter 8, we look at how nature harvests energy from the sun and the photovoltaic and photocatalytic applications of the lessons we learn from understanding those natural systems. We discuss why it is crucial that we develop inexpensive, environmentally friendly, and efficient sources of non-fossil fuel-based energy. We talk a bit about some of the solid-state and chemical approaches to harvesting solar energy and creating solar fuel, as well as a short discussion of the state-of-the-art in general. We touch on how quantum mechanics may be playing a role in the photosynthetic process. Looking primarily at the plant and bacterial mechanisms of solar capture, that is the initial conversion of an incident photon to an electron, we discuss some of the nanostructures being used to improve conversion efficiency of solar cells, and even how some animals are capable of using photosynthesis as an energy source. We delve a bit into chemistry and the approaches being taken to mimic photonic acceptor dye molecules found in photosynthetic organisms. Finally, we discuss solar fuel generation and nanoantenna-based approaches to artificial photosynthesis.

The final Chapter 9 examines a few emerging areas, such as biological fabrication, and the challenge of fabrication and scale. We discuss the limits of understanding and current capabilities, and some as yet unanswered questions to illustrate new areas for potential future examination. We talk about how this field is excellent for STEM education. Finally, we touch on the importance of multidisciplinary research and how basic research and application, particularly in the field of bioinspired photonics, are intimately linked.

Viktoria Greanya

Acknowledgments

For more than a decade, my career has involved assessing, leading, and nurturing government investment in scientific research. None of the research described within is my own, although I have been involved in some of it from the governmental funding and management side. Throughout the book, I have attempted to demonstrate my great respect for the researchers at the bench and in the lab doing the hard work. I have also attempted to offer my view and opinion on the key advances and challenges ahead. Writing a single-authored text has been an incredible challenge and I am thankful I did not know how much work it would be when I first agreed to do it. It would not have been possible without the help and encouragement of many friends and colleagues. I would like to thank a great many people who have helped me in the process of writing this book. First, I would thank my editor at Taylor & Francis Group, Luna Han, who first suggested and championed the idea for this book, provided her expertise and counsel, and has been extraordinarily patient with me as I figure out how to be a first-time author. I would like to extend my love to my friends and family who have provided sustained encouragement. I would also like to thank my colleagues on the planning committee for the "Bioinspired, Biointegrated, Bioengineered Photonic Devices" SPIE BIOS conference, and would encourage anyone interested in the field to attend that conference in the future. Finally, I would particularly like to thank my friends and colleagues: Professor Stephanie Palmer, Professor Caroline Schauer, Dr. Kim Sapsford-Medintz, Professor Luke Lee, Professor Peter Vukusic, Dr. Roger Hanlon, Dr. Rajesh Naik, Dr. Radislav Potyrailo, Professor Fiorenzo Omenetto, and Professor Monique Van Hoek who provided their considerable expertise in both discussion and review of the text as well as their support to me. I am very grateful to them; all mistakes are entirely mine.

Author

Viktoria Greanya, PhD, is the chief of basic research in the Chemical and Biological Technologies Department at the U.S. Defense Threat Reduction Agency and a research associate professor at George Mason University. She has over a decade of experience in leading government discovery and funding investment in revolutionary research and development. Prior to joining GMU and DTRA, she was a program manager at the Defense Advanced Research Projects Agency (DARPA), often described as the "Pentagon's Mad Scientists," where she created programs in bioinspired photonics and advanced nanostructured materials, as well as managing the DARPA Young Faculty Award. In 2011, she was awarded the Office of the Secretary of Defense Award for outstanding achievement for her portfolio of work at DARPA. Her background includes nanotherapeutics, bioinspired photonic systems, nanostructured functional materials, heterogeneous integration, high-power and vacuum electronics, flexible photonic and electronic systems, and liquid crystals, among other areas.

Dr. Greanya received her BS in physics with departmental honors from Guilford College in 1995, and her PhD in condensed matter physics from Michigan State University in 2001. She completed a prestigious National Research Council Postdoctoral Associateship at the U.S. Naval Research Laboratory Center for Bio/Molecular Science and Engineering. Her broad technical background, combined with the viewpoint of a Department of Defense (DoD) science and technology manager, provide a unique perspective on the evaluation of emerging science.

This book was written by Viktoria Greanya in her personal capacity. The opinions expressed in this article are the author's own and do not reflect any views of the Defense Threat Reduction Agency, U.S. Department of Defense, or the United States government.

1
Introduction to Bioinspired Photonic Systems

Fluttering around your garden on the wings of a butterfly are sophisticated nanostructured photonic crystals more complex than we can create in the laboratory. Swimming in our oceans are the potential future generations of high-speed, organic, color-changing display technology as well as creatures that were the genesis of a transformation in microbiology. In our flowerbeds and deep in the ocean live solar power generation technologies that may help us to eliminate our fossil fuel dependence. These examples, some seen in Figure 1.1, are the beginning of a technology revolution spurred on by discoveries made about the structures and systems that have evolved to produce, reflect, absorb, and manipulate light in nature. Lessons and inspiration we are gaining from these systems may enable entirely new classes of organic and inorganic devices, improve the ecological impact on technology development, and reduce our dependence on fossil fuels. There is an incredible diversity and complexity to these natural systems that we are only at the beginning of our ability to replicate in the laboratory. This book is about our discovery and understanding of these photonic systems and how we are using that knowledge to develop devices and applications that enable breakthrough capabilities. This is how we take inspiration from nature to create bioinspired photonic systems.

1.1 Biological and Bioinspired Photonics

Light is the basis of all life on the planet; it provides the energy, warmth, and sustenance on which all organisms depend. It is intimately a part of the story of the emergence and evolution of life on earth over the past ~3.5 billion years. Most organisms on earth have

Figure 1.1 Examples of animals and technologies that are making key contributions in bioinspired photonics. (a) Morpho butterfly. Photonic structures on its wing scales are the source of its iridescent blue coloration. (b) Cephalopods demonstrate dynamic, rapid color change to mimic their surroundings. (c) Plants and bacteria are being used as model systems to improve solar energy capture. (d) Genes found in bioluminescent jellyfish are used to make glowing proteins, allowing us to track cellular behavior.

evolved to utilize light in one way or another, and there is tremendous breadth and complexity to the way we (plants and animals) do so. Some organisms absorb light and convert it to energy for survival. Green sulfur bacteria, for example, can convert light to energy with nearly perfect efficiency.[1] Many creatures on earth have the ability to sense light in some fashion. Marine and land animals have developed complex and widely differing image-forming eye structures and other light-sensing systems.[2] For example, dragonflies (example in Figure 1.2) have highly functional compound eyes with more than 30,000 lenses each. They take up most of the space on the dragonfly head, which is typically less than an inch wide.[2] Eyes capture the necessary information from our environment to find food, or identify safety, and relate to others around us as predators, family, and mates. In tandem with the evolution of image-forming eyes, many organisms

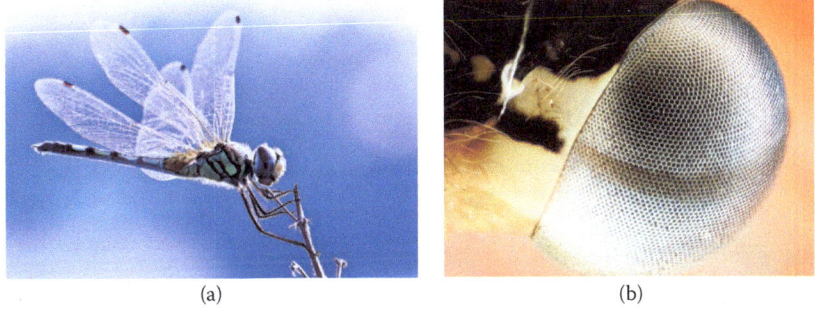

Figure 1.2 Dragonflies have complex image-forming compound eyes with excellent visual acuity. (a) Image of a dragon fly. (b) Close-up of a single dragonfly eye with the individual lenses of the compound eye visible.

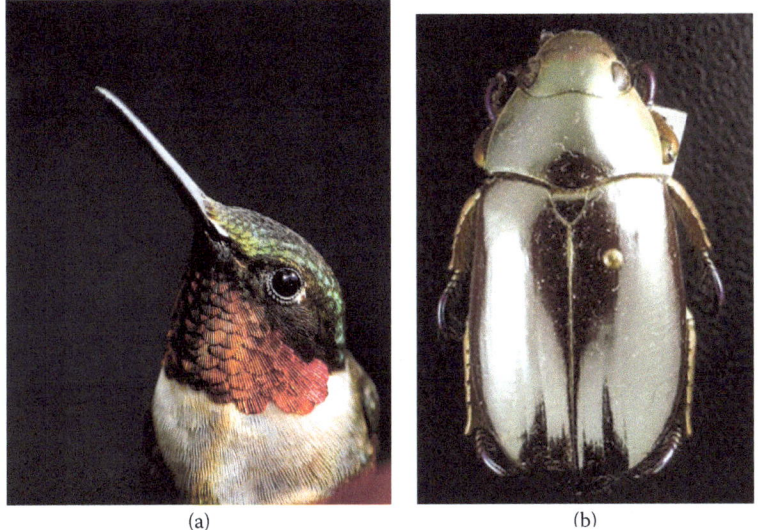

Figure 1.3 Examples of spectacular structural color. (a) Striking iridescent red hummingbird gorget and green iridescent crown. (b) Silver metallic looking elytra of the *Chrysina chrysagyrea*. (From Seago, A.E. et al., *J. R. Soc. Interface*, 6 Suppl 2, S165–S184, 2009. With permission.[5])

(both plants and animals) have developed complex color display capabilities.[3,4] Color enables an organism to communicate with other organisms. They may identify themselves to others of their species, announce fertility, provide warning, or entice certain behaviors. Color is also used for an organism's survival from predation, where it can be used as camouflage to hide from, confuse, or warn off the things that want to eat them. Many of these color displays are bright, colorful, and fascinating, such as the iridescent gorget and back feathers of some hummingbirds, as in Figure 1.3, or the beetle *Chrysina chrysagyrea*, which is a bright, shiny gold and yet has no metal in its body.[5] Within this book, we will discuss these and many more examples of how light interacts with nature.

When most people speak of light, they refer to visible light—electromagnetic radiation capable of being seen by the human eye. The photoreceptors in the human eye are sensitive to light at wavelengths from approximately 380 nanometers (nm) to 750 nm, which correspond to colors that range from violet to red, respectively. Visible wavelengths make up only a very small portion of the electromagnetic spectrum, as you can see in Figure 1.4, and many animals can see outside of the spectral range of human perception. Extending well past the limit of red wavelengths human eyes can discern, some bats, snakes, and beetles can sense mid-wave infrared (IR) wavelengths at 3–5 μm (3000–5000 nm).[6] They do this to find warm prey and locate appropriate breeding conditions. These animals utilize a different sensing mechanism than the photoreceptor in image-forming eyes, as the low energy of the IR photons is not sufficient to trigger biological effects. On the opposite side of the spectrum, at significantly shorter wavelengths some insects, arachnids, and birds can see into the ultraviolet (UV).[7-9] Mantis shrimp, bees, and butterflies, for example, can also see in the UV. Mantis shrimps have extremely complex, oblong compound eyes with multiple separate functional zones, which are each slightly different structurally. They have 12 photoreceptor channels, the most of any organism found thus far, four of which operate in the UV.[10,11] The term "light" as we will use it really encompasses wavelengths all of the way from IR through UV.

Photonics is the study of how we can manipulate light particularly in the regime where light acts as its quantized particle nature: the photon. Classically light behaves as a wave emitting from a source,

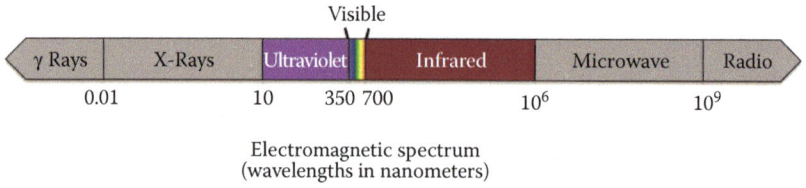

Figure 1.4 The electromagnetic spectrum. Light in the context of this book includes ultraviolet, human visible, and infrared wavelengths. Wavelengths are in nanometers.

and the study of optics encompasses how those waves travel through lenses, reflected and refracted at interfaces, and so on. The term "optics" is also used to describe the lenses, and other optical components themselves. In the early twentieth century, theory and evidence began to support the idea that light was quantized in particle (photon) form. Photonics in the modern sense arguably encompasses the entire study and manipulation of light and how it interacts with matter, as well as all of the related technology development. It is a broad field that describes nearly everything formerly described as optics, as well as newer technical areas such as nanophotonics, which deal with the interface of light and nanoscale structures.

Biophotonics is the study and manipulation of photons as they interact with biology. A good portion of work in this field relates to medicine and the development of imaging techniques for studying cells and tissues, as well as diagnostics, for example, the bioluminescent materials discussed in Chapter 6. Biophotonic bioluminescent proteins are having a huge impact on our ability to study and understand cellular behavior, for example, in areas such as cancer research, and how the neurons in our central nervous system work and interact with other neurons, as well as many other technology spaces. Biophotonics also describes the photonic materials and structures found in nature and the ways they are used by organisms to create, manipulate, and detect light. It describes the characterization of these structures and our attempts to understand the properties and functionality of these biological photonic systems. What we have discovered is that there are a wealth of interesting, complex phenomena and evidence of multi-scale, adaptable, and fault-tolerant functionality that we cannot yet match in our designs and fabrication capabilities. Although it is often tempting to believe that nature no longer holds any surprises for science, this is clearly not true as new species, new physics, and new capabilities are regularly discovered. We have only just scratched the surface of what exists and there are many things we still do not understand. The ability to delve into these complex systems has resulted in a rapidly developing field of science, which is taking inspiration in design, structure, function, fabrication, material, and other aspects and leveraging these advantageous properties for synthetic systems.

Bioinspired photonics is an applied subset of the biophotonics field. It is the design, fabrication, and utilization of photonic materials whose material, structure, or functionality is inspired in some way by what is found in nature. Bioinspired photonics is taking the leap from understanding how nature has evolved to respond to or produce light and creating a structure or system inspired by some aspect of that but altering to fit our own specific requirements. The biological photonic systems have evolved for purposes that most times do not match our technology needs. Bioinspired photonics examines what nature has created and then moves forward, taking design ideas and components from different organisms and combining them to invent completely novel designs. We may desire to use different materials than natural systems so we can access different optical properties (such as index of refraction), improved robustness, or ease of fabrication and integration into existing systems. We may take an aspect of design and alter it so as to achieve different wavelengths or selective activation to varying stimuli. Sometimes, inspiration comes simply from the fact that nature has proven something is achievable.

Fabrication and design of these systems is difficult. We either use the tools available to us from the commercial semiconductor industry or use lab-developed techniques not yet ready for scale-up. We are only beginning to be able to create novel synthetic systems similar to natural biophotonic structures with significant fidelity. Because of these difficult challenges, many times our first steps down the road to complete design and fabrication capability are biomimetic structures, where we simply attempt to replicate natural biophotonic structures. Soon however, hopefully in the not too distant future, we will be able to fabricate complex, integrated hierarchical systems with multiple organic and inorganic materials, and multiple length scales. We will be able to design from the ground up, and fabricate at ambient pressure and temperature conditions just as is done in nature, to achieve environmentally friendly and low-cost, high-performance devices. We can combine aspects of design from multiple organisms and concepts to create entirely novel systems. As the basic research uncovers more and more, we progress from the lab to innovative applications and this field will have a dramatic impact on science and technology development in the future.

1.2 Evolution

Of course, nature did not start out with fantastically complex optical systems. As the survival needs of biological organisms changed over time, so too did the organisms themselves. A discussion of the general principles of evolutionary processes can be found in Box 1.1. Most of what we know about the earliest beginnings of life is inferred from geological and geochemical evidence; soft tissue does not fossilize, so we have no direct fossil evidence of life forms from this time.[12,13] Photosynthesis has existed since the earliest glimmer of life on this planet (see Figure 1.5). From carbon signatures, we believe the earliest life began 3.8 Ga (giga-annum or billion years ago). The first life forms were photosynthetic anoxygenic bacteria, which absorbed solar radiation in the near IR and produced sulfur/sulfate compounds. Oxygen appeared in the atmosphere approximately 2.4 billion years ago after the first oxygen-producing photosynthetic bacteria, cyanobacteria, appeared just before this time. The eukaryotes, organisms with nuclei and other organelles within the cell membrane, emerged somewhere around 1.2 billion years ago. These eukaryotes are linked to the "great oxygenation event" that was a second increase in atmospheric oxygen that occurred around 850 million years ago. Algae appeared—first red and brown (1.2 Ga) and then green (~750 Ma) as chlorophyll-based systems emerged. Then around 500 million years ago, we begin to see evidence that green and vascular land plants evolved.

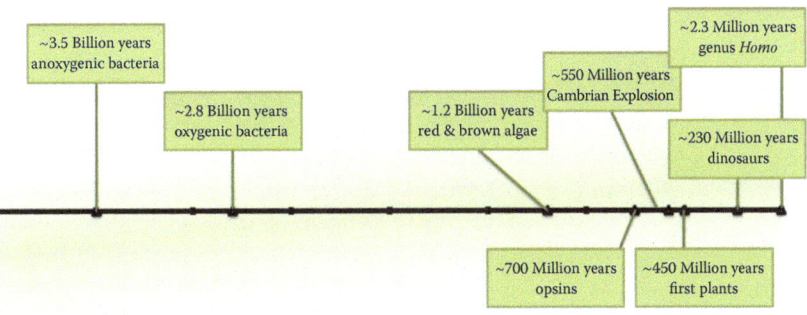

Figure 1.5 Illustration of the evolutionary time line of light-interacting structures in organisms from the formation of the planet through the rise of genus *Homo*.

BOX 1.1 EVOLUTIONARY PROCESSES

Evolution is the heritable genetic change of organisms descending from a common ancestor.[14] Changes to these traits within populations are driven by a combination of mutation, genetic drift, migration, and natural selection, illustrated in Figure 1.6. Mutation refers to the random change in a gene or genes that can occur in offspring relative to their parent's genetic makeup. These offspring mate and reproduce and the frequency of these new genes increases in the population compared to before. Migration is another way of new genes entering a population, where influx or efflux of community members locally changes the genetic mix. Genetic drift describes how, in a given population, over time the relative frequency of genes will change due to random events. Some individuals will have more offspring survival (for reasons unrelated to their fitness), or an event occurs that eliminates a portion of the population. Genetic drift describes how a

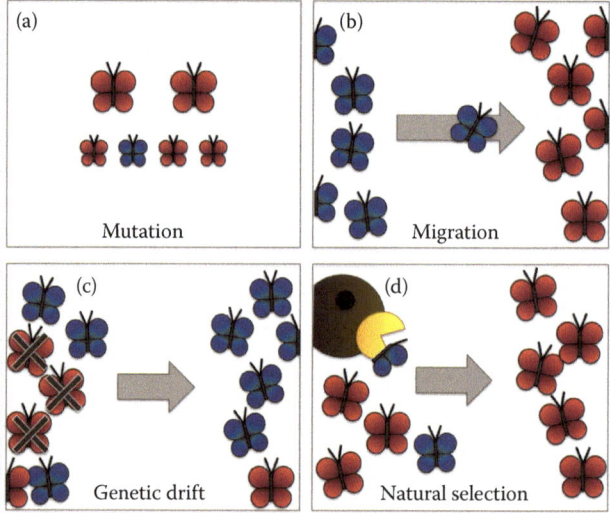

Figure 1.6 Illustration of the four primary mechanisms for evolutionary change. (a) Mutation—when an offspring introduces new genes to the population. (b) Migration—when new genes enter a population due to the influx or efflux of individuals. (c) Genetic drift—when random processes result in the increase in particular genetic frequency. (d) Natural selection—changes in a population due to the increase or decrease in reproductive success from characteristics that arise from changes in the genes in a population.

BOX 1.1 (*Continued*)   EVOLUTIONARY PROCESSES

population as a whole moves (genetically speaking) in a certain direction and is a random and unpredictable process. Natural selection describes the process by which genetic frequency for a given trait increases or decreases in a population as a result of reproductive success. Over time and through combinations of external pressure and survival requirements—such as pressure from predators, availability of food, changing environment, availability of mates, etc.—traits that reduce the survivability and reproductive success of a population decrease in frequency. Natural selection acts on the traits developed through the other evolutionary mechanisms. It occurs mostly by "selecting out" negative traits. Genetic frequency can only increase if an organism can reproduce. If a mutation results in an organism getting eaten before it can reproduce, then that was not a successful adaptation and that mutation is selected out of the population simply because no offspring occur to carry those genes forward. Natural selection does not result in traits that are beneficial (or at least neutral) to improving a population's reproductive successes that are carried forward into successive generations. Genes and traits that remain relatively unchanged over time between generations or species are said to be "conserved."

Evolution should not be thought of as a singular path. There are twists and turns, and occasionally dead ends and retreats. Sometimes, parallel or convergent evolution can happen as well. In parallel and convergent evolution, two or more species develop similar traits independently, even though they have different ancestors and lineages. (Parallel evolution occurs when the ancestors of both species were similar, and convergent evolution occurs when the ancestors of both species were not similar with regard to the specific trait.) The independent evolution of the cephalopod and vertebrate eyes, particularly the retina, is an example of convergent evolution. Coevolution is another particularly important aspect of evolution, where different traits of predator and prey evolve simultaneously, each influencing the other. For example, bees and other insects can see into the UV.

(*Continued*)

BOX 1.1 (*Continued*) EVOLUTIONARY PROCESSES

Some flowers have evolved coloration not only in the visible but also in the UV to take advantage of the bee's vision to entice a bee visit. Bees collect nectar from plants and in exchange carry pollen on their journey to other plants, but sometimes bees can demonstrate "nectar robbing" behavior and collect nectar without collecting any pollen. A variation on UV coloration in some plants has evolved in which a "pathway" to the nectar is highlighted by a UV patch called a "nectar guide" that guides the bee specifically to the nectar in a way that will lead the bee to pollen as well.[15] Coevolution is an important process, and we see repeatedly that codevelopment can shape multiple species of plant and animal simultaneously.

Modern biology categorizes organisms based on their common ancestry. Most are familiar with the biological classification ranks of domain and kingdom down through genus and species, with many subranks in between, which classify organisms by a hierarchical set of common physical features. Classification now looks at the evolutionary path as a prime determination of organizational taxonomy as they group organisms in units (taxa). A group that consists of an ancestor and its descendants is called a clade.

Something crucial occurred around 550 million years ago, which resulted in a tremendous evolutionary burst of life forms in a comparatively short amount of time (over only ~75 million years). This period, called the "Cambrian Explosion," is when the color and vision structures we will discuss throughout this book began to exist. The size and structure of organisms became amenable to the formation of these complex systems, and we find evidence thanks to their harder structural components, such as eyes and scales, that are more capable of fossilization. It is this time that we begin to see fossils of eye structures that resemble the image-forming eyes we are familiar with from extant (living) species, such as highly complex compound eyes in preserved arthropods (insects, crustaceans, arachnids, etc.).[16] *Anomalocaris*, for example (artist's renderings seen in Figure 1.7), was a giant apex predator of its time and sported two large compound

INTRODUCTION TO BIOINSPIRED PHOTONIC SYSTEMS 11

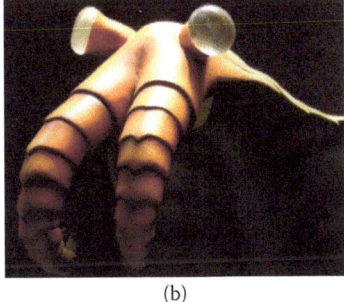

(a)   (b)

Figure 1.7 Artist's renderings of a 2-ft-long apex predator from the Cambrian era, *Anomalocaris*, which sported two large compound eyes. (From Wikimedia commons, With permission.[17,18])

eyes. It is not entirely clear what sort of optical systems the ancestral precursors to these image-forming eyes might have been. Their soft tissue structures left no record thus far discovered. The evolutionary path from simple photosensitive patch to an image-forming eye is not an all or nothing path.[19] We find evidence of organisms with eyes in the intermediate stages between patch and eye that exist as well. Modeling has shown that eyes can evolve from simple photosensitive patches through to an image-forming eye in a surprisingly small number of generations (as few as 2000) with just a small change in structure (~1%) per generation.[2]

We do not know for certain what evolved first—color display or image-forming eyes—but sometime around the Cambrian Explosion also come our first fossils with evidence of color. As we will delve into in Chapter 2, some colors are the result of light interacting with sophisticated nanostructures on the organism's outer surface, or integument. Fossil evidence of these nanostructures, such as on the scales (sclerites) of the hard outer shell of the Cambrian worm *Wiwaxia corrugate* (seen in Figure 1.8) and, *Canadia spinosa*, and the trilobite relative *Marrella splendens*, has been found from this time period.[4,20] In *Wiwaxia*, the scales have linear diffraction gratings that would have reflected color. The location of these gratings, on the outer hard scales of *Wiwaxia*, as well as the defensive parts of other fossil evidence, suggests that color was used primarily for defense. We now have many examples of photonic structures in fossils from this era and later, from the color in beetles, moths, and dinosaur feathers, to eyes, and other structures.[16,21–26]

Did primitive eyes have the ability to differentiate color before the image-forming eye structures evolved? Did color evolve specifically,

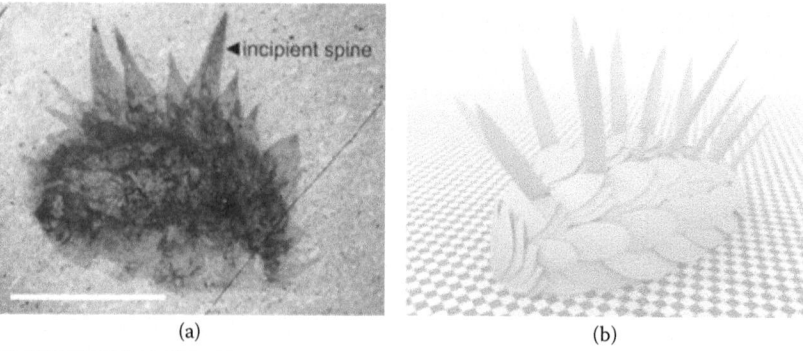

Figure 1.8 Image of (a) fossil and (b) artist's rendering of the Cambrian worm *Wiwaxia corrugate*. Grating structures were found on *Wiwaxia* fossil scales, indicating early structural color. (From Smith, M.R., *Palaeontology*, 2013. With permission.[20])

or was it initially a beneficial by-product of structures advantageous for other behaviors, like moving through the environment or antifouling, for example? We do not know. However, we can say that the development of an image-forming eye placed a tremendous amount of evolutionary pressure on physical appearance, and something like a coevolutionary "arms race" began.[2] These eye and color structures did not only evolve once. Opsins, the light-sensitive proteins found in photoreceptor cells, are believed to have existed for over 700 million years.[27] The opsins have a common ancestor, meaning that they evolved early in a single ancestor and that genetic feature was conserved through the evolutionary process. Although opsins may have evolved only once, eyes and color systems have evolved many times in many species with widely varying structures.

We often seem to think of evolution as a forward-only motion, moving us along a path to perfection. This is not truly the case. These evolutionary changes come with a fabrication and maintenance cost; optical and color systems require energy from the organism to grow and sustain. But from a survivability perspective, these systems do not have to reach some version of "perfect function." Most of nature's systems have evolved to be "good enough." There are many vestigial remnants from things that were at one time useful (or neutral) survival traits. Nature accepts compromise and creating new functionality out of existing materials and systems—such as the photoreceptor sensors in fire-seeking beetles seemingly adapted from the more common insect mechanoreceptors (we will discuss them more in Chapter 7).[28]

1.3 Historical Perspective and the Advent of Microscopy

Nature has been using these structures for over 500 million years. Just as these systems have evolved, so has our ability to investigate and understand them. Despite centuries of thought and experimentation, it is only in recent decades that we have begun to understand how these structures work and begin to use that knowledge to our own advantage. The early Greek scientist Anaximander (c. 610–546 BCE) can be said to be the progenitor of modern science and initiated the scientific study of nature. His student Anaximenes gave us the first written mention of bioluminescence (light produced by chemical reactions within living organisms), where he describes the glow seen when an oar strikes through water.[29] Aristotle was a prolific author of texts illustrating his study of nature, many of which survive today, such as his *History of Animals* (*Historia Animalium*).[30] Aristotle references bioluminescence in the text *Meteorologica* when he mentions the luminescence that can occur when water is struck by a rod, as well as in *De Coloribus* (though authorship of this text is disputed), where the author says, "For some things which are neither fire nor forms of fire seem to produce light by nature."[31,32] However, neither text explores the mechanism behind the effect. Pliny the Elder, several centuries later (23–79 CE), wrote detailed descriptions of color in animals and bioluminescence and described many bioluminescent organisms, including insects such as fireflies, and glowworms, and the bioluminescence of bacteria in dead fish, mollusks, and other marine life and fungus.[33] Nearly a thousand years later, Ibn al-Haytham (Alhazen) (965–c. 1040 CE) published some of the first studies of the eye and the physics of vision as well as other groundbreaking optics work in his *Kitab al-Manazir* (*Book of Optics*). The way we relate nature and light has clearly been of interest since the earliest days of science.

As in most other fields of science, the development and adoption of new tools is often the spark that drives new discovery. The development of optical devices such as the telescope and microscope is where the systematic scientific study of light and nature really begins. It is responsible for a huge movement forward in our understanding of the natural world. Natural philosophers of the seventeenth through nineteenth centuries became the forefathers of modern biophotonics. In 1669, Robert Hooke published his text *Micrographia*, illuminating a whole new world with descriptions of objects and insects viewed through the microscope.[34] In it, he describes many of the things that are still being investigated today such as the structure of fly eyes and refraction and reflection of light from thinly layered structures such as mica. Hooke's beautiful, hand-crafted microscope can be seen in Figure 1.9 (or at the National Museum of Health and Medicine in Washington, DC). Hooke describes the structures he saw in colorful bird feathers, the eyes and wings of flies, and the pearly scales of the "small Silver colour'd Book Worm," which we call today the silverfish. The silverfish is a small, nocturnal insect (only ½–1 inch long), as seen in Figure 1.10a. Hooke's drawing can be seen in Figure 1.10b. Of the color of the silverfish, Hooke referenced

Figure 1.9 Image of the hand-crafted microscope used by Robert Hooke, which was used to make observations noted in *Micrographia*. Currently on display at the National Museum of Health and Medicine in Washington, DC.

Figure 1.10 (a) Silverfish. (b) Drawing from Hooke's *Micrographia* illustrating the scales on a silverfish. (From Hooke, R., *Micrographia*, Project Gutenberg, http://www.gutenberg.org, 1669.[34])

the pearlescent color of seashells and said "… for they each of them consisting of an infinite number of very thin shells or laminated orbiculations, cause such multitudes of reflections, that the compositions of them together with the reflections of others that are so thin as to afford colours (of which I elsewhere give the reason) gives a very pleasant reflection of light." Newton, likewise, examined the structure of feathers (peacocks) to understand their color in his book *Opticks or A Treatise of the Reflections, Refractions, Inflections and Colours of Light* in 1704.[35] In it, he speculates about light reflection and transmission through thin and layered structures such as discussing peacock feathers, spider webs, colors of silk, and how immersion in water and oil might change their properties. Of peacock feathers specifically, he says "… their Colours arise from the thinness of the transparent parts of the Feathers … ." Both men speculated that structure drove the effect. The ability to see these and other structures with their own eyes, sparked intense scientific curiosity.

Hooke was also one of the many interested in the nature and cause of "shining" biological systems such as cat eyes and bioluminescence from rotting wood and fish.[34] He discusses the actively shining cat eyes, which he links to light emanating from the eyes particularly when the animal is hunting. Hooke ascribed many bioluminescent properties to motion and vibrational energy. Other scientists and authors, such as Kircher, Bartholin, and Boyle, in the 1600s experimented with and wrote detailed treatise on the subject of bioluminescence and theories on its mechanism.[36] Description of dinoflagellates and identification of their role in oceanic phosphorescence using the microscope occurred in the early 1800s (by Macartney in 1810 and Michaelis and Ehrenberg in the 1830s). Then, in 1887, Dubois discovered the organic molecules responsible for bioluminescence in beetles and mollusks (and also fireflies), which he named luciferin and luciferase.

The study of eyes, particularly compound eyes, also advanced significantly during this time. As mentioned earlier, Hooke describes the eyes of a drone fly in *Micrographia*.[34] In 1695, Antoni von Leeuwenhoek published his observations of images formed through a compound eye in a letter to the Royal Society of London.[2] The first to publish an account of looking through the insect eye optical array, he describes looking at a candle flame through the optical array and seeing the formation of hundreds of small, inverted images of the flame. Anatomical and histological (tissue) study led to some increased understanding of the photoreceptors and cell structures. But it was not until Exner published his monograph in 1891 that a mostly accurate description of how the structures of the compound eye focused images was outlined.[2,37] Particularly ground breaking for the time was the assertion that each facet of the compound eye was an individual focusing element or a lens cylinder.

To really understand how biophotonics works, a tremendous amount of understanding in many areas of science had to co-develop. The understanding of the quantization and propagation of light, cellular structure, and function and neural behavior, photoelectric transduction, and more had to be developed. All of this work and centuries of study were to understand how nature has been utilizing and manipulating light for hundreds of millions of years. As the study and disciplines of science matured during the eighteenth and nineteenth centuries, and its participants were increasingly driven to more focused study, the Natural Philosopher gave way to specialized scientists (biologist, chemist, physicist, etc.). For most of the twentieth century, biologists and zoologists performed much of the investigation of biophotonic structures and systems. They studied these structures as well as the evolutionary and behavioral implications of how and why these structures came to be. This was extremely important, fundamental work, but from a device perspective, to physicists, material scientists, and engineers, for example, excepting perhaps eye structures, it was an exotic curiosity.

In the 1940s, commercial electron microscopes entered the laboratory and revolutionized our ability to see smaller and smaller structures.

These devices use focused electron beams to create images. The electron beam illuminates a sample, and data are collected from one of a variety of modes depending on the type of electron microscope. In transmission electron microscopy, electrons travel through the sample and are then collected on the opposite side. In scanning electron microscopy (SEM), electrons strike the surface of the sample and are either absorbed and reemitted (secondary electrons) or reflected from the surface (backscatter). The data can then be processed to construct a high-resolution picture of the surface of the object being imaged. Because the wavelengths of the electrons are small compared to visible light wavelengths, this device produces a significantly higher resolution image compared to optical microscopy, on the order of tens of nanometers and below. Anderson and Richards (at RCA and the Zoology Laboratory at the University of Pennsylvania, respectively) used the device to perform the first detailed studies of the structures responsible for the vibrant and iridescent blue wing color in Morpho butterflies, as well as other insects.[38]

In the early 1980s, the field of nanotechnology was beginning to take off. Fabrication capabilities were reaching the point where very small structures could be made, and experimental and modeling capabilities were driving the investigation of smaller and smaller systems. Novel materials were being postulated, and in the late 1980s and early 1990s the first man-made photonic structures with dimensions on the order of the wavelength of light were described and fabricated. These "photonic crystals," described separately by Eli Yablonovich and Sajeev John in separate papers in 1987, were artificially structured heterogeneous materials that had multidimensional periodicity in index of refraction. Yablonovich's illustration of his structure and an illustration of some of the possible photonic crystal structures can be seen in Figure 1.11. They had the ability to controllably affect the transport of photons of different wavelengths through matter.[41,42] In other words, photonic crystals are artificially structured composites made of more than one material in some form of ordered architecture. This type of complex photonic structure also describes some of the optical systems seen in nature. These structures were one part of the nano revolution in photonic materials research, in fabrication, theoretical understanding, and technology component development. The tools being developed for study, fabrication, and application of

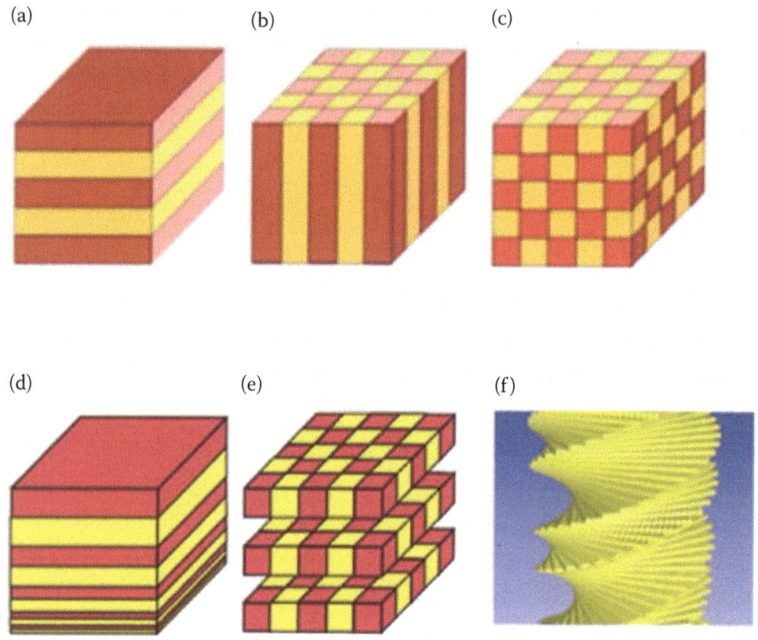

Figure 1.11 Illustrations of some possible photonic crystal structures. (a) One-dimensional (1D). (b) Two-dimensional (2D). (c) Three-dimensional (3D) structures. (d–f) Other structures seen in nature. (d) Chirped multilayer. (e) Perforated multilayer type structure. (f) Twisted plywood structure. (From Biro, L.P., and Vigneron, J.P., *Laser Photon. Rev.*, 5(1), 27–51, 2011 and Johnson, S.G., http://jdj.mit.edu/book/. With permission.[39,40])

nanomaterials and nanophotonics materials enabled the further study, fabrication, and now application of these biophotonic systems.

1.4 Tools of the Trade

The development of ever-improving tools is intimately related to our ability to make significant discoveries. Biological systems are incredibly complex from both a structure and a composition perspective. They are often composed of heterogeneous materials or cell types, each with their own morphology. The study of these systems requires a suite of techniques that can probe at, for example, different length scales, different material systems, and different wavelengths.[43,44] We need not only qualitative but also quantitative capability. Ultimately, our desire is to fully understand the composition and architecture of each system as well as fully characterize its functionality. Optical and electron-based imaging characterization techniques are one set of examples that enable us to probe the former, and spectroscopy and scatterometry are some techniques that enable us to probe the latter. Modeling and simulation tools help us to bridge the knowledge gap by intimately linking the two. Chemical, histological (tissue), and behavioral analyses also give us vital clues to their structure and performance. These techniques can provide us with insights into the relationship between structure and function in these systems.

1.4.1 Microscopy

Microscopy techniques are part of a primary set of tools that researchers use in their investigation of biological photonic structures and the characterization of their synthetic structures. Microscopy gave biological photonics its true start as a field of study. It gives us the ability to probe and understand the physical structure and link the structure to other measureable properties. Optical microscopy, fluorescence microscopy, and electron microscopy are all tools used to look at these systems across the multiple relevant length scales seen in natural systems, as seen in Figure 1.12.

Each technique has strengths and weaknesses. Optical microscopy and fluorescence microscopy are particularly useful as they allow us to view living cells and organisms. However, most standard techniques have comparatively low resolution. Confocal microscopy and superresolution fluorescence microscopy and other superresolution techniques overcome (to some level) this limitation. Modern electron microscopy techniques such as SEM and tunneling electron microscopy (TEM) are attractive because they enable us to image samples with sub-10 nm resolution. Indeed, high-resolution TEM (HRTEM) can reach

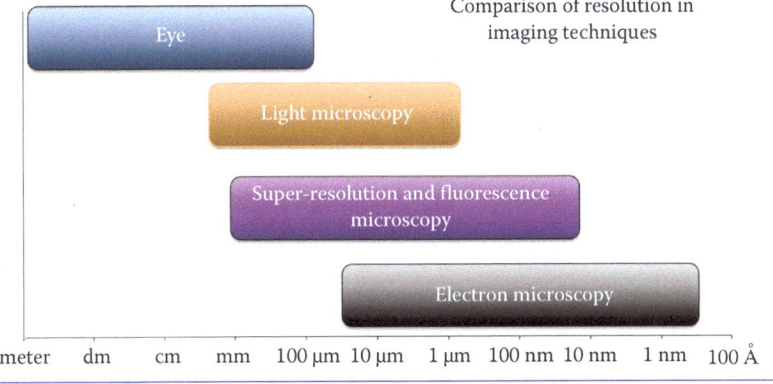

Figure 1.12 Comparison of resolutions of different imaging techniques.

resolutions of 0.05 nm, capable of imaging atoms in a crystalline structure. They can give us detailed views of the system ultrastructure, i.e., the specific, intricate structure of the biological organism. Other techniques such as TEM tomography (which has been used to more fully construct three-dimensional [3D] images of biophotonic crystals), atomic force microscopy, small-angle x-ray scattering (SAXS), and 3D nano-x-ray imaging can also be used to image biological samples.[45]

There are, however, challenges with most of these techniques. For both SEM and TEM, there is an extensive amount of sample preparation that must be done before imaging can occur.[43,44] Both techniques are done under ultra-high-vacuum conditions, which are incompatible with moist, squishy, and delicate biological samples. Samples must be fixed, either chemically (e.g., with glutaraldehyde) or physically (e.g., cryo-freezing). Wet samples must be dehydrated before they can enter the SEM/TEM chamber. Chemical processes attempt to preserve the natural structure of the system under study, but this is not always the case. In the case of dry arthropod specimens, whose hard exoskeletons can survive the imagine process, fixing and dehydration may not be required. SEM has good resolution and good depth of field, so images give some 3D information. SEM generally only probes the surface of a sample, however, so to see subsurface features inside the sample it must be cut or sectioned in such a way to expose the structure within. This can be done either with a very careful and steady hand or more accurately with commercial cryo-sectioning or focused ion beam to cleanly destroy a portion of the sample. The sample can then be plated. It is necessary that the sample conducts electrons to prevent charging under the high-voltage electron beam; so, the structures to be studied must be plated in a conducting material, such as gold.

TEM requires even more careful sample preparation. Because the electrons are transmitted through the material, the sample must be sectioned extremely thinly (>approximately 100 nm thick) to be transparent to the electrons. Typically, the sample is fixed, stained (with contrast agents of differing electron density, often heavy metals), and then embedded in resin for slicing. This technique requires a significant amount of experience, expertise, and careful sample preparation and handling. But, done correctly it can yield meaningful understanding of the structure. Many of these techniques are slow and only work with dry, dead specimens and are destructive to the samples. Development

INTRODUCTION TO BIOINSPIRED PHOTONIC SYSTEMS 21

in methodologies and techniques that enable ultrahigh resolution and nondestructive, 3D imaging of wet or nonconducting or non-vacuum-compatible samples will move the field forward significantly.

Helium ion microscopy (HIM) is a relatively new technique that uses a focused beam of helium to probe the surface of the structure to be imaged. HIM has several advantages over SEM such as charge reduction, minimized sample damage, increased depth of field, and 5 angstrom imaging resolution and high surface contrast without the need for metal coating.[46] However, as the technology is relatively new, there are very few in operation at this time. Figure 1.13 shows examples of image taking via optical microscopy, SEM, HIM, and TEM.

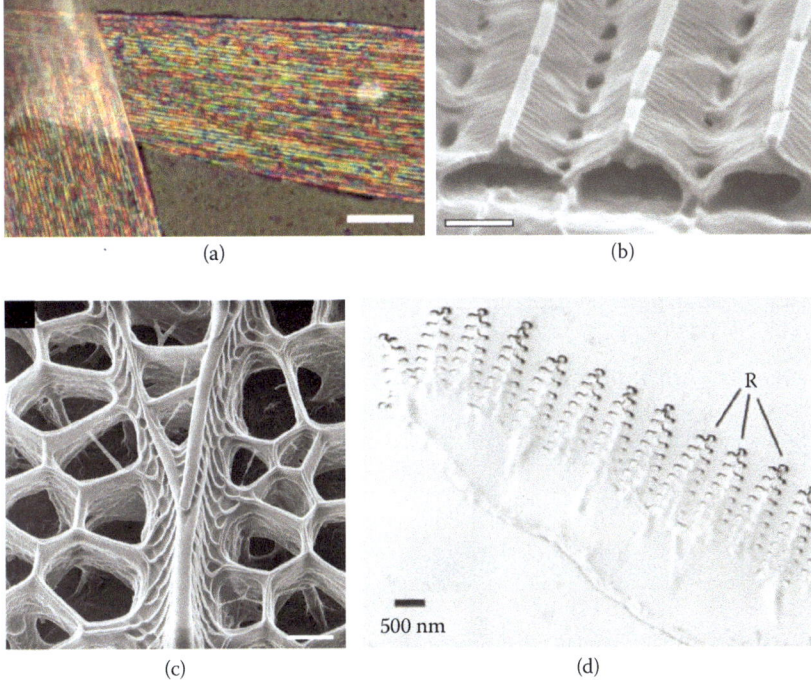

Figure 1.13 (a) Optical microscope epi-mode image of two single *A. argenteus* scales. Scale bars: 15 µm. (b) SEM taken at near grazing angle of open-ended structures on the *A. argenteus* wing scales. Scale bar: 0.8 µm. (c) Helium ion microscopy (HIM) images of *Papilio ulysses* black ground scale. Scale bar: 400 nm, 21° tilt. (d) Tunneling electron microscopy (TEM) view of a transverse section of the edge of a Morpho scale, showing the closely packed ridges (R) and their associated lamellae. The section also catches some of the pillars that connect the photonic structure with the relatively featureless bottom layer of the scale. Scale bar: 500 nm. (From Boden, S.A. et al., *Scanning*, 34, 107–120, 2012; Potyrailo, R.A. et al., *Nat. Photonics*, 1, 123–128, 2007. With permission.[47–49] Vukusic, P. et al., *J. R. Soc. Interface*, 6 Suppl 2, S193–201, 2009.)

1.4.2 Spectroscopy and Scatterometry

Spectroscopy and scatterometry are another toolset that researchers use in their study of these systems.[43,44] Spectroscopic techniques are used to probe the transmitted or reflected light intensity as a function of wavelength. Scatterometry captures the light scattered from the surface away from the angle of, or in the absence of, specularly reflected light. In both techniques, light from a source is focused on and illuminates the sample. Photodetectors placed in front of or behind the sample collect the available light, measuring reflection or transmission, respectively. Using a goniometer, for example, enables us to accurately rotate the detector or sample and measure the angular dependence of reflected and transmitted light at various wavelengths. Figure 1.14 shows a very simplistic schematic of a spectrometry setup, with light illuminating a sample and a rotating detector capturing the reflected and transmitted light as a function of angle. Key properties are the numerical aperture and the light collection capability of the system.[44] The light source may emit a very limited range of wavelengths, e.g., with a monochromatic laser or a broadband light source. Care must be taken to account for the nonuniformities of light source emission over the band of interest as a function of wavelength in the subsequent data. Cameras, of course, operate primarily in the visible and capture only a small fraction of light, and special light sources (such as Xenon or Deuterium lamps for UV) and detectors may be required to take

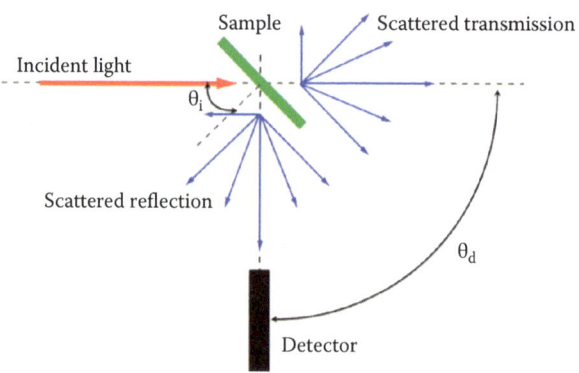

Figure 1.14 Simple diagram of spectrometry setup. Goniometer setup. A collimated beam of light macroscopically illuminates the sample at an angle of incidence θ_i, which can be varied from 0° (normal incidence) to about 90°. The scattered light in transmission and reflection is collected for different angles, θ_d. (From Vignolini, S. et al., *J. R. Soc. Interface*, 10, 87, 2013. With permission.[43])

images in the UV or IR.[43] Care should also be taken that neither the illumination intensity nor the exposure time damages the sample.

In addition to wavelength and intensity, polarization state is also of interest in many of these systems.[43] Several passive and active color systems have polarization information (both linear and circular), and some eyes, such as some spider eyes, can also detect light polarization. For samples that strongly scatter light, any polarization information of the scattered illuminating light will be lost. Also, some components of the instrument itself, such as the fibers of a fiber-optic probe, may lose the polarization information.[44]

1.4.3 Challenge of Working with Live Specimens

Carrying out experiments with small biological specimens sometimes requires creative thinking and great inventiveness. New and improved tools and techniques need to be invented constantly to see farther, better, and deeper. To model the biological nanostructures responsible for structural color, for example, we must develop new techniques to account for lack of perfection in the biophotonic nanostructure. This is particularly challenging as many modeling techniques rely on the repeatability of a structure to simplify the calculations. Studying live animals can be especially challenging. To tease out the specific pathways and channels that communicate with and control the photonic systems found in cephalopods such as squid, for example, researchers work with both skin samples and live animals. In addition to behavioral and observational studies, scientists study the mechanisms of physical, chemical, and neural control over the color and shape changes.[50,51] Some of the fluorescent materials and techniques described in Chapter 6 can also be used to elucidate the link between structure and function.

Another example is the special care and ingenuity needed when examining the Panamanian tortoise beetle *Charidotella egregia* to study color change.[52] The effect only happens in live specimens. The beetle changes from gold to red at the slightest disturbance. So in addition to studying the beetle in its natural habitat, the team had to develop a special technique to measure the reflectance spectrum in its undisturbed state. Because picking the beetle up to place it in the spectrometer would cause it to change color, they built small transparent housing boxes for each beetle with a net-covered hole into which a spectrophotometer probe could be placed without disturbing them and without allowing them to escape.

1.4.4 Fabrication Approaches

Fabrication is also extremely challenging. The scale of the hierarchical structures responsible for structural color, for example, can be on the order of only a few nanometers up to hundreds of nanometers and incorporated in even larger structures up to micron and mm scale. Developing fabrication methods capable of fabricating the intricate detail called for some of these systems is an ongoing desire. One method of faithful reproduction of a biological structure is biotemplating.[53] In this method, the biological structure a precursor is allowed to infiltrate the structure. After removal of the original biological structure, the subsequent inverse structure can be analyzed or used as a mold to grow a replica of the original structure. This technique only works with open architectures, such as the structural color systems, but it enables a relatively accurate replication of the nanostructures. This method does not allow any alteration to the biological structure, but scale-up can be achieved by, for example, stamping.

Another option is to fabricate at larger size scales. We have many fabrication techniques that can create complex 3D structures at millimeters and above, such as 3D printing. By using high-resolution TEM images, large-scale replicas several thousand times the size of natural systems can be made. This will enable us to explore the physics of the structure, though at much longer wavelengths. With lenses and wafer scale processes, macroscopic versions of these systems can be made for demonstration. However, true 3D fabrication of a man-made design at nanometer length scales is still a major challenge.

Other methods of fabricating nanostructures for near infra-red (NIR), visible, and UV wavelengths can be divided generally into "top–down" and "bottom–up" approaches. In top–down approaches you are etching something away from a substrate (the sometimes sacrificial base layer of the material), whereas with bottom–up approaches you are growing something onto the substrate. Current nanofabrication techniques such as electron beam (e-beam) lithography, soft lithography, nanoimprint lithography, and others can be expensive and slow and prone to defects and not yet capable of more than relatively simple 3D structures. Colloidal growth is often faster and cheaper but lacks any significant local control over structure and is also prone to defects. We cannot yet match the resolution and control needed to replicate these types of complex structures.

1.5 Bioinspired Photonics in the Twenty-First Century and the Challenge of Multidisciplinary Science

The applications-focused field of bioinspired photonics is a truly multidisciplinary field. However, it is not a single cohesive field. As you will see throughout this book, this field is more like a descriptor for a collection of many other scientific disciplines and communities. Developing the understanding of how nature's photonic systems work and applying that understanding to device design require expertise in diverse fields such as entomology, zoology, biology, neuroscience, photonics, physics, chemistry, microbiology, engineering, and more. Each of these communities approaches the topic with a different viewpoint and interest. It is no secret that research in multidisciplinary science and technology has tremendous potential for expanding our knowledge and seeding our new technology development. The National Science Foundation (NSF), National Institutes of Health (NIH), and Department of Defense (DOD), the major sources for funding of basic research in the United States, all place a high priority on multidisciplinary science and technology (S&T). Research at the interfaces between fields requires innovative thinking, and biophotonics is a good example of this.

Bioinspired photonics is a field of rapidly advancing technology with great potential. Applications range from compact, low-cost, and low power consumption products such as displays, and imagers, to lightweight, energy efficient, and environmentally safe energy-harvesting technologies to advances in nanomedicine, neuroscience, and biomedical imaging. Bioinspired photonics can impact all of these areas. We will achieve major technology breakthroughs by the fusion of different disciplines. The emergence of nanophotonics as a field of study, development and continuing maturation of tools for nanoscale fabrication and characterization, as well as development of tools and techniques for studying the microbiology, neuroscience, and chemistry of natural biophotonic structures, among other things, are dramatically increasing the number of researchers and publications in the field. Since the 1980s, the biophotonics arena has expanded from only a few diligent investigators in the early years to now hundreds of publications a year from all disciplines of science. Figure 1.15 shows an analysis of search results for a variety of biophotonic key words, and the cumulative publication in these areas has increased nearly every year.

The field is extremely challenging and yet widely fascinating, attracting students and researchers from many disciplines. Bioinspired materials in general and photonics specifically seem to capture the imagination, and science news articles about novel discoveries or approaches can be found popping up regularly online. One challenge is to increase the number of knowledgeable researchers and trained personnel in the field. Multidisciplinary training for future generations is a vital need. These future scientists and technologists must balance depth and breadth of technical knowledge with the ability to speak the language of multiple disciplines. Opportunities now abound for interactions among traditionally disparate disciplines in S&T. Popular science books as well as technical books and articles

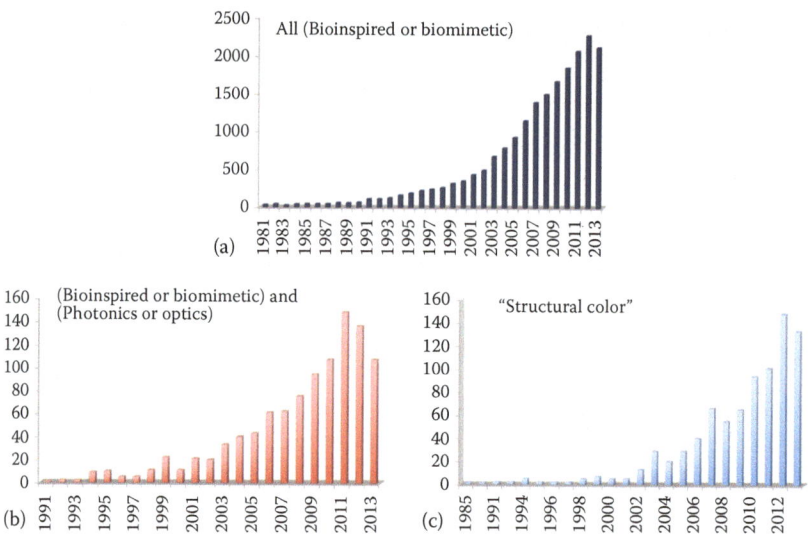

Figure 1.15 Examples of the increasing interest in bioinspired and biomimetic science and technology. Number of journal articles per year in (a) all bioinspired and biomimetic articles. Search term = (bioinsp* or biomim*). (b) Structural color. Search term = "structural color*." (c) Bioinspired or biomimetic optics and photonics articles. Search term = (bioinsp* or biomim*) and (optic* or photon*). Search and analysis via Thomson Reuters Web of Knowledge as of January 2014.

capture the imagination; new journals such as the IOP's *Bioinspiration and Biomimetics* and dedicated conferences and symposia such as the Bioinspired, Biointegrated, Bioengineered Photonic Devices conference of SPIE Photonics West, the Bioinspiration Conference, or topical materials symposia at the Materials Research Society meetings increase the visibility of the field. Research in this area firmly straddles the basic research domain of trying to understand these systems, their structures, and properties and the applied research domain of building devices and systems based, in part, on nature. We have begun to identify the most interesting and challenging aspects of natural systems. Indeed, it is becoming widely recognized that science at the intersection of biology and materials is poised to have a dramatic impact on S&T development in the coming years. The enticing combination of our burgeoning understanding of diverse natural photonic systems and tools developed for nanomaterials' study and fabrication brings this field to a critical point where innovative application can begin.

References

1. N. Lambert, Y. N. Chen, Y. C. Cheng, C. M. Li, G. Y. Chen, and F. Nori, *Nat. Phys.* **9** (1), 10–18 (2013).
2. M. F. Land and D. Nilsson, *Animal Eyes*. (Oxford University Press, Oxford, 2002).
3. B. J. Glover and H. M. Whitney, *Ann. Bot.* **105** (4), 505–511 (2010).
4. A. R. Parker, *J. Opt. a-Pure Appl. Opt.* (2000).
5. A. E. Seago, P. Brady, J. P. Vigneron, and T. D. Schultz, *J. R. Soc. Interface* **6 Suppl 2**, S165–S184 (2009).
6. A. L. Campbell, R. R. Naik, L. Sowards, and M. O. Stone, *Micron* **33** (2), 211–225 (2002).
7. A. D. Briscoe, *Integr. Comp. Biol.* **49**, E21–E21 (2009).
8. R. DeVoe, *J. Gen. Physiol.* **59**, 247–269 (1972).
9. I. C. Cuthill, J. C. Partridge, A. T. D. Bennett, S. C. Church, N. S. Hart, and S. Hunt, *Adv. Study Behav.*, **29**, 159–214 (2000).
10. D. L. Cowles, J. R. Van Dolson, L. R. Hainey, and D. M. Dick, *J. Exp. Mar. Biol. Ecol.* **330** (2), 528–534 (2006).
11. J. Marshall, T. W. Cronin, and S. Kleinlogel, *Arthropod Struct. Dev.* **36** (4), 420–448 (2007).
12. R. E. Blankenship, *Plant Physiol.* **154** (2), 434–438 (2010).
13. J. Xiong, *Genome Biol.* **7** (245) (2006).
14. Much of the information in this section and other excellent resources on evolution can be found: http://evolution.berkeley.edu/evolibrary/home.php.

15. D. M. Hansen, T. Van der Niet, and S. D. Johnson, *Proc. R. Soc. B* **279** (1729), 634–639 (2012).
16. J. R. Paterson, D. C. Garcia-Bellido, M. S. Y. Lee, G. A. Brock, J. B. Jago, and G. D. Edgecombe, *Nat.* **480** (7376), 237–240 (2011).
17. *Anomalocaris model by Gaetan Lee. Licensed under Creative Commons Attribution 2.5 via Wikimedia Commons*, http://commons.wikimedia.org/wiki/File:Anomalocaris_model.jpg.
18. *Anomalocaris predator model by Yinan Chen. Licensed under the Creative Commons Public Domain Dedication*, http://commons.wikimedia.org/wiki/File:Gfp-anomalocaris-predator.jpg.
19. M. F. Land and R. D. Fernald, *Ann. Rev. Neurosci.* **15**, 1–29 (1992).
20. M. R. Smith, *Palaeontology*, n/a-n/a (2013).
21. M. E. McNamara, D. E. Briggs, P. J. Orr, H. Noh, and H. Cao, *Proc. Biol. Sci.* (2011).
22. M. E. McNamara, D. E. Briggs, P. J. Orr, S. Wedmann, H. Noh, and H. Cao, *PLoS Biol.* **9** (11), e1001200 (2011).
23. J. Vinther, D. E. Briggs, J. Clarke, G. Mayr, and R. O. Prum, *Biol. Lett.* **6** (1), 128–131 (2010).
24. W. Wichard, A. Gras, H. Gras, and D. Dreesmann, *Entomol. Gen.* **27** (3–4), 223–238 (2005).
25. Q. G. Li, K. Q. Gao, J. Vinther, M. D. Shawkey, J. A. Clarke, L. D'Alba, Q. J. Meng, D. E. G. Briggs, and R. O. Prum, *Sci.* **327** (5971), 1369–1372 (2010).
26. G. Kuhl, D. E. G. Briggs, and J. Rust, *Sci.* **323** (5915), 771–773 (2009).
27. R. Feuda, S. C. Hamilton, J. O. McInerney, and D. Pisani, *Proc. Natl. Acad. Sci. USA* **109** (46), 18868–18872 (2012).
28. G. Siebke, P. Holik, S. Schmitz, M. Lacher, and S. Steltenkamp, in *Bioinspiration, Biomimetics, and Bioreplication 2013*, edited by R. J. Martin Palma and A. Lakhtakia (Spie-Int Soc Optical Engineering, Bellingham, 2013), Vol. 8686.
29. A. Rees, *The Cyclopaedia or Universal Dictionary of Arts, Sciences, and Literature*. (Longman, Hurst, Rees, Orme and Brown, London, 1819).
30. Aristotle, *Historia Animalium*. (http://classics.mit.edu/Aristotle/history_anim.mb.txt, 350 BCE).
31. Authorship of text contested. http://penelope.uchicago.edu/Thayer/E/Roman/Texts/Aristotle/de_Coloribus*.html.
32. Aristotle, http://classics.mit.edu/Aristotle/meteorology.mb.txt (350 bce).
33. P. T. Elder, Naturalis Historia. (c. 77–79 CE).
34. R. Hooke, *Micrographia*. (Project Gutenberg, http://www.gutenberg.org, 1669).
35. I. Newton, *Opticks*. (Project Gutenberg, http://www.gutenberg.org, 1704).
36. J. Lee, *J. Sib. Fed. U. Biology* **3**, 194–205 (2008).
37. A. Parker, *In the Blink of an Eye: How Vision Sparked the Big Bang of Evolution*. (Basic Books, New York, 2004).
38. T. F. Anderson and A. G. J. Richards, *J. Appl. Phys.* **13**, 748–758 (1942).
39. L. P. Biro and J. P. Vigneron, *Laser Photon. Rev.* **5** (1), 27–51 (2011).

40. J. D. Joannopoulos, S. G. Johnson, J. N. Winn and R. D. Meade, *Photonic Crystals: Molding the Flow of Light 2nd Edition*. (Princeton University Press, Princeton, 2008, accessed, 2014) http://jdj.mit.edu/book/.
41. E. Yablonovitch, *Phys. Rev. Lett.* **58** (20), 2059–2062 (1987).
42. S. John, *Phys. Rev. Lett.* **58** (23), 2486–2489 (1987).
43. S. Vignolini, E. Moyroud, B. J. Glover, and U. Steiner, *J. R. Soc. Interface* **10** (87) (2013).
44. P. Vukusic and D. G. Stavenga, *J. R. Soc. Interface* **6**, S133–S148 (2009).
45. Xradia, http://www.xradia.com.
46. M. S. Joens, C. Huynh, J. M. Kasuboski, D. Ferranti, Y. J. Sigal, F. Zeitvogel, M. Obst, C. J. Burkhardt, K. P. Curran, S. H. Chalasani, L. A. Stern, B. Goetze, and J. A. J. Fitzpatrick, *Sci. Rep.* **3**, 7 (2013).
47. S. A. Boden, A. Asadollahbaik, H. N. Rutt, and D. M. Bagnall, *Scanning* **34** (2), 107–120 (2012).
48. R. A. Potyrailo, H. Ghiradella, A. Vertiatchikh, K. Dovidenko, J. R. Cournoyer, and E. Olson, *Nat. Photonics* **1** (2), 123–128 (2007).
49. P. Vukusic, R. Kelly, and I. Hooper, *J. R. Soc. Interface 6 Suppl 2*, S193–201 (2009).
50. L. M. Mathger, E. J. Denton, N. J. Marshall, and R. T. Hanlon, *J. R. Soc. Interface* **6 Suppl 2**, S149–163 (2009).
51. L. M. Mathger and R. T. Hanlon, *Cell Tissue Res.* **329** (1), 179–186 (2007).
52. J. P. Vigneron, J. M. Pasteels, D. M. Windsor, Z. Vertesy, M. Rassart, T. Seldrum, J. Dumont, O. Deparis, V. Lousse, L. P. Biro, D. Ertz, and V. Welch, *Phys. Rev. E* **76** (3) (2007).
53. K. L. Yu, T. X. Fan, S. Lou, and D. Zhang, *Prog. Mater. Sci.* **58** (6), 825–873 (2013).

2

STRUCTURAL COLOR I

Low-Dimensional Structures

2.1 Next Generation Applications Inspired by Ancient Structures

There is no denying that the colors seen in nature—the saturated greens of summer, the riot of color in a field of wildflowers, and warm browns and oranges of leaves in the fall—can be breathtakingly beautiful. Animals, too, can display a wide variety of colors; land animals, for example, display numerous shades of browns, yellows, oranges, and reds. Many of these colors come from biological pigments such as melanin, which produces browns, blacks, and some reds; chlorophyll, for greens; and carotenoid, giving red, orange, and yellow.[1] These pigments each selectively absorb specific wavelengths of light and selectively reflect others, resulting in the colors we see. But among the sea of greens and browns of nature, we find examples of different and more startling colors: electric blues, ultravivid greens, and purples surprising in their intensity; metallic colors in organisms that contain no metal; and iridescent colors that shimmer and shift as the organism or observer moves. Examples, such as those in Figure 2.1, are found not only in one species, or even in one class of organisms (such as insects), but across a wide variety of animals and plants, from mammals to fruit, from insects to flowers, vertebrates and invertebrates, in marine environments, and on land, indicating that nature has come to this solution over and over again.[2–7] The optical effects are the result of sophisticated micro- and nanostructures comprising or embedded in the outer cells and protective outer coverings (i.e., integument, e.g., skin) of these organisms. This is structural color, and the nanostructures responsible are stunning in their ability to produce a wide range of vivid optical effects.

Biological photonic structures are fascinating to many scientists and technologists because these structures are ancient, having first evolved millions of years ago, and yet they are directly analogous to

Figure 2.1 Various examples of structural color. (a) *Pollia condensata* fruit. (b) *Morpho didius* butterfly. (c) White *Cyphochilus* beetle. (d) *Chrysina aurigans* beetle. (e) Jumping spider (Salticidae). (f) *Troides magellanus* butterfly. (From Vignolini, S. et al., *Proc. Natl. Acad. Sci. USA*, 109(39), 15712–15715, 2012; Vukusic, P. et al., *Sci.*, 315(5810), 348, 2007; Campos-Fernandez, C. et al., *Opt. Mater. Express*, 1(1), 85–100, 2011; and Van Hooijdonk, E. et al., *J. Appl. Phys.*, 112(7), 8, 2012.[12,13,16,77])

the synthetic photonic crystals we create in the laboratory. The optical properties of photonic crystals emerge from the combination of length-scale, periodicity, and dimensionality of the structure as well as the refractive indices of the constituent materials. By controlling the fabrication and material properties of the nanophotonic structures, we can completely control the optical properties, making them highly tunable for application requirements. In addition, we can create materials with new properties and novel functionalities that the constituent materials may never have otherwise possessed. Nanophotonics will play a key role in the development and advancement of technologies in the coming decades, and synthetic photonic nanostructures are being explored for a wide range of applications from paints and cosmetics to the building blocks for optical computers.[8] The direct analogy between synthetic and biological photonic structures allows us to learn critical lessons from natural systems and apply those lessons in the laboratory. Natural biological photonic structures are sophisticated and complex composite optical systems, often combining multiple structures and materials. The intricacy of these systems, complexity of their optical response, and replication of the details of their design are something we are not yet able to achieve with current fabrication techniques for visible wavelengths. A great deal of research is being performed to understand, design, and fabricate synthetic photonic crystals. Natural structures can provide us with design blueprints for future synthetic optical structures. So far, these structures have inspired novel approaches and applications in lighting, displays and sensors, solar energy harvesting, and components for optical computing, to name just a few. Structural color is a very attractive approach for paints, displays, and cosmetics, for example, as it can create intense colors with interesting optical effects that are photostable for longer periods than many dyes and paints (they would not bleach). That biological photonic crystals are constructed from organic materials offers other avenues of inspiration, as organic materials can be flexible, and biodegradable, and the fabrication techniques are often less costly than for inorganic materials. Medical biophotonics is one area where these features might be extremely useful as the biological materials from which these are made can also be nonimmunogenic (biocompatible), opening the door for a host of possible applications in in vivo sensing and diagnostics. As our knowledge of biological photonic systems becomes increasingly sophisticated, and our ability to fabricate nanophotonics systems at the resolutions required becomes more capable and scalable, we begin to apply our lessons learned from structural color in bioinspired photonic systems.

2.2 Sparkly, Vibrant, Bright, and Shiny—Light and Biology in Action

The bright, saturated, and sparkling colors that emerge from structural color arise from refraction, scattering, diffraction, and interference, or a combination of these effects, as light travels through and reflects from these biological photonic crystals. Evolutionary pressure has resulted again, and again, in systems that achieve a tremendous variety of optical properties and effects (in Figure 2.1) from a comparatively limited palate of materials.[9] Structural color is most often associated with iridescence, in which the color changes with viewing angle. Iridescence can be seen in the rainbow-like shimmer on the feathers of some birds, in flowers such as the Queen of the Night tulip, or in the shifting color of butterfly wings such as those of the *Morpho* genus.[10,11] Some biological photonic structures, such as those seen in the gold and silver beetles *Chrysina aurigans* and *Chrysina limbata*, create metallic broadband reflection.[12] Other structures are specifically designed to produce a matte reflection, such as the white *Cyphochilus* beetle.[13] These structures can produce highly directional signals so that the effect can only be seen when at a specific viewing angle, such as the *Pierella luna* butterfly.[14] Polarization effects found in biological systems are very interesting to us because we cannot perceive them with the human eye. Many other organisms, such as some insects, marine animals, and arachnids, can utilize polarization as another communication channel. Polarization effects have been found in fruits such as the *Pollia condensate* and in beetles such as the *Chrysina gloriosa*.[15,16] Some structures exist to enhance the brightness of existing pigmentary or fluorescent color, such as the *Troides magellanus* butterfly and *Phelsuma* lizards.[17,18] These effects and their production are all achieved with biological photonic structures and evolutionarily adapted for habitat, behavior, and biological cost. In this chapter, as well as in Chapter 3, we will be focusing on static (passive) color structures. Some organisms are able to actively control their optical display, and we will be discussing those in Chapter 4.

The function of these systems is inextricably linked to the organism's behavior as well as the interplay between the organism and its environment. Optical effects play a very important role in animal communication to enhance reproduction, and to avoid predation.[19] Structural color displays can be found over the entire visible wavelength range as well as the ultraviolet (UV) and are used in both intra-specific (within species) and inter-specific (between species) signaling. The optical effects that arise from structural color can function as a way for conspecifics (two members of the same species) to recognize each other. They can function as sexual communication where the color and intensity are good indicators of, for example, the sex of the organism, its health, and its suitability as mate. Many of the color markings on animals are sexually dimorphic; members of each sex have distinctive gender-specific markings. The intensity of the color may relate to the organism's development, diet, or even wear and tear to the structures. Because these are very complex biological systems, there is a significant biological "cost" to an organism's development and maintenance of these color displays; it is thought that structural color functions as an "honest indicator" of health, as only healthy individuals would be able to pay the cost for high-intensity signaling. Structural color effects can provide camouflage functionality (called crypsis) either allowing the organism to blend in to its environment, hide its size by blurring its shape and outline, or enable mimicry of other organisms. Optical effects can serve to make the organism more visible (aposematic displays) and can function as a way to startle or warn away predators (e.g., warn of toxicity), or provide a channel of communication that enables group behavior such as flock or school formation. They can play a role in agonistic displays (aggressive displays) during mating to fight off competitors, or indicate territoriality.

In some cases, the optical function of the structure may be a beneficial side effect of structures evolved for other purposes. The functions of some structures on beetles (order Coleoptera) and butterflies (order Lepidoptera), for example, may be more specifically evolved for thermal regulation, rather than display. Diffraction gratings prevalent among species of snakes, lizards, and certain beetles may contribute to improved movement through swamp or mud-based habitats, either through friction reduction or by providing additional support for forward motion in dense or compressed environments.[2] Iridescent structures over dark pigments are often seen in bird wings near the tips of the flight feathers, contributing significantly to their robustness, whereas the blue iridescent multilayer structures in certain shade-growing plants, such as *Begonia pavonina*, seem to enable resistance to photodamage.[4,20–22] Structural color can also be used by plants to entice pollinators to visit.[4]

Structural color has been more widely studied in some organisms than in others and there are many excellent review articles to be found. Beetles and butterflies are the most widely studied, due in part to their flashy, conspicuous color, particularly specimens with seemingly metallic or iridescent colors.[2,5] Birds and fish have been somewhat well studied, while building our knowledge of many other organisms such as spiders, plants, and other insects seems to have barely begun.[4,23–25] Exploration of structural color in fossils is another area that is growing in publications recently and is a potentially rich area of study.[26]

2.3 Describing Biological Photonic Structures

The photonic systems found in biological organisms range from simple, quasi-ordered structures to highly ordered complex architectures. They are typically constructed of a biomaterial, which acts as a high index of refraction material, and air, which acts as the low index of refraction material. The biomaterials are typically keratin or chitin, although the specific composition is often unknown. Identifying the individual component structures, what fundamental physics is involved in generating the optical response, and how the structure's functions combine to create the observed optical properties is challenging. Biological photonic structures are complex and often formed of multiple integrated coincident structures. The structures may consist of components on several length-scales from micrometer to nanometer sizes, each component contributing to some aspect of the spectral response. Scattering elements often exist within an interference structure to alter the reflection from being highly directional to being much more diffuse. They can act in combination with pigments, either to enhance the pigmentary color or the pigment will enhance the structural color by absorbing stray light at undesirable wavelengths. Synthetic photonic crystals strive to create perfect, idealized structures with well-known, predictable physics. However, biology is anything but pristine, idealized, and predictable, and the disorder in the structures creates added complexity in the optical response, as well as complicating our ability to characterize them.[5,27] In fact, disorder may be a great lesson to learn from biological structures, and learning to harness that disorder for synthetic systems may offer potential advantages.[5]

When describing the physical characteristics of these systems, coming up with a common language has been a slowly evolving task, due in large part to the diverse fields of the scientists investigating and describing these systems. Over the last few decades, however, as the pace of discovery has increased, more examples have been found and several different potential classification schemes have been described. Ghiradella proposed a classification scheme from a biology perspective, and much of the language we use to describe the physical structures on the surface of the butterfly scales, specifically, comes from her work (see Figure 2.2).[5,28–30] Beetle classification was proposed by Seago et al.[2] How we describe and quantify iridescence has also been evolving, with a recent proposal by Meadows et al.[31] Given the structural similarity of the biological photonic structures to synthetic photonic crystals, it is natural to describe them in common terms. Photonic crystals themselves borrow their classification from electronic materials and classical crystallography. We describe the biological structures as photonic crystals in the broad sense, in that they are ordered and most often periodic optical structures.[5] We will describe them by the dimensionality of the structure (1D, 2D, and 3D), the periodicity and degree of order within the structure, and the primary physical effect driving the optical response.

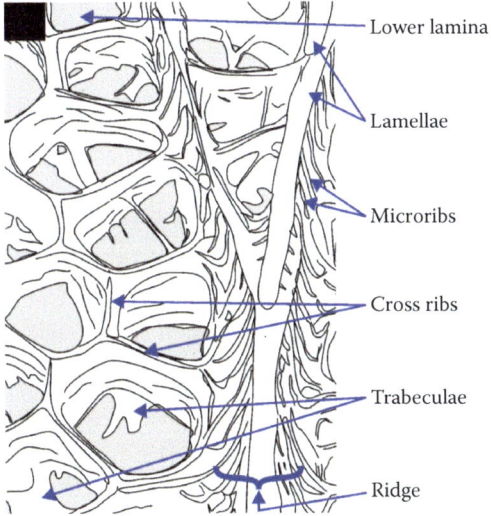

Figure 2.2 Annotated sketch of *Papilio ulysses* structures with identifying features on the wing scale using nomenclature from the work of Ghiradella (1985, 1989). (From Boden, S.A. et al., *Scanning*, 34(2), 107–120, 2012. With permission.[78])

2.4 One-Dimensional Layered Structures

One-dimensional (1D) photonic structures are the simplest photonic structures and are biologically comparatively easy for organisms to form. Idealized 1D structures only vary periodically in a single direction, either as a single layer thin film or as a periodic multilayer (essentially a stack of planar layers). These layered structures can be found widely in nature. The reflective properties of the film depend on the optical path the light takes from source to observer, as described by the Bragg conditions for constructive and destructive interference.[32] The thickness of the planes, the refractive indices of the different material layers, and the angle of the incoming light are the key properties responsible for the spectral properties of the reflectance.

2.4.1 Single Layer Thin Films

Most people are familiar with the optical behavior of thin films having seen the interference patterns in oil on water or in the changing colors on a soap bubble. Iridescence arising from single layer thin films can be found on the wings of many flying insects, such as wasps, sawflies, ants, and bees (order Hymenoptera). Unlike some of the organisms we will be discussing further on in this chapter, structural color in insects of this order has not yet been comprehensively studied. What has been examined, however, is the iridescence found on their wings, which is primarily driven by thin-film interference.

The giant wasp *Megascolia procer javanensis* is a large Javanese insect, approximately 5 cm long.[33] This distressingly large wasp is parasitic and reproduces by laying its eggs onto Scarabaeid larvae. The wasp's wings (seen in Figure 2.3) are opaque and darkly colored with long narrow ridges along the length of the wing. The wings reflect iridescent blue and green, particularly at large viewing angles. Scanning electron microscopy (SEM) images taken of the wing cross section show that the wing is made up of many layers. Most of these layers likely provide mechanical stability and structure to the wing. The body of the wing contains melanin, which acts as an optical absorber and gives the wing its black coloring. However, the thickness of the layers in the multilayer stack varies across the wing (from 400 nm to 1 μm). These layers are likely to play little role in the iridescence of the wing given the variation in layer thickness and lack of spectral peaks

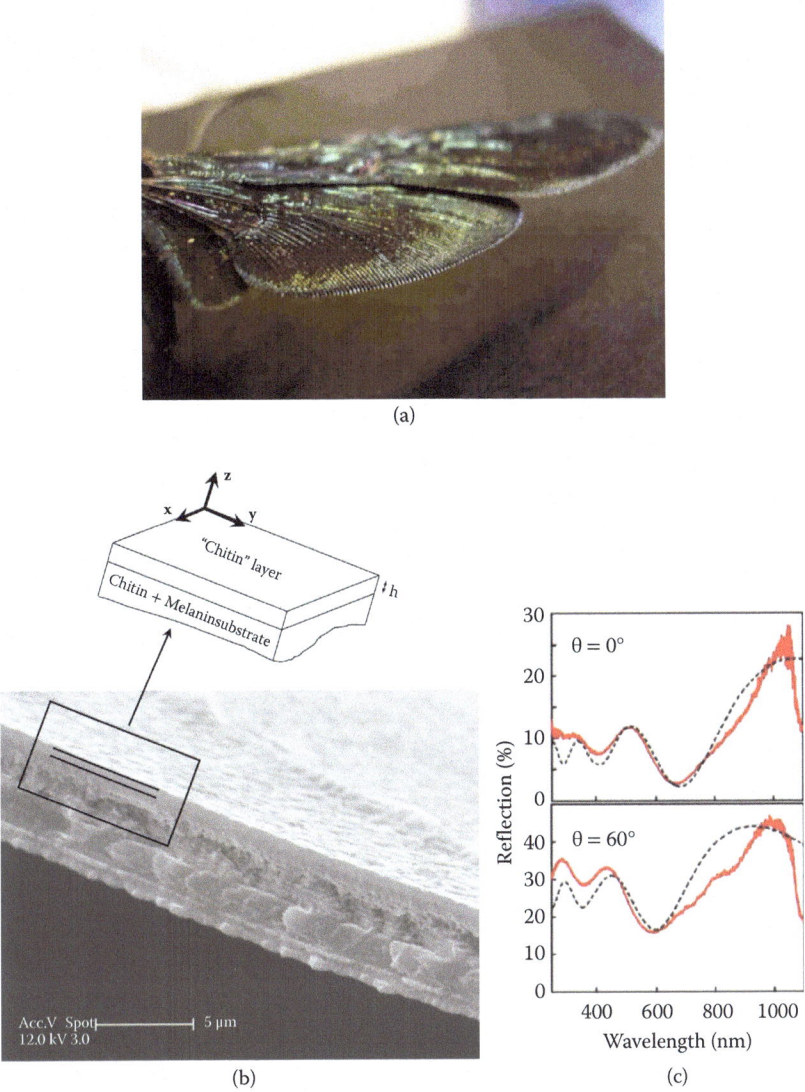

Figure 2.3 (a) Wing of a male *Megascolia procer javanensis* (Hymenoptera) where blue/green/purple iridescence can be seen. (b) Cross section of *M. p. javanensis* wing via SEM and model used for simulation. (c) Wing reflection as a function of the wavelength for θ = 0° and 60°. Experimental data (solid line) and simulations results (dashed line). (From Sarrazin, M. et al., *Phys. Rev. E*, 78(5), 5, 2008. With permission.[33])

that would be expected. The structure that plays the primary role is the topmost layer. This top layer is thin, approximately 300 nm thick, and acts as a thin-film reflector. Reflectance spectra indicate that the wing reflects over three relatively broad wavelength ranges centered at 325 nm, 505 nm, and 1015 nm, corresponding to UV, blue/green, and infrared wavelengths, respectively. Modeling this system as a thin biopolymer layer over a larger mechanical layer stack resulted in a calculated reflectance spectrum where the peak reflectance wavelengths match experimental data fairly well. This suggests that the top layer acts as an optical filter. The dark coloring that results from the melanin makes it easier to see the iridescent shimmer on the wings. The model was not a perfect match for the experimental data, but it is possible that other physical features, such as the large-scale roughness of the surface of the wing (as can be seen in the SEM image), may also play a role. Thin films like this, which can also be found in birds, for example, are a relatively low biological cost way for nature to add iridescence to a pigment-based or transparent color system.

Small changes in the film thickness result in variations in the wavelengths reflected from the film. Much like the iridescence of a soap bubble or oil layer, it was assumed that the color of an iridescent wing would be variable and random for any given individual. However, that does not seem to be the case. A comprehensive study out of Lund University in Sweden has demonstrated that the wing interference patterns (WIP) in transparent winged insects (Hymenoptera and Diptera [flies]) are stable and taxon specific.[34] The wing interference patterns are heritable traits and are similar in all members of a given taxon. The wings in these insects primarily consist of two layers of transparent chitin compressed into a single membrane (perhaps combined with pigment, such as in the wasp discussed previously). The wing membrane varies in thickness from approximately 300 to 600 nm, on the same size scale as the wavelengths of visible light. The resulting wing interference patterns show a complexity driven by film thickness, corrugation of the surface, veination, hair placement, and pigmentation. Examples for Diptera are shown in Figure 2.4 and are fascinating in the variety of color and pattern.[34] The specific patterns are hard to see in bright white light, but when seen on the wing against a dark, opaque background, such as a dark green leaf, the iridescent pattern becomes vibrantly apparent and may play a role in species differentiation and mate selection.

Similar reflective patterning can be seen on the transparent sections of the wing of the glasswing butterfly, *Greta Oto*, but the specific patterns, and their conservation on different specimens has yet to be studied (Prof. Caroline Schauer, personal correspondence).

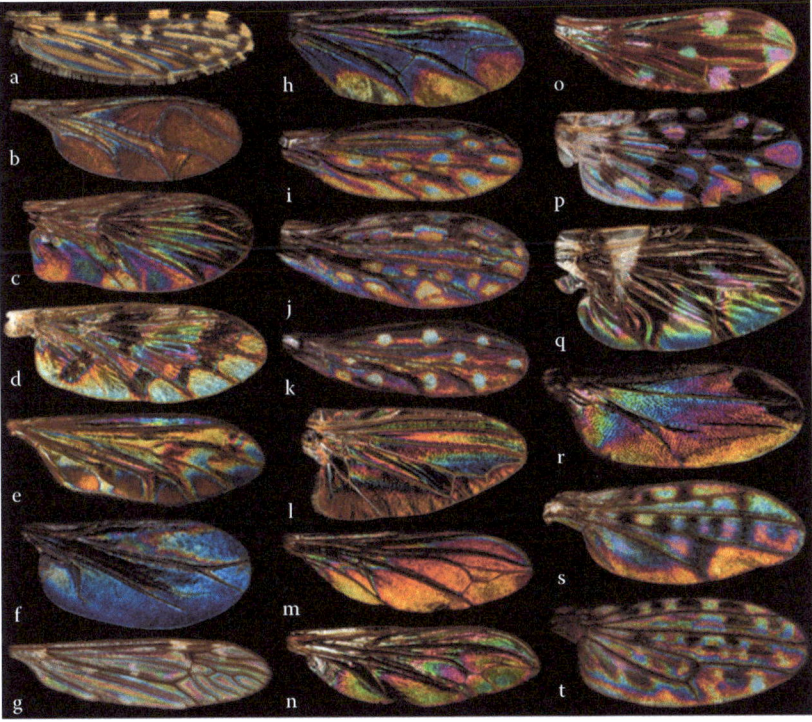

Figure 2.4 WIP diversity across Diptera. (a–g) displays lower Diptera ("Nematocera"), (h–n) displays lower flies. (o–t) displays higher flies (Acalyptrata). (a) Culicidae, *Anopheles melas* (female, Ghana). (b) Sciaridae, *Zygoneura* sp. (male, Japan). (c) Keroplatidae, *Macrocera fascipennis* (male, Sweden). (d) Keroplatidae, *Proceroplatus* sp. (male, Honduras). (e) Lygistorrhinidae, *Lygistorrhina pictipennis* (male, Japan) (f) Scatopsidae, *Swammerdamella brevicorne* (female, Sweden). (g) Tipulidae, *Tipula confusa* (male, Sweden). (h) Dolichopodidae, *Condylostylus* sp. (female, Canada). (i) Empididae, *Dolichocephala guttata* (female, Sweden). (j) Empididae, *Dolichocephala irrorata* (female, Sweden). (k) Empididae, *Dolichocephala ocellata* (male, Sweden). (l) Platypezidae, *Paraplatypeza atra* (male, Sweden). (m) Pipunculidae, *Eudorylas obscurus* (male, Sweden). (n) Pipunculidae, *Nephrocerus scutellatus* (male, Sweden). (o) Diopsidae, *Teleopsis rubicunda* (male, Philippines). (p) Tephritidae, *Actinoptera discoidea* (male, Sweden). (q) Tephritidae, *Rhagoletis pomonella* (male, USA). (r) Chloropidae, *Chloropsina* sp. (female, Malaysia). (s) Ephydridae, *Paralimna* sp. (female, Ghana). (t) Ephydridae, *Limnellia quadrata* (male, Sweden). (From Shevtsova, E. et al., *Proc. Natl. Acad. Sci. USA*, 108(2), 668–673, 2011. With permission.[34])

2.4.2 Simple Multilayers

At their most basic, multilayer structures can be described as a series of stacked layers, where the layers alternate between materials with comparatively high refractive index (RI) and low refractive index (e.g., chitin and air). Multilayers in various forms are relatively common in structural color, found in insects, fish, birds, plants, and fruits.[6] One example of animal-based structural color from a basic multilayer stack can be found in the jumping spider family (Salticidae).[35] Structural color has been found in spiders at wavelengths ranging from UV to yellow, although relatively few structural color studies have been performed so far.

In both sexes of the North American red-backed jumping spider (*Phidippus johnsoni* [Peckham and Peckham, 1883]), the long, protruding mouthparts called chelicerae are vibrant iridescent green, as seen in Figure 2.5.[35] The chelicerae reflect green light at around 520 nm, and the green tends toward bronze as the surface curves. (This 520-nm peak is actually the first order [$m = 1$] optical maximum of the stack as the zeroth order [$m = 0$] resonance peak actually occurs in the infrared.) The cuticle surfaces of the chelicerae are smooth and relatively featureless. The cross section seen by tunneling electron microscopy (TEM) in Figure 2.5b, however, shows an 86-layer multilayer stack made of (most likely) alternating chitin and air. The thickness of the individual layers shows large variation across the chelicera (236 ± 75 nm and 135 ± 38 nm for the dark and light layers in the TEM image, respectively). Modeling indicated that primarily the first 10 or so layers contribute to the bulk of the optical reflectance properties.

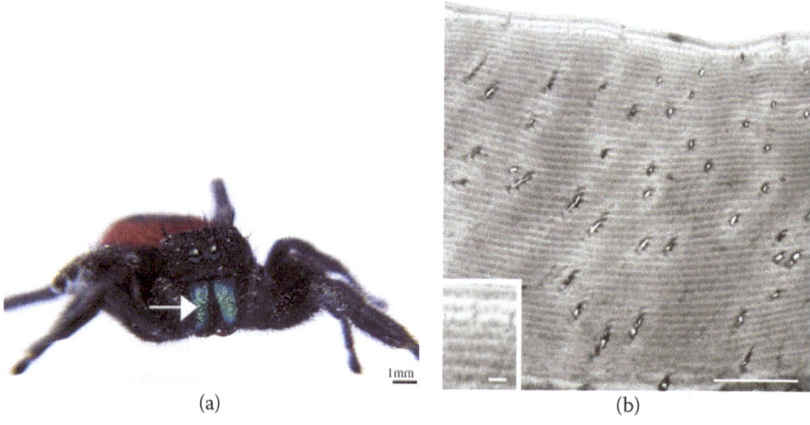

Figure 2.5 (a) Front view of adult female *Phiddipus johnsoni* (Salticidae). Arrow indicates green chelicera. (b) Cross-sectional view of chelicera showing upper layers of a large multilayer stack with 86 total layers (scale bar: 1 μm). Inset shows magnified section of the outermost region of the basal chelicera segment (scale bar: 0.1 μm). (From Ingram, A.L. et al., *Arthropod Struct. Dev.*, 40(1), 21–25, 2011. With permission.[35])

Multilayers may produce not only iridescent color, but they can also enhance pigment-based color. Tropical *Phelsuma* geckos (seen in Figure 2.6) have multilayer structures that supplement the reflective properties of pigment cells in their skins.[17] The multilayers in these geckos contain either ordered or disordered guanine nanocrystals. Blue and sometime green color in these animals is due to multilayer interference from the ordered nanocrystal multilayers. In some cases, the green skin color may be due to color mixing from yellow pigmentary cells combining with blue multilayer reflection. The disordered nanocrystal multilayers act as broadband scatterers and are responsible for both creating white coloring spots and enhancing the reflectivity of red spots. It has been noted that the developmental mechanisms behind the specific patterning of organized versus disorganized multilayer as well as the tremendous colocalization with specific pigment cells in these geckos are unknown. Guanine inclusions in structural color elements are not that rare and have also been found as a component in spider and fish structural color.[23]

For layered films (both single layer and multilayer reflectors), the simplicity and relative ease of fabrication makes basic layered structures attractive and already widely used in industrial, laboratory, and commercial applications. 1D photonic films find use as optical filters; sensors; coatings on solar cells, automobiles, and windows; and other high-quality optical coatings. There are many fabrication methods used to create these films, depending on the materials from which

Figure 2.6 (a) Gecko of genus *Phelsuma*. (b) Example SEM of guanine multilayers in skin of different colors. Ordered guanine crystals found in green/blue skin, disordered guanine crystals in red and white skin. Scale bar: 1 μm. (From Saenko, S.V. et al., *BMC Biol.*, 11, 12, 2013. With permission.[17])

they are being made, such as physical or chemical vapor deposition techniques, spin-coating, and more. Multilayers grown in the laboratory tend to have nearly pristine, perfectly ordered structure, with highly controlled dimensions, very even layer thickness, and precise composition. They result in highly specular (i.e., mirrorlike) reflections, but will be so in only a small range of wavelengths (i.e., narrowband), and the color will be highly incident angle dependent, as the optical path length will change with angle.

One feature that is interesting in natural multilayered films found in nature is the repeatable nonperiodic variations in thin-film thickness that construct the wing interference patterns in flies and wasps. Nature creates these patterns repeatedly generation after generation. Our current fabrication techniques find it difficult to handle nonperiodic but nonrandom structures in any scalable fashion. Applications for bioinspired multilayers may evolve over time as our fabrication techniques catch up to nature. These applications may be as simple as novel packaging features, such as barcodes, or time-dependent degradable markings, or as coatings for decorative patterned iridescence, and as complex as security features and others we can only imagine.

Organic multilayers have many of the same benefits and challenges as biological multilayers. The range of indices of refraction achievable is also more limited than with inorganic materials. However, the processing is comparatively low temperature and they can be biocompatible, flexible, and soft. The deformability of the layers in a multilayer stack can therefore be used to create mechanochromic photonic gel systems, where mechanical deformation changes the optical path.[36] Researchers are investigating ways to include other materials into organic multilayers, much like the guanine crystal multilayers of the *Phelsuma* gecko. Hybrid multilayers were fabricated incorporating inorganic materials, such as titania (TiO_2) alternating with layers of organic polymers.[37] The polymer layers can be reversibly swollen without affecting the thickness of the TiO_2 layer, creating a controllable color change.

Rolled multilayer fibers were inspired by similar structures found in the seed coats of the *Margaritaria nobilis* fruit.[38] The photonic structures in this fruit are arrays of cylindrical cells, each of which is made of a multilayer structure that forms concentric rings around a central axis, as seen in Figure 2.7. This inspired the development of a photonic fiber that is formed by a multilayer rolled around a central core.

STRUCTURAL COLOR I

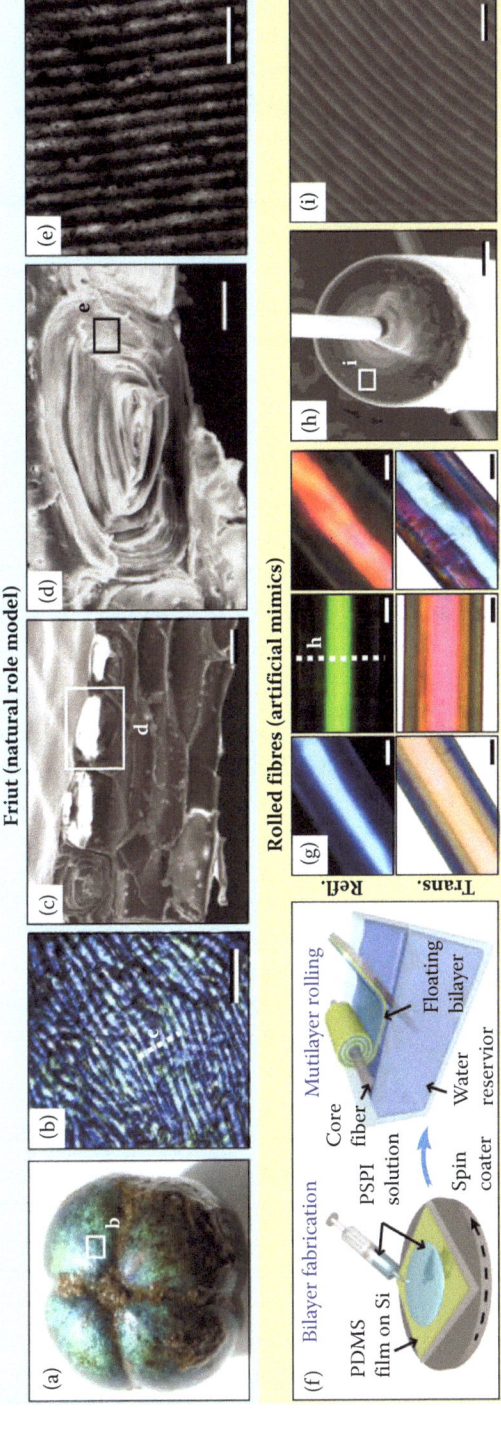

Figure 2.7 (a) *Margaritaria nobilis* fruit without its capsule (~10 mm diameter). (b) Optical image of the fruit surface with elongated blue cells. Scale bar: 200 µm. (c) SEM cross section through the outer layers of the fruit's endocarp. Scale bar: 20 µm. (d) Cross section through a single tissue cell showing a concentric flattened cylindrical layered structure. Scale bar: 10 µm. (e) TEM of a section layered structure within a single cell. Scale bar: 500 nm. (f) Illustration of the photonic fiber fabrication process. (g) Optical image of three rolled-up multilayer fibers with different layer thicknesses and colors in reflection (top) and transmission (bottom). Scale bar: 20 µm. (h) SEM cross section showing multilayer cladding around the glass fiber core. Scale bar: 20 µm. (i) SEM of the individual layers in the cladding. Scale bar: 1 µm. (From Kolle, M. et al., *Adv. Mater.*, 25(15), 2239–2245, 2013. With permission.[38])

BOX 2.1 CHALLENGE OF MATERIAL COMPOSITION AND REFRACTIVE INDICES

One way to test whether an organism's color is structural in nature is to dip or drop liquid onto the section under study. If the structure has voids, or layers of air, the infiltration of a liquid will cause the color to change as the air is replaced with the liquid, which has a different index of refraction (see Figure 2.8). Identifying the composition of the biological materials and their properties is key to understanding the photonic response. Unfortunately, our knowledge of the exact material composition of the photonic nanostructure (and their refractive indices) is not yet sufficient to truly accurately model their performance. Literature often states that a given structure is made out of keratin (such as in bird feathers) or chitin (such as in insect cuticles), but the reality is that the exact composition of these materials is often not known. They are likely heterogeneous composites of the comparatively small palate of materials with which nature tends to work: biopolymers, proteins, lipids, etc. The composition may not necessarily even be uniform across a single structure. The structure may incorporate inclusions or other materials such as guanine, uric acid, or silica.[17,23,39,40] Many of these structures also include pigments, either in a separate layer or incorporated into the material of the photonic structure itself. In the example

Figure 2.8 Image of Morpho butterfly with natural color on right. Introduction of ethanol ($n = 1.362$) (top left) and toluene ($n = 1.497$) (bottom left) to the structure resulted in reflective color change. (From Potyrailo, R. and Naik, R.R, In D.R. Clarke [ed.], *Annual Review of Materials Research,* Palo Alto, Annual Reviews, 2013. With permission.[79])

of the jumping spider, the exact composition of the multilayer stack was unknown, although modeling results implied that it is chitin.[35] We also assume that the composition of each layer or discrete structural unit does not change across the system.

Identifying the RI of the materials for use in modeling of the optical properties is a critical issue.[5] The true RI of a system is difficult to accurately measure experimentally. Most calculations make assumptions about the RI of the materials in question. By using liquids of known RI, infiltration can be used to give an estimate of the real RI of the material. Once a liquid with matching index infiltrates the structure, the structural color component of the color display ceases to be visible (assuming that the structure's composition is uniform). Biological materials, such as keratin and chitin, have a very narrow range of refractive indices (~1.4 to ~1.8 for the real component). This means that a relatively small differential between large and small refractive indices can be achieved for any given biological photonic structure. Keratin and chitin have an RI = 1.33 and 1.56, respectively; chitin with uric acid has an RI = 1.684, and guanine crystals have higher RI compared to most other biological materials, with an RI = 1.8.[5,39]

This novel fiber achieves both the high reflectivity of a planar multilayer, but also the large angular range of reflection due to curvature seen in the biological structure. In addition, the fiber can be fabricated out of soft polymer materials and the rigid core removed to create stretchable fibers that now change color with strain.

2.4.3 Chirped Multilayers

As was said above, regular multilayers can result in highly specular reflections, but these reflections will typically be narrowband, and highly angle dependent. Nature creates variations on the basic multilayer structure to achieve other optical effects, such as broadband reflection, by varying the uniformity, periodicity, thickness, and composition of the multilayer stacks. Figure 2.9 shows three examples of alternate multilayer structures

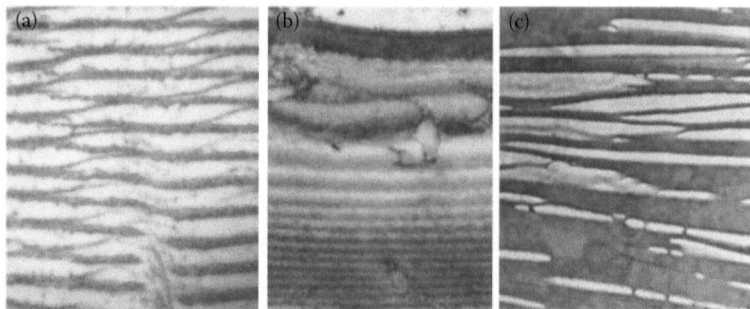

Figure 2.9 Examples of (internal) multilayer reflectors (transmission electron micrographs). (a) A quarter-wave stack with corrugations in the swimming crab *Ovalipes molleri*. The corrugations, not considered in physics, provide interesting variations to the reflector. The branching of the layers are also interesting, introducing stability and robustness to the stack. (b) A "chirped" broadband reflector, or mirror, in an amphipod crustacean (*Danaella* sp.). This mirror is positioned on the antenna and is employed to redirect bioluminescence produced from the body. (c) A "chaotic" broadband reflector in fish skin. In (a) all high-index layers (dark) are equal in thickness; in (b) and (c) the high-index layers (dark and light, respectively) vary in thickness. All layers are of the order 100 nm thick. (From Parker, A.R., *Philos. Trans. R. Soc., A*, 362(1825), 2709–2720, 2004. With permission.[80])

found in animals: a corrugated structure found in the crab *Ovalipes molleri*, a chirped structure found in an amphipod crustacean, and a chaotic multilayer found in fish skin.[39] The king penguin *Aptenodytes patagonicus* has folded multilayer polycrystals in its beak, which create UV reflection.[41,42] Some of the structures, such as corrugation, provide simultaneous optical effect and structural function, where the cross-linking of the layers provides extra physical robustness and flexibility.

Complex multilayers can create fascinating broadband metallic effects as seen in the beetles *Chrysina aurigans* and *Chrysina limbata*, seen in Figure 2.10.[12] Beetles are structurally wildly diverse, found in both living and fossil examples, and are some of the most broadly studied organisms when it comes to structural color.[2] In *C. aurigans* and *C. limbata*, the structure that results in the gold and silver metallic color is called a chirped multilayer.[12] In chirped multilayer reflectors, the physical thickness of each layer decreases with depth in the structure. As the physical thickness changes, the optical path-length changes and so different sections of the chirped structure are effectively tuned to different wavelengths of light, with the deeper layers reflecting smaller wavelengths as the thickness decreases. As the light leaves the surface of the beetle's integument, the wavelengths combine to form broadband color reflection. The multilayer that creates

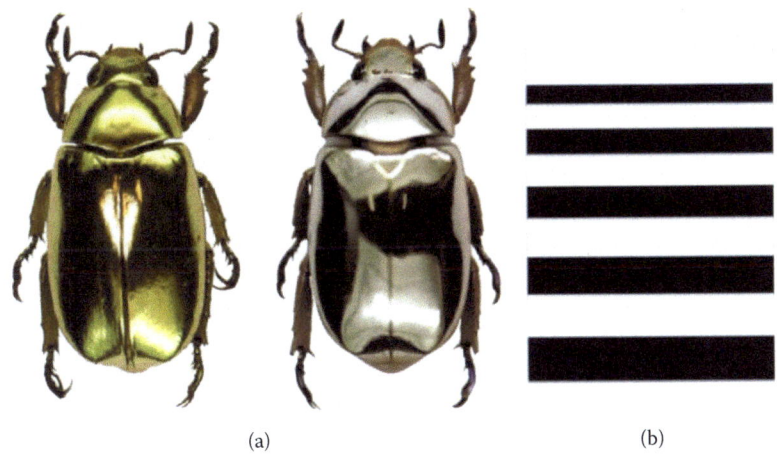

Figure 2.10 (a) Pictures of the golden *Chrysina aurigans* and silver *Chrysina limbata*. Broadband reflection emerges from chirped multilayers in the beetle carapace. (b) Illustration of a chirped multilayer. (From Campos-Fernandes, C. et al., *Opt. Mater. Express*, 1(1), 85–100, 2011. With permission.[12])

the silver color seen in *C. limbata* reflects almost 97% of light across the entire visible wavelength range. The multilayer that creates the gold color, such as with *C. aurigans*, reflects all visible wavelengths except the shortest: blue and violet.

Broadband reflection, specifically angle-independent highly specular reflection, is very attractive for applications. One area in which these reflectors are of interest is in high-efficiency lighting, particularly for white light light-emitting diodes (LEDs).[6] A reflector placed at the back of the LED redirects light that would otherwise be lost, essentially controlling the light emission direction. The laboratory-grown Bragg reflectors used for these applications tend to be high-quality reflectors but reflect only over a narrow range of wavelengths. White light sources would require reflectors able to operate over a much broader wavelength range. White light in light-emitting devices is often created by combining red, green, and blue light. A single multilayer that is capable of reflecting the three specific source wavelengths would also be very attractive. Chirped and other complex multilayer structures may be able to provide alternative inspiration for a host of architectures creating precisely shaped reflection spectra, enabling highly controlled light emission over multiple wavelengths.

Following the need to be able to create reflectors with directly designed spectral properties, researchers were inspired by the disorder and design flexibility found in natural structures and attempted to develop designs for dielectric nanostructures that can display whatever color effect they desire.[43] Using a technique called topology optimization, they specify the desired spectral outcome, and the analytical technique derives the notional reflector structure for given material parameters. Novel structural architectures can be designed without geometrical restrictions, although their lack may sometimes result in structures that are difficult to fabricate using current technologies. The methodology can also introduce manufacturing limitations in the form of geometrical restrictions. Figure 2.11 shows some of the

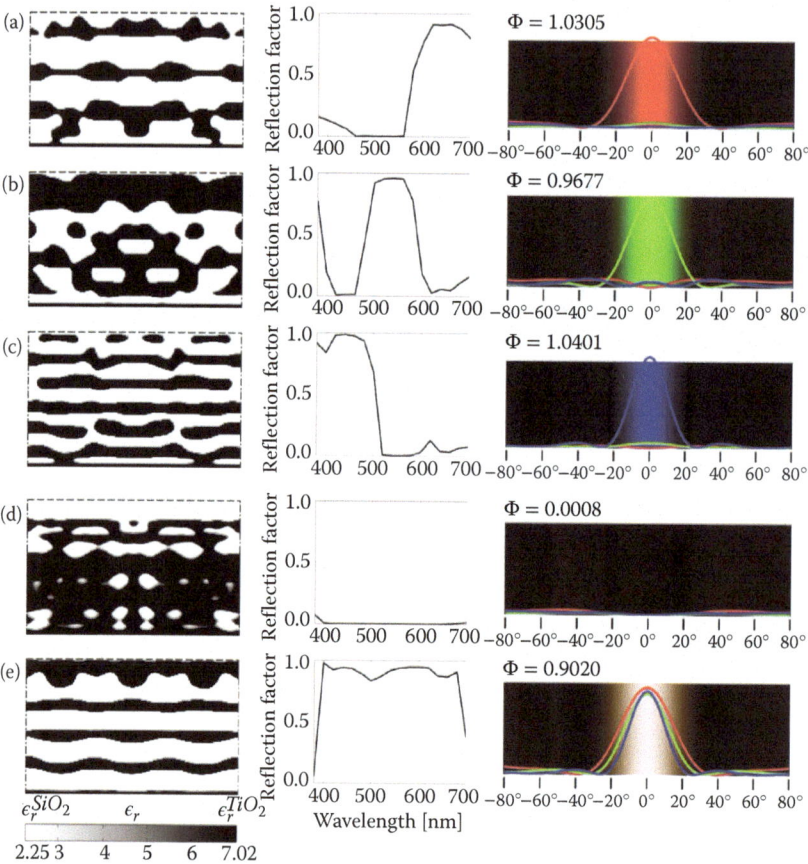

Figure 2.11 Material design for color optimization for 0° observation angle. Material design for (a) red (RGBr = [1; 0; 0]), (b) green 0,1,0, (c) blue 0,0,1, (d) black 0,0,0, and (e) white 1,1,1, respectively (left); simulated frequency response as seen from the observer (center); simulated angular color spectrum (right). (From Andkjaer, J. et al., *J. Opt. Soc. Am. B-Opt. Phys.*, 31(1), 164–174, 2014. With permission.[43])

multilayer structure designs that are capable of creating several different colors; the methodology is also able to create any desired spectral shape.[38] The results provide some interesting insight into the design space of these systems. Complex structures can generate more intense colors compared to simple layer structures and wide-angle reflection can be achieved, but the peak intensity will be reduced. Finally, the performance and optimized surface topologies of these structures do not depend significantly on light polarization.

2.4.4 Sculpted Multilayers

Some butterflies from the *Papilio* genus have another very interesting variation on the periodic multilayer structure.[44,45] The dorsal wings of *P. ulysses* butterflies (in Figure 2.12), a strikingly beautiful butterfly from Indonesia, are black and an almost glowing, emerald green. These butterflies are also known as emerald swallowtails. In general, butterfly wings are made of a double cuticle membrane. Lepidoptera rest with their wings closed, and in this pose the ventral side of their wings (the side closest to the belly of the butterfly) is visible. When opened and viewed from above, the dorsal side of the wings, the side closest to the back of the butterfly, can be seen. Both sides of the wing may show drastically different coloration, and it is often on the dorsal side of the wings that the vibrant structural colors can be found. On the surfaces of the wings are overlapping arrays of scales approximately 50–200 μm × 15–50 μm in size.[5,46] The scales are typically chitin and are long, flattened, and oval in shape.[46] There are usually two layers of scales, with an upper layer of scales, cover scales, lying atop a lower layer of scales, called ground scales, that are closer to the wing membrane. These scales lie very densely packed on the wing (around 200–500/mm^2). The cover and ground scales often will have different size, pigment content, structure, and biological function.[47] The photonic structures may be on or within either of the scale types. The form and structure of either scale type may change from one area on a butterfly's wing to another. In *P. blumei*, for example, the structures that create the green color are found on the smaller cover scales.[48] Larger ground scales beneath the bright cover scales absorb light, reducing back-scattering and increasing contrast and purity of the green coloring in the cover scale above.

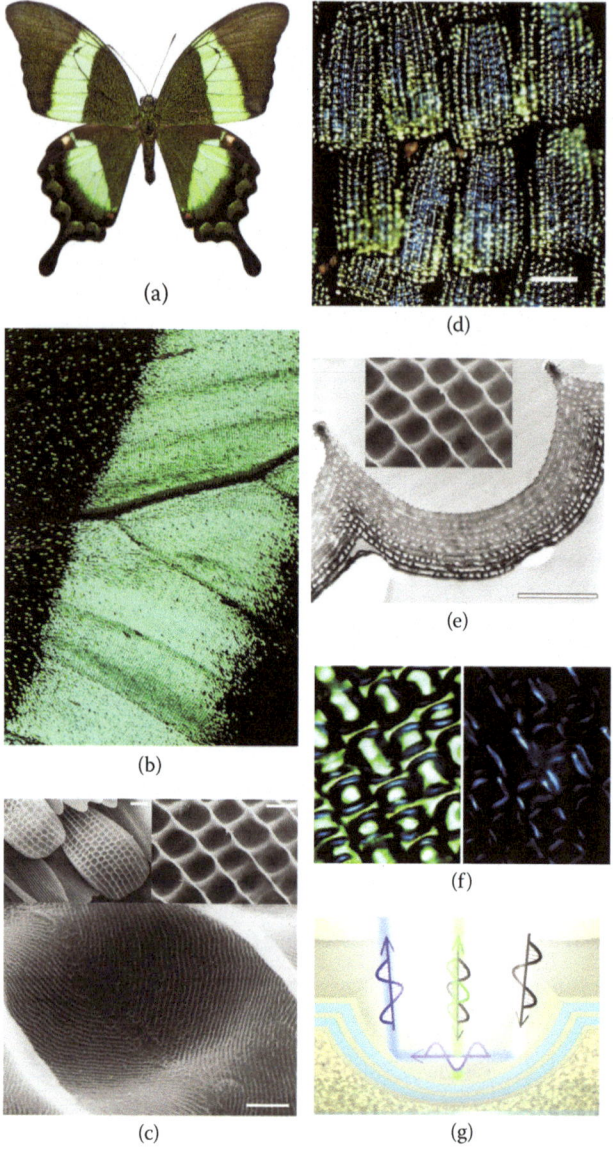

Figure 2.12 (a) Image of a *Papilio palinurus*. (b) A close-up of a wing with visible scales. (c) SEM of *P. palinurus*: whole iridescent scale on the wing, a small region of iridescent scale, and single dimple. Scale bars: 10 μm, 5 μm, and 1 μm. (d) Close-up of *P. blumei's* brightly colored wing scales. Scale bar: 100 μm. (e) TEM image of dimple cross section on a *P. palinurus* iridescent scale. Inset: SEM image of the surface of an iridescent scale. Scale bar: 1 μm (inset, Scale bar: 7 μm). (f) Dimple multilayers in *P. blumei* reflect yellow-green light at the centers and blue at the edges (left). Viewed through crossed polarizers, yellow-green light is eliminated, but the blue light still visible along each edge (right). (g) Double reflection inside the dimple results in polarization rotation. (From Vukusic, P. et al., *Nat.*, 404(6777), 457, 2000; Vukusic, P. et al., *Appl. Opt.*, 40(7), 1116–1125, 2001; and Kolle, M. et al., *Nat. Nanotechnol.*, 5(7), 511–515, 2010.[44,45,48])

On the scales of both the males of *P. palinurus* and *P. blumei*, the surface structures look almost like the surface of a golf ball, with closely packed arrays of somewhat regularly ordered concave dimples, each several μm across.[44] An SEM image of the cross section of one of the dimples, seen in Figure 2.13, shows a concave structure comprising a bowled or sculpted perforated chitin–air multilayer. The perforations form the air layers and are potentially interconnected air pockets.[49] Each dimple can be deep on the scale of the wavelengths of light, ranging from 0.5 to 3 μm, and the sides and bottoms are not perfectly round, but somewhat flattened.[48] The sculpted multilayers do not specifically reflect green light. Rather, the sculpted multilayer creates an overall green appearance (when viewed from a distance) via color mixing. Optical microscopy at normal incidence reveals that the flattish bottom of the dimple reflects a bright yellow color, whereas the walls of the structure reflect blue light due to the different angle of incidence of the illuminating light hitting the sides versus the bottom. This blue reflection is directed back out to the observer via retroreflection. At the resolution of the human eye, especially at a distance, these colors combine and we perceive the wing scale to be bright green. This structure also causes polarization effects, as the double reflection of the blue off of the walls of the dimples causes the polarization of the retroreflected light to change. Sculpted multilayer structures have also been found in the *Chlorophila obscuripennis* beetle; its bluish green coloring is also due to color mixing of a blue color reflected from the basin of the structure, and a green color reflected from the walls.[50] Physically, these structures are technically two-dimensional (2D), but the physics of each color reflection is due primarily to a 1D structure (the multilayer).[5] This is not the only structure that does not fit neatly into a single dimension category.

Multiple efforts have been made to replicate this structure in the laboratory. In one approach, atomic layer deposition (ALD) of thin TiO_2 layers was used to infiltrate and coat the sculpted multilayer structure on an actual butterfly wing scale.[49] A second approach, seen in Figure 2.13 fabricated synthetic versions of the *Papilio* biophotonic structures in a clever combination of bottom-up and top-down techniques.[48] Polystyrene (PS) beads approximately 5 μm in diameter were self-assembled onto a gold-coated silicon substrate. A thin layer of gold or platinum was then deposited to fill in the interstitial gaps between the colloids, which were then removed ultrasonically. This created a

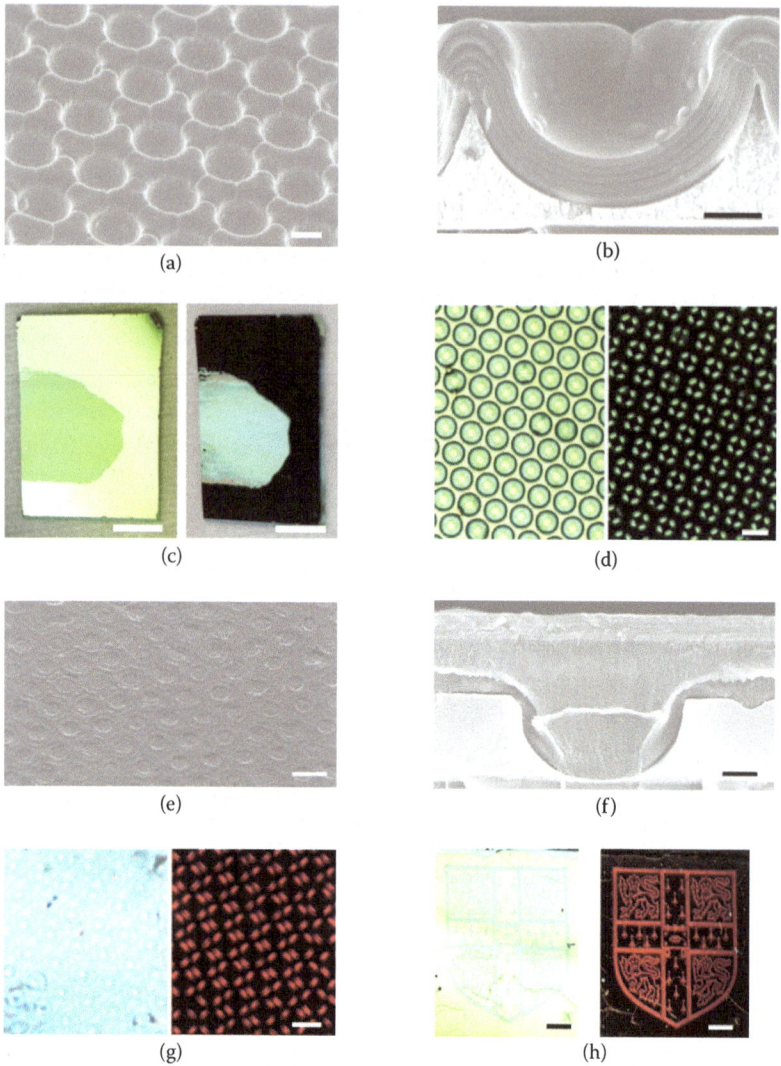

Figure 2.13 (a) SEM images of synthetic dimples covered by a conformal multilayer stack (scale bar: 2 μm). (b) SEM cross section of a dimple showing 11 alternating layers of titania and alumina. (scale bar: 1 μm). (c) Artificial replica appears green when illuminated by normally incident light and blue at grazing incidence (scale bars: 5 μm). (d) In optical microscope image, the dimple edges appear turquoise, and the centers and interstitial regions are yellow (left). Between crossed polarizers only the green dimple edges are visible (scale bar: 5 μm). (e–h) Modified mimic with enhanced optical performance. (e) SEM image of melted colloidal multilayer stacks (scale bar: 5 μm). (f) SEM of cross section (scale bar: 1 μm). (g) Edges of the dimples appear blue under unpolarized light, while the centers and interstitial regions are reflecting in a broad-spectral range (left). Only reflected red light is detected between crossed polarizers (right). Scale bar: 2 μm. (h) Sculpted multilayer logo viewed in direct specular reflection and in retroreflection. Scale bars: 5 μm. (From Kolle, M. et al., *Nat. Nanotechnol.*, 5(7), 511–515, 2010. With permission.[48])

negative replica of the bead layer—an array of hexagonally arranged metal dimples. A very thin layer of carbon was sputtered on the gold to act as an absorbing layer, much like a melanin-absorbing layer. Finally, an alternating (11 layer) stack of TiO_2 and alumina (Al_2O_3) was deposited by ALD creating the sculpted multilayer. The indices of refraction for TiO_2 and Al_2O_3 were $n_{TiO_2} \approx 2.5$ and $n_{Al_2O_3} \approx 1.7$, respectively, significantly higher than the refractive indices of biological materials, so the layer thicknesses could be smaller (only ~57 nm thick for the TiO_2 and ~82 nm thick for Al_2O_3). The resulting structure demonstrated a peak reflectivity of nearly 95% of the light, with a peak wavelength around 550 nm.

The same team created a modified structure based on these techniques by simplifying the fabrication process. In this process, the colloids were instead left in place and annealed, causing the beads to melt into a film, on top of which the TiO_2–Al_2O_3 multilayer was deposited. This created arrays of polymer-filled resonant cavities. SEM images of top view and cross section of this modified structure are shown in Figure 2.13. The optical properties of this new structure also show color separation and polarization effects, although the mechanism for color production is slightly different. In this structure, the flat multilayer reflects blue light. Red light travels through the multilayer and into the polymer-filled cavity where it is reflected from the gold layer (which acts as a mirror). Patterns in the dimple arrays can be created by blocking the initial gold deposition. This technique is then interesting for potential applications in static displays, paints and other coatings, as well as security labeling.

2.5 Two-Dimensional Structures

2D structures vary periodically in two directions (and not in the third). They might be ordered arrays of ridges, grooves, corrugations, fibers, or tubules arranged in parallel, periodic structures. Compared to multilayers, they are somewhat less common in nature though they can be found in, for example, bird feathers and on flower petals.[4,24,51,52] Arrays of keratin tubules, for instance, are responsible for the coloration seen in peacock feathers. Larger 2D optical structures that influence color display are found in fireflies where they improve light transmission from luminescence organs. Applications abound for these structures in astronomy and space applications, lighting, laser optics, optoelectronics, telecommunications, spectroscopy, solar, holography, and more.[53–56]

2.5.1 Arrays in Peacock Feathers

Simple arrays of aligned ridges, tubules, fibers, etc. can be found producing color in many different organisms. Many of these 2D structures cause color via diffraction and are generally categorized as diffraction gratings. They create iridescence in flowers (thought to enhance pollinator attraction), fish, bird feathers, beetles, snakes, and butterflies, often to enhance the mechanism primarily responsible for color.[4–6] In bird primary feathers, the external portion typically consists of a semirigid shaft called the rachis.[57] Attached roughly perpendicular to the shaft length (as seen in Figure 2.14) are long barbs and thin barbules and hooks formed between these barbs. Structural color elements such as melanin rods, keratin, and air holes are incorporated into the barbule structure.[6] Different color effects are achieved by variations in the structure of the barbule nanostructure, such as its crystal lattice parameters (the dimensions of its repeating units), the number of periods, and the size and radius of the melanin rods or air holes. Male peacocks, for example, have incredibly beautiful tail feathers. The iridescent feathered train is used in mating displays and the iridescence seems to play a role in mating success.[58] Peacock feathers possess 2D photonic crystal structures that create the iridescence color in their blue, green, yellow, and even their brown feathers.[59,60] The reflectance spectra of peacock feather barbules show relatively narrowband reflection in blue and green, broader reflection in yellow, and a very broad, low reflection peak in brown. Structural color in brown feather barbules is actually quite surprising, as brown is most often assumed to be due to pigments.

The outer surface of the peacock feather barbule is made of a thin layer of keratin. Below this layer is an array of square or rectangular melanin rods connected by keratin, and separated by holes. The nanorod/air array for a brown barbule can be seen in Figure 2.14 (the keratin surface layer is on the right) along with an illustration of the array structure. The melanin nanorods run parallel to the keratin surface layer. What is quite intriguing about these structures is, again, that while we might expect that structural color plays a major role in

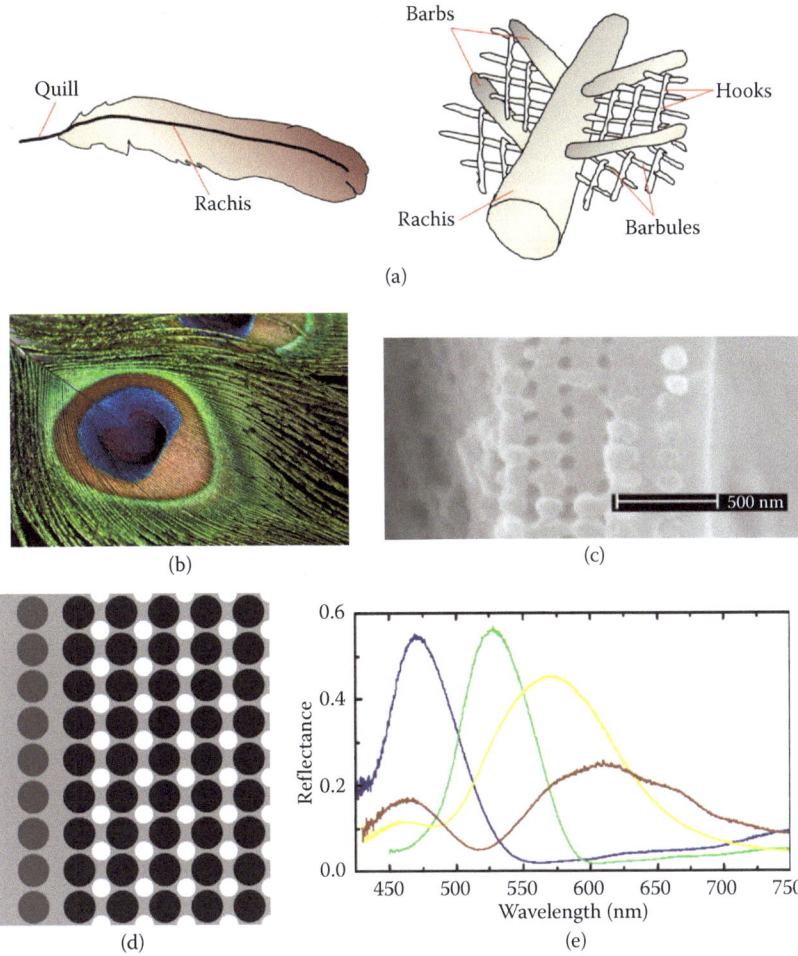

Figure 2.14 (a) General structure of primary feather and feather central section in birds. (b) Eye feather marking from a brightly colored peacock tail feather. (c) Cross section of the cortex for brown barbules. The 2D photonic crystal-like structure (melanin rods and air) is visible beneath the surface keratin layer. Melanin rods are parallel to the cortex surface and to the page. (d) Illustration of the microstructure cross section of the cortex layer of brown barbules. Dark gray dots represent melanin rods nearest to the cortex surface. Black and white dots are other melanin rods and air holes of the 2D photonic-crystal structure, respectively. Light gray indicates keratin matrix. (e) Measured reflectance of blue, green, yellow, and brown barbules for E polarization at normal incidence. (From De la Torre-Ibarra, M.H. and Santoyo, F.M., *J. Biomed. Opt.*, 18(5), 8, 2013; Li, Y.Z. et al., *Phys. Rev. E*, 72(1), 4, 2005; and Zi, J. et al., *Proc. Natl. Acad. Sci. USA*, 100(22), 12576–12578, 2003.[57,59,60])

the case of green, blue, and yellow, we do not expect to see structural color play a role in brown reflectance and yet it seems to. In the blue, green, and yellow 2D photonic crystals, the nanorods are squareish, and the number of melanin nanorod layers plays an important role in the color these structures produce. Brown colored barbules have rectangular nanorods and they have fewer nanorod layers than the other colors. The thickness of the top layer in these barbules is smaller than in the subsequent layers and, importantly, does not have air gaps between the two melanin layers nearest the surface. Band structure calculations of the 2D lattice portion of the brown barbule structures (excluding those top melanin layers) indicated a partial band gap exists that seems to correlate with orange to red. It is thought that the two melanin layers closest to the surface form a blue interferometric reflector. The peacock feather structure creates the brown coloring in the barbs in its iridescent feathers via color mixing.

2.5.2 Templated Growth and Replication

Several groups have created inorganic nanorod structures from those found in the peacock feather, but in a manner that does not require lithographic techniques. Using the biological photonic structure as the template enables the reproduction of the complex photonic structures, with some control over the resulting material properties. Cadmium sulfide (CdS) nanocrystals were incorporated into an actual feather structure, both on the surface and within the 2D array, as seen in Figure 2.15.[61] This effort created a novel hybrid photonic crystal demonstrating that materials could be incorporated into the structure. As we might expect, the incorporation of CdS into the peacock structure changed the structure's reflectance compared to the original feather. In another approach, zinc oxide (ZnO) nanorod arrays were created also by using the actual feather structure as a fabrication scaffold.[62] In this effort, the ZnO was incorporated into the keratin and melanin structure. The organic material was then removed in a process that also enabled the formation of ZnO nanorods in the melanin rod alignment. Unlike the CdS inclusion, the resulting ZnO structure was not a hybrid structure, as all of the organic matter was removed.

The 2D biological structures found in flowers, such as tulips and roses, have also been used to create replicas of the floral photonic structures. Many tulips have ordered striations on their petals creating surface diffraction grating structures.[4,63] Epoxy casts of the diffraction gratings found on the tulip *T. kolpakowskiana* replicating the 2D structures. SEM images of the epoxy casts can be seen in Figure 2.16,

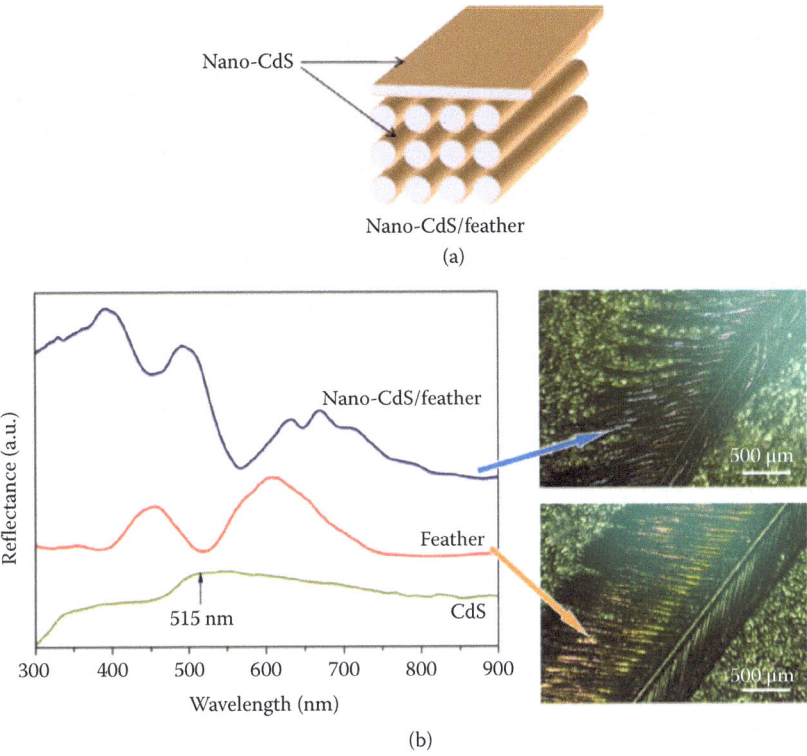

Figure 2.15 (a) Illustration of nano-CdS incorporation onto keratin/melanin structure. (b) Reflection spectra of CdS on glass substrate, the original red feather barb, and the nano-CdS/feather hybrid. Optical images of the nano-CdS/red barb hybrid (top right) and original red barb (bottom right). (Reprinted with permission from Han, J. et al., *Langmuir: ACS J. Surf. Colloids*, 25(5), 3207–3211. Copyright 2014 American Chemical Society.[61])

and the larger undulations are due to the epidermal cells of the petal. It is believed that iridescence in plants acts as a cue to lead pollinators to nectar and these replicas were used to create flower mockups to test this theory with bees. The diffraction grating on the Queen of the Night tulip was mimicked using sheared nanocrystalline cellulose composites that form trapped liquid crystalline phases.[10] The semi-aligned 2D grating structures found on rose petals were created in Poly (vinyl alcohol) (PVA) using a rose petal as a template and the inverse replica was then used as a mold to create a PS replica.[64] Although these techniques that use biological structures to create replicas or act as fabrication scaffolds may eliminate expensive lithographic approaches to complex 2D arrays, they also suffer with a challenge to scalability due to the use of actual biological structures as templates.

Figure 2.16 (a) Picture of Queen of the Night tulip. (b) SEM image shows the grating structure on the surface of the flower petal; the periodicity is ~ 2 μm. (c) Transmission polarizing optical microscopy image of sheered grating, taken between cross polarizers. (From Fernandes, S.N. et al., *Macromol. Chem. Phys.*, 214(1), 25–32, 2013. With permission.[10])

2.5.3 Quasi-Ordered 2D Structures and Quasicrystals

Most of the structures that are responsible for the production of color are highly ordered and periodic. However, examples have been found of structures that are only quasi-ordered but are still capable of producing color. There are not many blue pigments found in vertebrates, and those that have been found exist primarily in fish.[3,7,65] When noniridescent blue color is found, it is often assumed to be due to a disordered structure that produces incoherent (Rayleigh) scattering.[3] However, several examples of blue skin in mammals, such as the face and rump of the primate mandrill (*Mandrillus sphinx*) seen in Figure 2.17, and other primates and marsupials, are actually due to coherent scattering, i.e., coherent interference. When sections of blue skin were examined, arrays of closely packed collagen fibers were found, some with layers of melanin underneath. Fourier analysis was performed on the TEM images of these collagen arrays and instead of complete disorder they found that the parallel collagen fibers were actually quasi-ordered. The weevil *Eupholus magnificus* is a beetle with an elytra banded by yellow and blue.[66] The scales on the blue banded area have a semidiffuse blue reflection. The structures there show little obvious sign of periodicity and are actually aperiodic quasi-ordered crystals.

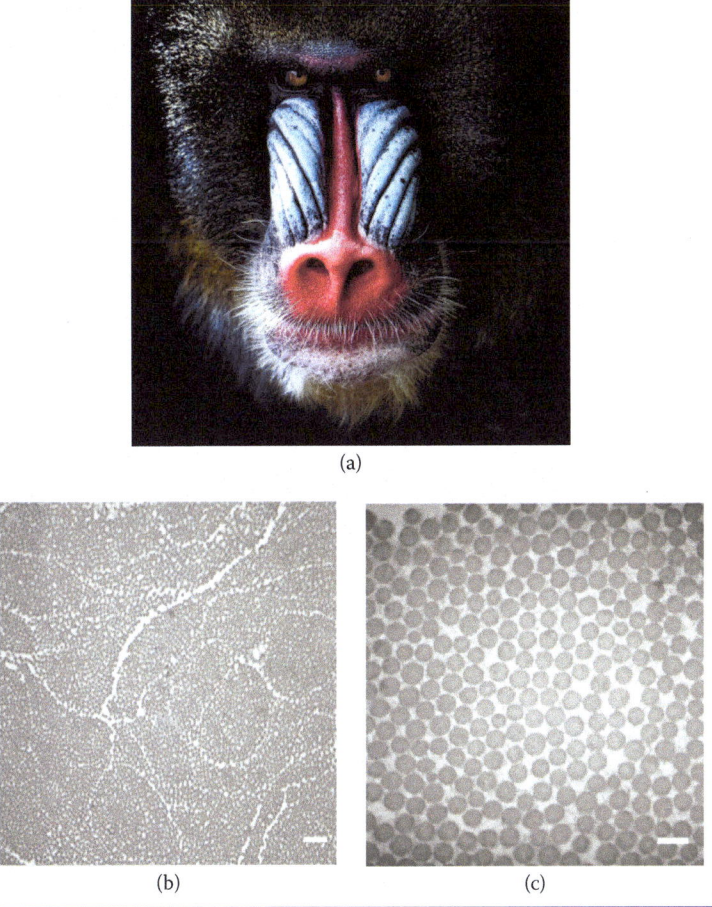

Figure 2.17 (a) Male adult Mandrill face. (b) TEM images of collagen arrays from a female mandrill (*Mandrillus sphinx*) facial skin at 7500×. Scale bar: 500 nm. (c) TEM of collagen array in male mandrill facial skin at 40,000×. Scale bar: 250 nm. (From Prum, R.O., *J. Exp. Biol.*, 207(12), 2157–2172, 2004. With permission.[3])

Quasicrystals and quasi-ordered crystals have long-range order, but are aperiodic (meaning that they do not have translational symmetry, i.e., a minimum unit cell that repeats) and can be 1D, 2D, or three-dimensional (3D). Quasicrystals are relatively new to material science, having only been discovered in 1982, and their discovery won the 2011 Chemistry Nobel Prize.[67,68] Quasi-ordered crystals have been seen in other biological photonic structures in birds and beetles, with primarily blue and green coloring.[2] Aperiodic photonic structures are capable of complex physical phenomena, including band gaps. For example, aperiodic nanoimprinted structures were fabricated from silk films and their performance as colorimetric sensors have been explored.[69–71] There remain many challenges when it comes to fabrication, as well as understanding the underlying physics, and therefore tapping into the potential application space. One potential benefit of these structures is the ability to create complete band gaps without the requirement of a high index of refraction difference. It is anticipated that aperiodic quasicrystals have the potential to provide significant control of near and far-field spectral properties for applications involving transmission and reflection, photoluminescence, and light transport, such as optical computing, sensors, plasmonics, and lasers.

2.5.4 Improving Light Extraction

Most biological photonic structures are responsible for the direct production of color. Some structures enhance pigmentary color production as in the *Phelsuma* geckos. Other structures direct and enhance the emission of light created by the organism itself, bioluminescent light. The snail shell of *Hinea brasiliana* acts as a color-specific optical diffuser with excellent diffusion and transmission efficiency compared to a commercial diffuser.[72] However, it is unknown at this point what structural mechanism within the snail shell enables this function.

Fireflies produce bioluminescent yellow-green or orange-red light from an organ in their abdomen called a lantern.[73] The lanterns of *Photuris* species of fireflies have complex architectures, as seen in Figure 2.18. The topmost layer is made of a combination of nano- and microstructures that sit atop the light production layers. These top-layer structures contribute to the firefly's ability to emit light efficiently. The microscale structure is a closely packed array of scales, each about 10 microns long on average. Each scale is tilted so that one edge of the scale protrudes from the body plane by about 3 μm. These tilted scales form a large grating structure. The protruding tips of the scales scatter light and the transmission intensity and light extraction are higher at these tips, perhaps nearly 50% more light than a flat surface. 2D arrays of nanoscale corrugations lie on the surface of these tilted scales, which act like diffraction gratings and antireflection structures.

High-efficiency optical emitters (such as LEDs, organic light emitting diode etc.) can benefit from any structures that can improve their

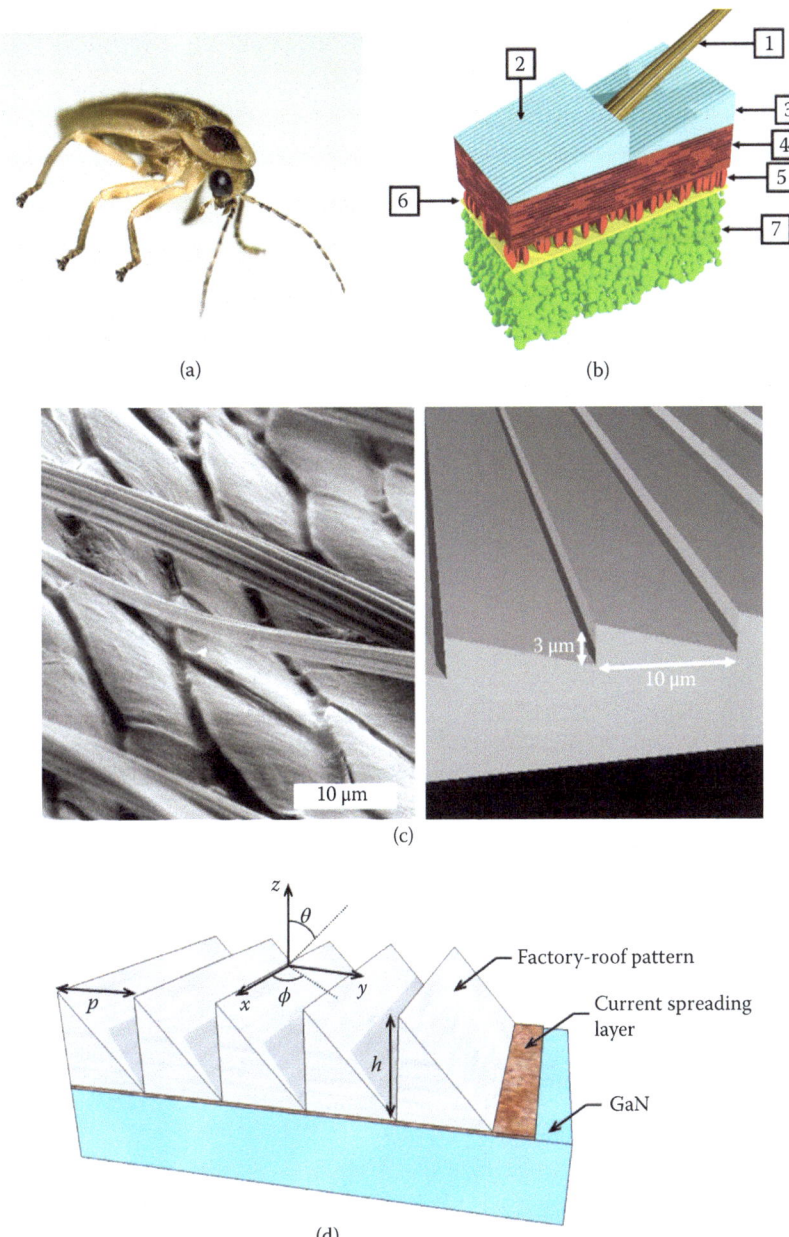

Figure 2.18 (a) The *Photuris* sp. firefly. (b) Structures found in the *Photuris* firefly light organ. (1) Setae; (2) ribs separated by 250 nm; (3) misfit scales, 10 μm long and 3 μm high; (4) textured cuticle volume 2.4 μm thick; (5) muscular layer, variable thickness; (6) photocyte cluster membrane, 60 nm thick; (7) photocytes layer, containing spherical peroxisomes. (c) SEM image of the "misfit scales" cuticle structure and model used for simulations of the light propagation. Period = 10 μm. Height = 3 μm. (d) Illustration of the "factory-roof pattern" over the gallium nitride device.

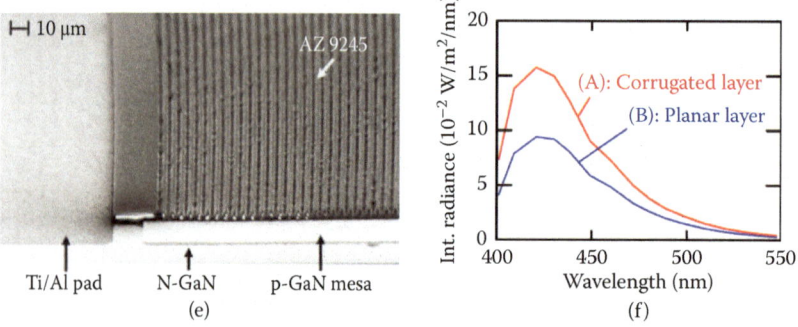

Figure 2.18 (*Continued*) (e) Factory-roof patterns in AZ 9245®photoresist coating on a gallium nitride-based light emitting diodes, by direct-writing laser lithography. (f) Measured radiance integrated over the full 2π hemispheric solid angle for two identical gallium nitride diodes covered by a photoresist layer terminated. (A) By a factory-roof pattern and (b) by a flat surface. The patterned surface causes a 68% increase of the emitted power. (From Bay, A. et al., *Opt. Express*, 21(1), 764–780, 2013 and Bay, A. et al., *Opt. Express*, 21(1), A179–A189, 2013. With permission.[73,74])

extraction efficiency (how much light you can get out of the structure for a given power input), which is a challenge due to internal reflection.[6] Approaches to improve LED performance have included both optical reflectors below and nanostructured air/device interfaces above the light-emitting structures, which is very similar to the structure of natural light organs.[75] 2D nanostructured photonic crystals have been fabricated into LED and OLED structures to enhance the extraction efficiency. Lithographic techniques, such as e-beam lithography and nanoimprint lithography, as well as etching are common techniques to fabricate these types of structures.

Nature cannot afford to waste light, so these natural structures can serve as inspiration for novel optical designs that can be incorporated into optically efficient emitters. Returning to the *Photuris* firefly, the large tilted scale structure is interesting because the size and mechanism are not a standard antireflection surface, and the size of the features makes it attractive for fabrication. Structures inspired by the tilted scale structure have been incorporated into a gallium nitride (GaN) LED device to increase in light extraction due to the scattering structure.[74] The structure, as seen in Figure 2.18, which the authors called a "factory-roof pattern," was fabricated out of photoresist as an overlayer on top of an existing GaN LED device. The photoresist was simply spin-coated onto the completed devices and

BOX 2.2 MODELING OF STRUCTURAL COLOR

Modeling of these systems is crucial to our ability to fully understand the observed reflectance spectra caused by structural color. Simple systems taken in isolation, such as a two-component multilayer or simple thin-film interference, are most accessible and successful at matching theoretical reflectance spectrum to observation.[76] For multilayers, for example, the multilayer parameters (the dimensions of the layers, and the complex refractive indices of the materials from which the structure is made) can be adjusted until prediction and observation match, as long as the parameters chosen are realistic. For more complex structures, such as 3D structures or systems that combine multiple structures, modeling is significantly more complicated. The finite-difference time-domain and finite-element methods are two of the commonly used techniques. Several commercial and homegrown software packages have been developed and are widely available.

Accurately modeling these structures is a significant challenge. Modeled systems most often start with idealized structures that are perfect and continuously periodic, due to the reduced theoretical complexity and computational ease. Biological photonic systems, however, are not perfectly periodic, idealized structures. They often contain additional structural complexity and disorder, which significantly complicates the theoretical model. They have finite thickness and may contain domain boundaries at which the photonic structure reorients relative to its neighbors. Refractive indices may be based on assumptions or estimates (which may not be accurate) and may include only the real and not imaginary component. Both our understating of the exact architecture and material composition of these biological photonic structures as well as the ability of modeling methods that can handle the challenges discussed above must improve before we will be able to quantitatively predict the complex optical spectra of these systems with sufficient accuracy.

then patterned using direct writing laser lithography. The dimensions of the microstructured grating were designed for the GaN device's optical properties, optimized for the wavelength of the GaN LED (425 nm compared to 560 nm for the firefly). The 2D triangular structures had 5-μm separation distance and 5-μm height. This additional structure had a significant impact on the LED's performance. Compared to an LED with flat photoresist top-layer, the structured photoresist layer was able to increase the emitted power by a tremendous 68%. This relatively simple fabrication approach and bioinspired design is very attractive for applications. The structure is potentially optimizable for any emission wavelength, the process is scalable to large areas, and as the structure can be fabricated in place as a supplemental overlayer it does not require any changes to the device's original architecture.

References

1. F. Delgado-Vargas, A. R. Jimenez, and O. Paredes-Lopez, *Crit. Rev. Food Sci. Nutr.* **40** (3), 173–289 (2000).
2. A. E. Seago, P. Brady, J. P. Vigneron, and T. D. Schultz, *J. R.. Soc. Interface* **6 Suppl 2**, S165–S184 (2009).
3. R. O. Prum, *J. Exp. Biol.* **207** (12), 2157–2172 (2004).
4. B. J. Glover and H. M. Whitney, *Ann. Bot.* **105** (4), 505–511 (2010).
5. L. P. Biro and J. P. Vigneron, *Laser Photon. Rev.* **5** (1), 27–51 (2011).
6. K. L. Yu, T. X. Fan, S. Lou, and D. Zhang, *Prog. Mater. Sci.* **58** (6), 825–873 (2013).
7. J. T. Bagnara, P. J. Fernandez, and R. Fujii, *Pigm. Cell Res.* **20** (1), 14–26 (2007).
8. *National Research Council, Optics and Photonics: Essential Technologies for Our Nation.* (The National Academies Press, Washington DC, 2013).
9. M. D. Shawkey, N. I. Morehouse, and P. Vukusic, *J. R. Soc. Interface* **6 Suppl 2**, S221–S231 (2009).
10. S. N. Fernandes, Y. Geng, S. Vignolini, B. J. Glover, A. C. Trindade, J. P. Canejo, P. L. Almeida, P. Brogueira, and M. H. Godinho, *Macromol. Chem. Phys.* **214** (1), 25–32 (2013).
11. P. Vukusic, J. R. Sambles, C. R. Lawrence, and R. J. Wootton, *Proc. R. Soc., B* **266** (1427), 1403–1411 (1999).
12. C. Campos-Fernandez, D. E. Azofeifa, M. Hernandez-Jimenez, A. Ruiz-Ruiz, and W. E. Vargas, *Opt. Mater. Express* **1** (1), 85–100 (2011).
13. P. Vukusic, B. Hallam, and J. Noyes, *Sci.* **315** (5810), 348 (2007).
14. J. P. Vigneron, P. Simonis, A. Aiello, A. Bay, D. M. Windsor, J. F. Colomer, and M. Rassart, *Phys. Rev. E* **82** (2), 8 (2010).

15. V. Sharma, M. Crne, J. O. Park, and M. Srinivasarao, *Sci.* **325** (5939), 449–451 (2009).
16. S. Vignolini, P. J. Rudall, A. V. Rowland, A. Reed, E. Moyroud, R. B. Faden, J. J. Baumberg, B. J. Glover, and U. Steiner, *Proc. Natl. Acad. Sci. USA* **109** (39), 15712–15715 (2012).
17. S. V. Saenko, J. Teyssier, D. van der Marel, and M. C. Milinkovitch, *BMC Biol.* **11**, 12 (2013).
18. J. P. Vigneron, K. Kertesz, Z. Vertesy, M. Rassart, V. Lousse, Z. Balint, and L. P. Biro, *Phys. Rev. E* **78** (2), 9 (2008).
19. S. M. Doucet and M. G. Meadows, *J. R. Soc. Interface* **6**, S115–S132 (2009).
20. K. S. Gould and D. W. Lee, *Am. J. Bot.* **83** (1), 45–50 (1996).
21. M. Butler and A. S. Johnson, *J. Exp. Biol.* **207** (2), 285–293 (2004).
22. R. H. C. Bonser, *Condor* **97** (2), 590–591 (1995).
23. A. Levy-Lior, E. Shimoni, O. Schwartz, E. Gavish-Regev, D. Oron, G. Oxford, S. Weiner, and L. Addadi, *Adv. Funct. Mater.* **20** (2), 320–329 (2010).
24. S. Vignolini, E. Moyroud, B. J. Glover, and U. Steiner, *J. R. Soc. Interface* **10** (87) (2013).
25. H. T. Ghiradella and M. W. Butler, *J. R. Soc. Interface* **6 Suppl 2**, S243–S251 (2009).
26. M. E. McNamara, *Palaeontology* **56** (3), 557–575 (2013).
27. S. Kinoshita and S. Yoshioka, *Chemphyschem: Eur. J. Chem. Phys. Phys. Chem.* **6** (8), 1442–1459 (2005).
28. H. Ghiradella, *Ann. Entomol. Soc. Am.* **77** (6), 637–645 (1984).
29. H. Ghiradella, *J. Morphol.* **202** (1), 69–88 (1989).
30. H. Ghiradella, *Ann. Entomol. Soc. Am.* **78** (2), 252–264 (1985).
31. M. G. Meadows, N. I. Morehouse, R. L. Rutowski, J. M. Douglas, and K. J. McGraw, *Behav. Ecol. Sociobiol.* **65** (6), 1317–1327 (2011).
32. S. Kinoshita, S. Yoshioka, and J. Miyazaki, *Rep. Prog. Phys.* **71** (7) (2008).
33. M. Sarrazin, J. P. Vigneron, V. Welch, and M. Rassart, *Phys. Rev. E* **78** (5), 5 (2008).
34. E. Shevtsova, C. Hansson, D. H. Janzen, and J. Kjaerandsen, *Proc. Natl. Acad. Sci. USA* **108** (2), 668–673 (2011).
35. A. L. Ingram, O. Deparis, J. Boulenguez, G. Kennaway, S. Berthier, and A. R. Parker, *Arthropod Struct. Dev.* **40** (1), 21–25 (2011).
36. E. P. Chan, J. J. Walish, A. M. Urbas, and E. L. Thomas, *Adv. Mater.* **25** (29), 3934–3947 (2013).
37. M. Karaman, S. E. Kooi, and K. K. Gleason, *Chem. Mater.* **20** (6), 2262–2267 (2008).
38. M. Kolle, A. Lethbridge, M. Kreysing, J. J. Baumberg, J. Aizenberg, and P. Vukusic, *Adv. Mater.* **25** (15), 2239–2245 (2013).
39. A. R. Parker, D. R. McKenzie, and M. C. J. Large, *J. Exp. Biol.* **201** (9), 1307–1313 (1998).
40. G. Strout, S. D. Russell, D. P. Pulsifer, S. Erten, A. Lakhtakia, and D. W. Lee, *Ann. Bot.* **112** (6), 1141–1148 (2013).
41. B. Dresp, P. Jouventin, and K. Langley, *Biol. Lett.* **1** (3), 310–313 (2005).

42. P. Jouventin, P. M. Nolan, J. Ornborg, and F. S. Dobson, *Condor* **107** (1), 144–150 (2005).
43. J. Andkjaer, V. E. Johansen, K. S. Friis, and O. Sigmund, *J. Opt. Soc. Am. B-Opt. Phys.* **31** (1), 164–174 (2014).
44. P. Vukusic, J. R. Sambles, and C. R. Lawrence, *Nat.* **404** (6777), 457 (2000).
45. P. Vukusic, R. Sambles, C. Lawrence, and G. Wakely, *Appl. Opt.* **40** (7), 1116–1125 (2001).
46. M. Srinivasarao, *Chem. Rev.* **99** (7), 1935–1961 (1999).
47. P. Vukusic, J. R. Sambles, and C. R. Lawrence, *Proc. R. Soc., B* **271**, S237–S239 (2004).
48. M. Kolle, P. M. Salgard-Cunha, M. R. J. Scherer, F. M. Huang, P. Vukusic, S. Mahajan, J. J. Baumberg, and U. Steiner, *Nat. Nanotechnol.* **5** (7), 511–515 (2010).
49. C. J. Summers, D. P. Gaillot, M. Crne, J. Blair, J. O. Park, M. Srinivasarao, O. Deparis, V. Welch, and J. P. Vigneron, *J. Nonlinear Opt. Phys. Mater.* **19** (3), 489–501 (2010).
50. F. Liu, H. W. Yin, B. Q. Dong, Y. H. Qing, L. Zhao, S. Meyer, X. H. Liu, J. Zi, and B. Chen, *Phys. Rev. E* **77** (1), 4 (2008).
51. D. Osorio and A. D. Ham, *J. Exp. Biol.* **205** (14), 2017–2027 (2002).
52. R. O. Prum, E. R. Dufresne, T. Quinn, and K. Waters, *J. R. Soc. Interface* **6 Suppl 2**, S253–S265 (2009).
53. S. Singh, *Opt. Laser Technol.* **31** (3), 195–218 (1999).
54. J. Barbe, A. F. Thomson, E. C. Wang, K. McIntosh, and K. Catchpole, *Prog. Photovoltaics* **20** (2), 143–148 (2012).
55. U. D. Zeitner, F. Fuchs, and E. B. Kley, in *Modern Technologies in Space- and Ground-Based Telescopes and Instrumentation Ii*, edited by R. Navarro, C. R. Cunningham and E. Prieto (Spie-Int Soc Optical Engineering, Bellingham, 2012), Vol. 8450.
56. H. J. Cornelissen, D. K. G. de Boer, and T. Tukker, in *Led-Based Illumination Systems*, edited by J. Jiao (Spie-Int Soc Optical Engineering, Bellingham, 2013), Vol. 8835.
57. M. H. De la Torre-Ibarra and F. M. Santoyo, *J. Biomed. Opt.* **18** (5), 8 (2013).
58. A. Loyau, D. Gomez, B. T. Moureau, M. Thery, N. S. Hart, M. Saint Jalme, A. T. D. Bennett, and G. Sorci, *Behav. Ecol.* **18** (6), 1123–1131 (2007).
59. Y. Z. Li, Z. H. Lu, H. W. Yin, X. D. Yu, X. H. Liu, and J. Zi, *Phys. Rev. E* **72** (1), 4 (2005).
60. J. Zi, X. D. Yu, Y. Z. Li, X. H. Hu, C. Xu, X. J. Wang, X. H. Liu, and R. T. Fu, *Proc. Natl. Acad. Sci. USA* **100** (22), 12576–12578 (2003).
61. J. Han, H. L. Su, F. Song, J. J. Gu, D. Zhang, and L. M. Hang, *Langmuir: ACS J. Surf. Colloids* **25** (5), 3207–3211 (2009).
62. J. Cao, H. L. Su, J. J. Chen, J. Han, W. J. Moon, and D. Zhang, *Crystengcomm* **14** (16), 5262–5266 (2012).
63. H. M. Whitney, M. Kolle, P. Andrew, L. Chittka, U. Steiner, and B. J. Glover, *Sci.* **323** (5910), 130–133 (2009).

64. L. Feng, Y. A. Zhang, M. Z. Li, Y. M. Zheng, W. Z. Shen, and L. Jiang, *Langmuir: ACS J. Surf. Colloids* **26** (18), 14885–14888 (2010).
65. M. H. Amiri and H. M. Shaheen, *Micron* **43** (2–3), 159–169 (2012).
66. C. Pouya, D. G. Stavenga, and P. Vukusic, *Opt. Express* **19** (12), 11355–11364 (2011).
67. Z. V. Vardeny, A. Nahata, and A. Agrawal, *Nat. Photonics* **7** (3), 177–187 (2013).
68. R. Lifshitz, in *Silicon Versus Carbon: Fundamental Nanoprocesses, Nanobiotechnology and Risks Assessment*, edited by Y. Magarshak, S. Kozyrev, and A. K. Vaseashta (Springer, Dordrecht, 2009), pp. 119–136.
69. S. V. Boriskina, S. Y. K. Lee, J. J. Amsden, F. G. Omenetto, and L. Dal Negro, *Opt. Express* **18** (14), 14568–14576 (2010).
70. S. Y. Lee, J. J. Amsden, S. V. Boriskina, A. Gopinath, A. Mitropolous, D. L. Kaplan, F. G. Omenetto, and L. Dal Negro, *Proc. Natl. Acad. Sci. USA* **107** (27), 12086–12090 (2010).
71. H. Tao, D. L. Kaplan, and F. G. Omenetto, *Adv. Mater.* **24** (21), 2824–2837 (2012).
72. D. D. Deheyn and N. G. Wilson, *Proc. R. Soc., B* **278** (1715), 2112–2121 (2011).
73. A. Bay, P. Cloetens, H. Suhonen, and J. P. Vigneron, *Opt. Express* **21** (1), 764–780 (2013).
74. A. Bay, N. Andre, M. Sarrazin, A. Belarouci, V. Aimez, L. A. Francis, and J. P. Vigneron, *Opt. Express* **21** (1), A179–A189 (2013).
75. J. J. Kim, Y. Lee, H. G. Kim, K. J. Choi, H. S. Kweon, S. Park, and K. H. Jeong, *Proceedings of the National Academy of Sciences of the United States of America* **109** (46), 18674–18678 (2012).
76. P. Vukusic and D. G. Stavenga, *J. R. Soc. Interface* **6**, S133–S148 (2009).
77. E. Van Hooijdonk, S. Berthier, and J. P. Vigneron, *J. Appl. Phys.* **112** (7), 8 (2012).
78. S. A. Boden, A. Asadollahbaik, H. N. Rutt, and D. M. Bagnall, *Scanning* **34** (2), 107–120 (2012).
79. R. Potyrailo and R. R. Naik, in *Annual Review of Materials Research*, edited by D. R. Clarke (Annual Reviews, Palo Alto, 2013), Vol. 43, pp. 307–334.
80. A. R. Parker, *Philos. Trans. R. Soc., A* **362** (1825), 2709–2720 (2004).

3
STRUCTURAL COLOR II
Complex Structures

Continuing the discussion of Structural Color from Chapter 2 we move to even more complicated structures. Beyond the one- and two-dimensional (1D and 2D) materials seen in Chapter 2, there are many examples of complex higher dimensional structures found in nature.

3.1 Quasi Two-/Three-Dimensional Structures

Many biological structures do not fit neatly into the definition of a 1D, 2D, or three-dimensional (3D) structure. In fact, none can likely be defined neatly by the strictest definition. Some structures, more than others, ride the cusp, for example, between the definition of 2D and 3D structures. To simplify the problem for analysis, we might categorize a structure based on its primary photonic effect. However, nature has shown its willingness to combine multiple effects into the same hierarchical heterostructure. These quasi-(X)-dimensional structures look (or act) a different dimensionality than how they might otherwise be categorized.

3.1.1 Tilted Structures and Narrow Angle Reflectance

Some butterflies, such as the *Ancyluris meliboeus*, *Lamprolenis nitida*, and *Pierella luna*, have evolved to have wing reflectance properties that shift only over a very narrow range of angles.[1-3] Males of *L. nitida* have two integrated diffraction gratings in their hind wing cover scales.[2,4] These two gratings work independently to create separate colors at different illumination and observation angles. In *L. nitida*, red to green iridescence can be seen from the front, while blue to ultraviolet reflection is visible from behind. The first grating is formed by ribs and cross-ribs sitting at a 30° angle with respect to the scale surface. Flutes on the sides of the ridges form the second diffraction grating. A second interesting example is the bent gratings found in *P. luna*.[3] In *P. luna*, the grating forms in scales bent at their tips nearly perpendicular to the wing surface. The optical effect occurs because of transmission-based diffraction instead of reflective diffraction. These tilted, or blazed, gratings enhance the diffraction efficiency at a given wavelength and angle at the expense of other diffractive orders. The structures are physically 3D, as the structure varies in all three directions. However, the physics can be described as a combination of 2D effects.

The *Troides magellanus* butterfly also has multiple components working in concert to enhance reflection color over a limited angle. *T. magellanus* is a threatened species of Asian butterfly that can have a wingspan of about 20 cm.[5] Its wings are deep black and bright yellow, as seen in Figure 3.1. However, when viewed at just the right angle, a flash of pale blue/green/white can be seen on the hindwings. The yellow scales of *T. magellanus* have ridges that run the length of the scales. Looking at their cross section, the ridges are actually triangular and are filled with a yellow pigment called papilliochrome II. Papilliochrome II is primarily responsible for the yellow color seen on the wings. The pigment is fluorescent, absorbing light in the ultraviolet (UV) range and emitting light in the yellow, reinforcing the color saturation and brightness of the yellow color. The triangular ridges are separated from each other by approximately 1.5 μm and are each about 4 μm tall. On the side of the ridges, running at a 60° angle are smaller lamellae each separated by ~250 nm. These structures have been modeled and described as both a blazed grating and alternately a longitudinal tilted air/chitin multilayer stack.[5-8] The reflectance spectra of the hindwing show that when the illumination and viewing angles are far from grazing, the pigmentary yellow dominates, as seen by the plateau at longer wavelengths. But as the viewing angle moves toward grazing angles, where the viewer is looking nearly parallel with the wing surface, a bright flash of blue/green/white color can be seen. At these angles, the reflectance spectrum shifts to dual-peak reflectance at shorter wavelengths; one peak is attributed to reflection from the tilted multilayer and the other to fluorescence emission. The larger scale ridges do not seem to contribute to the UV-vis optical spectra, due to their large size with respect to visible wavelengths.

Multilayer blazed gratings are important components of ultra-high-resolution extreme UV (EUV) and soft-x-ray spectrometry

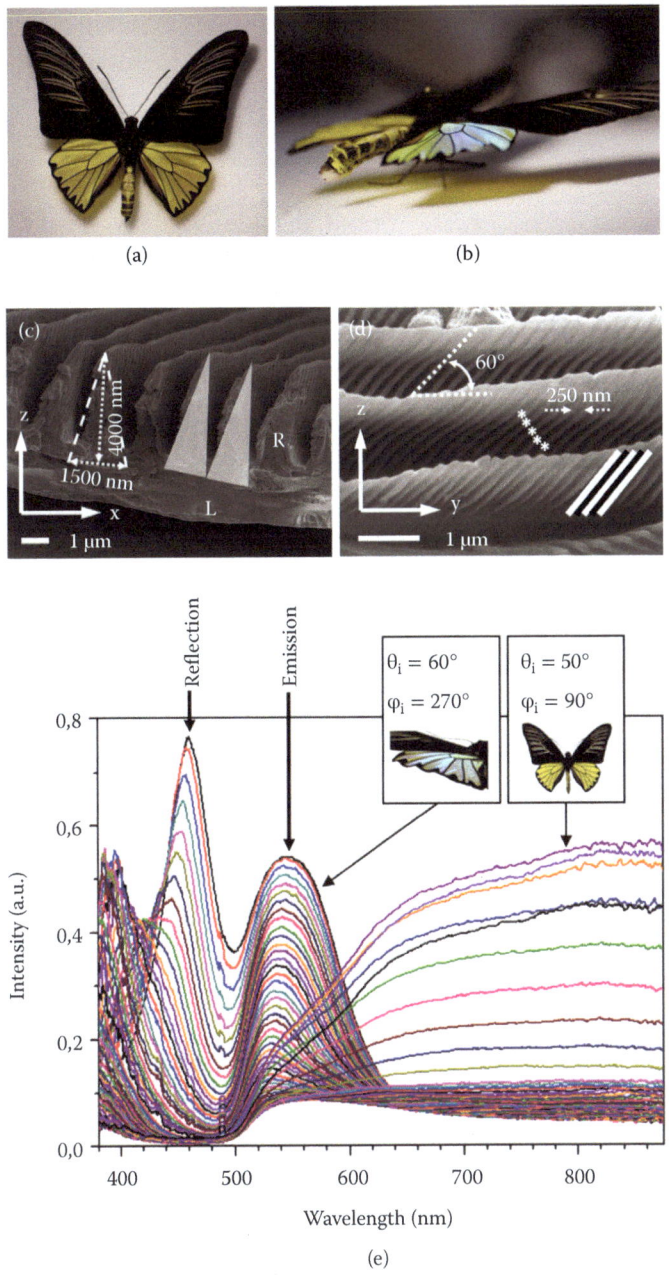

Figure 3.1 (a) Dorsal view of the male *Troides magellanus*. (b) A bluish or greenish flash is visible on the hindwings when the illumination and the observation are grazing. (c) Cross-sectional view of a fractured scale and (d) lateral view of a fractured scale. (e) Ultraviolet-visible spectra for a whole hindwing of the *T. magellanus* according to the incident direction (belonging to the *yz* plane). The two arrows indicate the reflection and emission peaks. (From Hooijdonk et al., *J. Appl. Phys.*, 112 (7), 8, 2012.[5])

for applications in astrophysics, plasma diagnostics, pulse compression of chirped x-ray beams, and more.[9–11] These structures can be optimized to achieve desired optical properties, such as high dispersion, spectral resolution, and diffraction efficiency, at a specific diffractive order. However, the resolution control required at these very short wavelengths makes fabrication of the grating structures extremely challenging.[10] Efforts have been made to overcome some of these challenges; high-quality periodic multilayer-coated blazed gratings have been fabricated (seen in Figure 3.2) and have demonstrated an absolute diffraction efficiency of 37.6% at 13.6-nm wavelengths. The structures, which were fabricated using scanning beam interference lithography followed by wet anisotropic etching on silicon, had a 200-nm period grating (~5000 lines/mm) that was coated with a 30 bilayer Mo/Si multilayer. Scanning electron microscopy (SEM) of the grating structure and subsequent multilayer structure show that there can be some difficulty in fabricating the blazed grating on the nonplanar surface. This grating was fabricated on a patterned surface to create the tilted structure. Fabrication can be challenging as the taller portions of the patterned substrates may shadow other portions preventing or reducing deposition of the sputtered material and creating nonuniform layer thicknesses. Deposition conditions can cause deformation of the underlying grating substrate structure. The measured diffraction efficiency of the structure is compared to simulated efficiencies in Figure 3.2d for the ideal grating structure, as well as the deformed grating profiles seen in the structure by atomic force microscopy. Many of these structures operate over a very narrow range of wavelengths (perhaps 1 nm FWHM [full width at half maximum]). To achieve broadband diffraction, strategies very similar to those seen in chirped or aperiodic biological structures have been successfully investigated.[9] By fabricating aperiodic multilayers onto blazed grating substrates, the resulting structures operate at much broader wavelengths over the desired spectral region. Genetic algorithm-based design codes have been used to design the optimal Mo/Si coating structure. The designed coating was expected to have an ~16% reflection across the 13- to 16-nm wavelength region. Differences in the fabricated 40 Mo/Si layer coating from the design reduced the overall efficiency to ~10% in the second order, but the wavelength range over which the structure reflected light was 13–16 nm, which is large for the EUV regime.

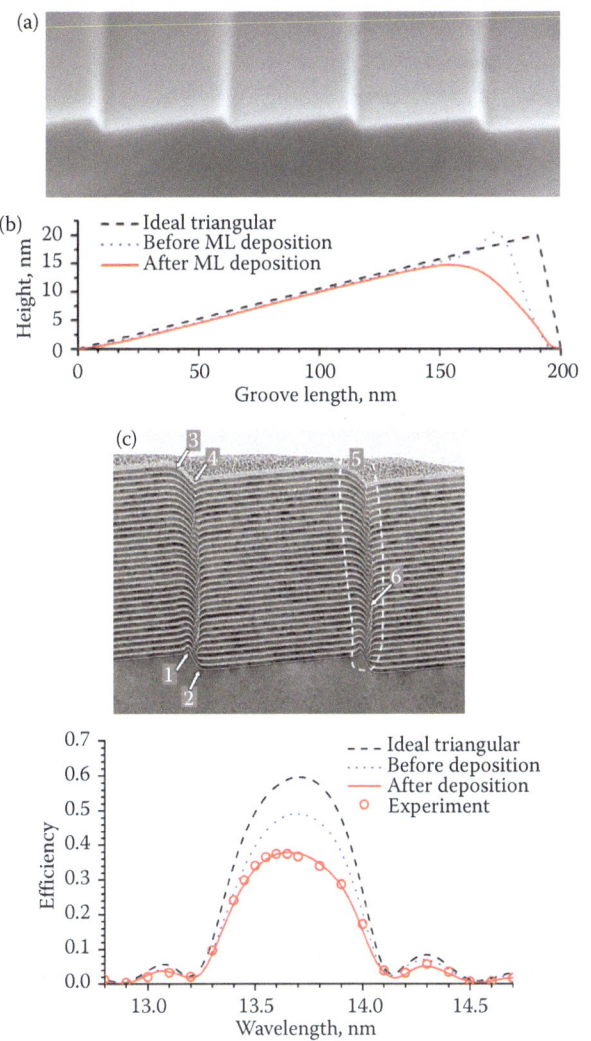

Figure 3.2 Example of an inorganic blazed grating. (a) SEM images of the silicon blazed grating used as a substrate for molybdenum/silicon (Mo/Si) multilayer growth. (b) Atomic force microscopy (AFM)-based groove profiles showing deformation of the blazed grating before (dotted curve) and after (solid curve) multilayer deposition. The dashed curve shows an ideal triangular profile. (c) Transmission electron microscopy cross section of a blazed grating coated with 30 Mo/Si bilayers, with a 6-degree blaze angle and 64-degree antiblaze angle. (1 and 2) Indicate small bumps in the upper part of the grooves and a rounding of the bottom junction of the facets. Deposition changes the grating profile. (3 and 4) Indicate there are rounded vertices and a trough on the top surface of the multilayer. (5 and 6) The area of perturbed multilayer stack and areas with collapsed multilayer structure. (d) Diffraction efficiency versus wavelength of the third diffraction order of a Mo/Si-coated blazed grating (period of 200 nm). Curves show simulated efficiency calculated for ideal triangle groove profile (dashed curve), and the AFM-measured profiles before (dotted curve) and after (solid line) multilayer deposition. Circles show experimental data. (From Voronov et al., *Adv. X-Ray/Euv Optics Components V*, 7802, 2010.[10])

3.1.2 Wings of the Morpho Butterfly

Arguably, the most iconic of all representatives of structural color in the wild is the *Morpho* butterfly genus. Various species of this genus are some of the most extensively studied examples of structural color. In many *Morpho* butterflies (seen in Figure 3.3), the ventral wing color is shades and patterns of (typically) brown with white spots. However, when opened, the dorsal side exhibits beautiful highly visible bright blue or green coloring that is intense and slightly iridescent. Native to South and Central America, their large wings (ranging up to nearly 20 cm) and conspicuous coloring made them obvious targets for early naturalists, and prized by scientists and collectors today.

The quasi-3D structure found on their wing scales has been described as "Christmas tree-like," with structured ridges running parallel with the long axis of the scale. The ridges consist of a tall central pillar

Figure 3.3 (a) Dorsal wings of a *Morpho peleides* butterfly (ventral wings inset). (b) A SEM image of the photonic *Morpho rhetenor* nanostructure. (From Potyrailo et al., *Annual Review of Materials Research*, Annual Reviews, Palo Alto, 2013.[12])

transected by multiple lamellae forming an effective multilayer supported by interstitial vertical microribs. The entire structure is held aloft from the surface of the scale. There are generally large spaces through which the interior of the scale can be seen. The iridescent coloration arises from a combination of interference, diffraction, scattering, and absorption due to the entire multiscale, hierarchical structure on the wing.

Several different strategies have been used to recreate the *Morpho* structure. Replication of the *Morpho* structure using the butterfly scales as templates has been performed by several different methods. In one example, an inverse replica of a male *Morpho peleides* scale was created by depositing a layer of Al_2O_3 directly onto a scale using atomic layer deposition (ALD).[13] The organic template was then removed. The inverse structure was able to replicate some of the reflectance properties seen in the original structure. The researchers were also able to demonstrate the ability to shift the reflected wavelengths by changing the thickness of the alumina layer. Other techniques include, among others, a modified sol–gel process used to deposit rutile TiO_2 onto a *Morpho* wing scale and polydimethylsiloxane (PDMS) replicas made via soft lithography.[14,15] Approaches to create the lamellar structures without a biological template, such as focused ion beam (FIB) and lithographic processes, have had good success creating structural color.[4,16,17]

The primary reflective color of the *Morpho* wing can be attributed to multilayer-like behavior of the lamellar ridge structure, and many researchers who approach modeling and fabrication of the *Morpho* structure begin their approximations with that assumption.[18] However, the *Morpho*'s optical behavior is much more complex. The *Morpho* color is comparatively angle independent, which is not typical of an ideal multilayer. In some species, the color can change from blue to deep blue to deep violet depending on the angle of incidence.[18,19] *Morpho rhetenor* has been found to show retroreflection in two inverse directions under oblique incidence. *Morpho*-like structures have been shown to be efficient low-pass filters for all angles of incidence and polarizations, and nearly completely transmissive above about 550 nm. Under certain conditions, backscattered light can be completely extinguished: when the wing is viewed parallel to the wing veins, the blue color disappears and the wing color appears black. Modeling efforts to explain these phenomena, and create new designs inspired by these butterflies, become increasingly complex.

The combination of high reflectance and wide viewing angle found in biological structures like those found on the *Morpho* provides powerful inspiration for color-related applications, such as low-power reflective displays, paints, signs, etc. Researchers from Osaka University and their collaborators have fabricated multi-layer structures inspired by the *Morpho* using a nanoimprinting technique. This approach has the benefit of scalability, which is necessary for large-scale application. First, a quartz substrate is patterned using e-beam lithography and dry etching to create randomly arranged 110-nm deep steps (seen in Figure 3.4). Next, a seven-pair stack of TiO_2/SiO_2 layers was deposited, each having a thickness of 40 and 75 nm, respectively. The TiO_2 layers have high refractive index, while the SiO_2 layers act as low-index materials. This process creates a large-area film in which the color is derived from the multilayer stacks and is highly reflective. The non-uniformity of the height provides scattering, helping the blue color

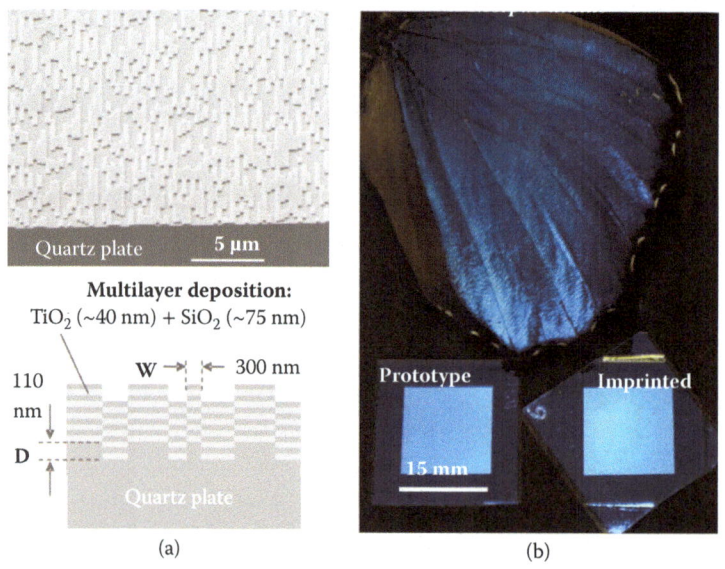

Figure 3.4 (a) SEM image of a quasi-1D pattern on a quartz substrate (top) and illustration of the *Morpho*-inspired structure (bottom). (b) Photograph of *Morpho didius* (upper) and replicated *Morpho*-blue prototype (left) and plate fabricated by nanoimprinting (right). (From Saito et al., *Adv. Fabrication Technol. Micro/Nano Optics and Photon.li*, 7205, 9, 2009.[16])

to reflect over a wider angular range than it might otherwise. The same substrate pattern, created in resin, can be used as a stamp for large-area, high-throughput nanoimprint lithographic fabrication. Figure 3.4 shows a *Morpho didius* wing, the e-beam fabricated structure and the nanoimprinted structure, both of which show intense blue color over a wide viewing angle. However, these structures do not have as wide a viewing angle as the natural wing structure and do not match some of the more complex spectral features.

Inspired by the complex optical properties of *Morpho* but also attempting to reduce fabrication complexity, researchers from KAIST in Korea (and colleagues) created a disordered multilayer structure using silica microspheres to create a flexible substrate multilayer stack.[20] An approximate monolayer of silica microspheres with wide size dispersion were spin-coated onto a Si wafer. A chromium layer 300 nm thick was then sputter-deposited onto the microspheres, mimicking an absorbing melanin layer, to reduce the back-reflection of transmitted light and achieve higher color purity (Box 3.1). An eight-pair stack of TiO_2/SiO_2 layers was sputter-deposited atop this dark background layer. Because the microspheres are not strongly attached to the Si, the entire structure can be backfilled with a transparent polymer, such as PDMS, and pulled from the Si wafer to create a thin, flexible film. The structure produced an angle-independent saturated blue color as well as other colors depending on the multilayer layer thicknesses. They show a bright color reflection with a small change in brightness and color with angle, in a flexible form.

3.1.3 Helicoidal Multilayers

Helicoidal multilayers are also called twisted multilayers. Each layer in the stack is made of fibers that are all generally aligned within the same direction. The fiber layers are stacked in such a way that the average direction of the fibers rotates slightly (in the same direction) with each subsequent layer. The rotation direction forms a helix, which is periodic and can be seen in Figure 3.5. The distance it takes for the average fiber direction to rotate all the way around to the starting orientation is called the pitch (p). These twisted multilayer structures have been found in some fruit and beetles, for example. Importantly, these structures enable the organisms to reflect polarized light, which

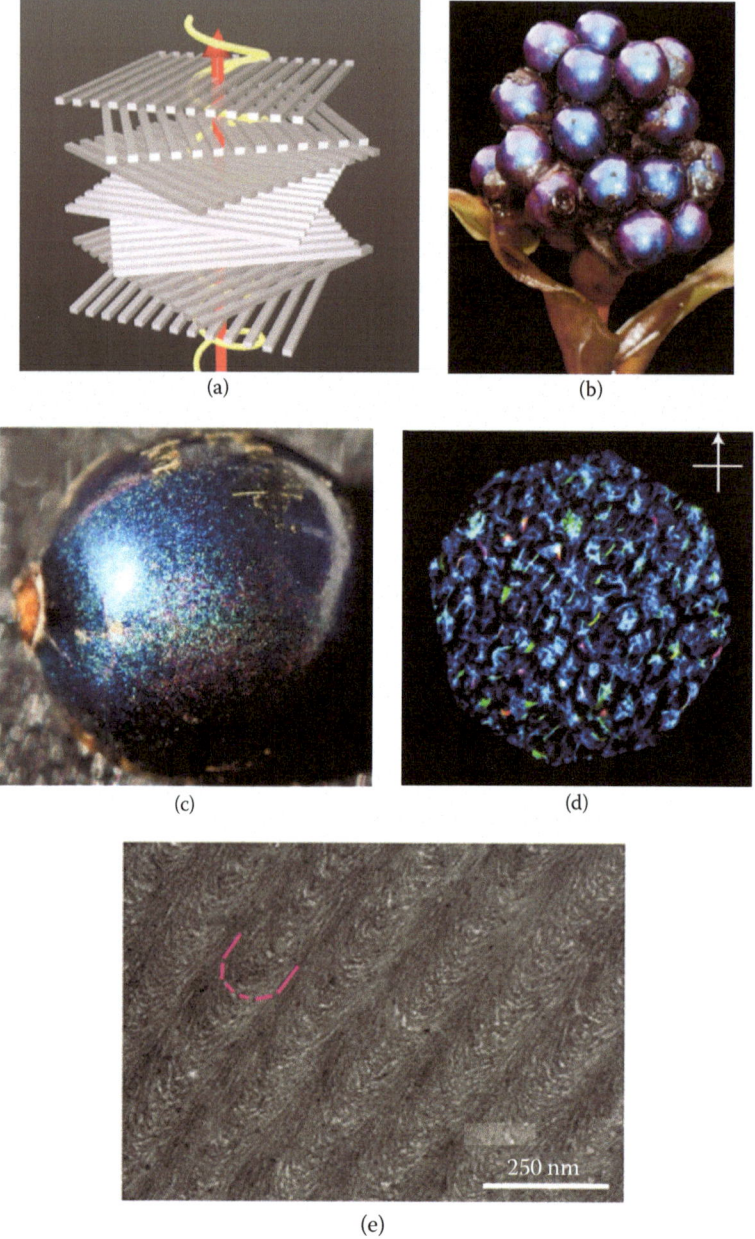

Figure 3.5 (a) 3D representation of the orientation of cellulose microfibril assembly and a transmitted circularly polarized beam. (b) Fruit cluster from preserved specimen. (c) Single dried *Pollia condensata* fruit (fruit averages ~5 mm diameter). (d) Reflection of *P. condensata* fruit surface imaged between crossed polarizers. (e) Transmission electron microscopy of the cellulose microfibrils that constitute the thick cell wall. The red lines highlight the twisting direction of the microfibrils. (From Vignolini et al., *Proc. Natl Acad. Sci. U S A*, 109 (39), 15712–15715, 2012.[21])

is visible to many marine animals and insects.[4] The particular direction of rotation of the structure is called the handedness, or chirality, of the structure, which determines the handedness of the polarization. Structures like these have been compared to cholesteric liquid crystals, which also have layers where the average alignment of molecules in those layers (called the director) twists in the z direction.

The plant *Pollia condensata* fruits in clusters of up to 40 small, metallic blue berries.[21] Each berry is about 5 mm in diameter. Looking closely at the surface of the *Pollia* fruit, the surface has "pixelated" iridescent blue coloring with green and purple/red speckles. The color forms from structures in the outermost layer of the tissue surrounding the seeds (called the epicarp). The twisted multilayers are made of helically stacked layers of cellulose fibers; each fibril is approximately 5 nm wide. They reside in the cell walls of the topmost cells in this tissue. Below these cells are layers of optically absorbent cells that contain brown tannin pigment and thin cells that scatter transmitted light. The reflected color arises from Bragg reflection from the fiber stacks. Polarization effects occur due to the continuous rotation of the orientation of the multilayer fiber planes resulting in the circular polarization of the reflected and transmitted light. In *Pollia* fruit, each helicoid stack can have either handedness, so the handedness of the polarized reflection varies from cell to cell.

Beetles, such as *Plusiotis resplendens* and other scarabaeid beetles, have also been found that can reflect polarized light.[22] *Plusiotis resplendens* reflects both left and right circularly polarized as well as broadband reflection.[4] Polarized light carries a tremendous amount of information and is often used as a navigation and communication channel. It provides information on surface shape, material contents, and local curvature of objects and can enhance object recognition, contrast enhancement, camouflage breaking, signal detection and discrimination.[23] This structure may actually have originally emerged as a mechanism to enhance mechanical resistance.[18]

These chiral structures and devices made using them are very attractive for many applications. Cholesteric liquid crystals are a commercially available display technology for reflective, low-power requirement displays, e-paper, and writing tablets.[24] Helicoidally twisted thin films and fibers are also commercially available in applications in telecommunications, lasers, biological and chemical sensing, displays, and lighting.[25] Organic liquid crystal lasers use (dye-doped) liquid crystals as a resonator cavity. Three-dimensional omnidirectional microlasers have been reported, which are created by optically pumping microdroplets of dye-doped cholesteric liquid crystals.[26] The helical pitch of the liquid crystal determined the lasing wavelength. Bioinspired approaches have also included using inspiration from the helicoidal structures found in *Plusiotis resplendens* to fabricate a heterojunction structure for lasing and optical diode behavior.[4,27,28] The structure layers a polymer cholesteric liquid crystal layer, an anisotropic dye-doped nematic liquid crystal layer, and a second polymer cholesteric layer. The cholesteric layers have different pitches from each other and the middle doped layer behaves like a half-wave retarder and resonator cavity between the reflective layers.

A similar structure (an untwisted layer sandwiched between two twisted layers) is found in *Chrysina resplendens*, which results in beetle with a broadband metallic gold reflection.[29] In this beetle, the middle layer is birefringent (its refractive index depends on light polarization) and the pitch of the helical layers varies, which contributes to the broadband reflection. The synthetic structure was fabricated by first depositing a nematic alignment layer, which controls the initial direction of the cholesteric liquid crystal at that interface. Two layers of cholesteric liquid crystals were then deposited, the first with a reflection peak of 639 nm and the second with a reflection peak at 562 nm. An untwisted layer, which operated as a half-wave retarder at 595 nm (a wavelength between the wavelengths of the twisted layers), was deposited. Finally, two more cholesteric layers (562 and 639 nm wavelengths) were deposited on top of the untwisted layer. A scanning electron microscopy (SEM) of the synthetic layered structure is shown in Figure 3.6. The *Chrysina resplendens* beetle can be seen in that figure, next to the similarly colored structured film. Both show metallic gold coloring. This technology may be attractive as a low-cost manufacturing route for reflectors in displays, and the group is exploring roll-to-roll processing and scale-up of the process.

3.1.4 Intercalated Structures

Biological photonic structures often combine multiple individual structures (intimately collocated) to create complex optical phenomena. Nature also creates structures where more than one type of photonic structure is actually completely intermingled, or intercalated, with another structure. The Taiwanese beetles *Trigonophorus rothschildi varians* are very colorful, as seen in Figure 3.7.[30] They can be found (in the same population and habitat) in colors ranging from black, orange, purple, and green (green being most prevalent).

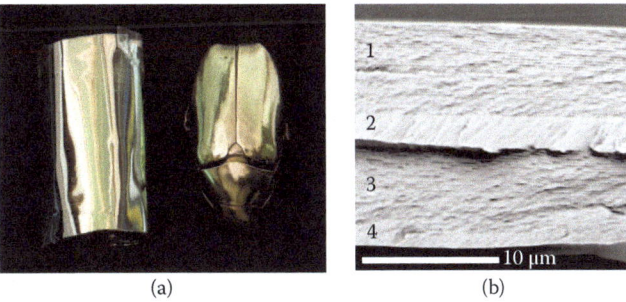

Figure 3.6 (a) Photograph of a composite 562- and 639-nm cholesteric film (left) and a *Chrysina resplendens* beetle (right). (b) SEM image of the composite cholesteric film. Layers labeled as (1) cholesteric film, (2) untwisted retarder layer, (3) cholesteric film, and (4) untwisted alignment layer. Layers 1 and 3 are bilayer films (not distinguishable in the image) with two cholesteric sublayers (pitches 562 and 639 nm). (From Matranga et al., *Adv. Mater.*, 25 (4), 520–523, 2013.[29])

Figure 3.7 (a) Group of *T. rothschildi varians* in their natural habitat showing the three color variations. (b) Reflectance versus wavelength for the three-colored elytra samples, as measured with an integrating sphere (circle, orange; square, violet; and triangle, green). (c) SEM images of cross section of elytra of purple beetle. Scale bar: 1 μm. (From Biro et al., *J. Roy. Soc., Interface/Roy.l Soc.*, 7 (47), 887–894, 2010.[30])

SEM images taken of the differently colored beetle elytra show a multilayer structure of chitin and air with a period of 200, 180, and 150 nm for the orange, green, and violet beetles, respectively. Within the multilayer, rodlike structures pierce the layers roughly perpendicular to the plane of the chitin/air layers. The rods are randomly distributed and each may cross many layers in the multilayer stack (as many as 10–20 periods). The multilayer material also seems to contain pigment, likely melanin, potentially acting as an optical absorbing layer. In this structure, the peak reflectance wavelengths are the result of the multilayer reflector, but the presence of the rods broadens the angular range over which the reflection occurs for non-normal incidences.

An intercalated multilayer structure inspired by *T. rothschildi varians* was fabricated to explore the role the secondary structures play in the optical reflectance. To do so, an amorphous silicon oxide (SiO) and silicon germanium (SiGe) multilayer was fabricated using physical vapor deposition. The total stack consisted of five layers each of alternating SiO and SiGe, 50 and 45 nm thick, respectively. The multilayer was then patterned with holes using an FIB, with the holes either randomly distributed or ordered in a regular square lattice. The random holes were spaced at four different distances: 743, 743, 609, and 580 nm apart (average distance). The regular square holes were spaced 746 nm apart. The patterns were each tiled in a 3 × 3 array to increase the size of the probe-able nanostructure. The structures, seen in Figure 3.8, were studied to identify changes in reflectance of the patterned multilayers at varying angles.[30] In the figure, the black background is the unpatterned multilayer, which showed no reflection under these conditions. The leftmost four squares show the randomly patterned structures at the varying spacing, while the rightmost square is the regular square lattice pattern. As the angle between incidence and observation varies, the color of the square lattice changes significantly, while the random pattern changes only slightly. When the plane of illumination is rotated 45° with respect to the rows of holes in the square lattice, the lattice pattern ceases to reflect, whereas the random pattern still reflects light. Much like the biological inspiration, the synthetic structure shows colored reflection due to the multilayer structure and broad angle independence due to the intercalated structure.

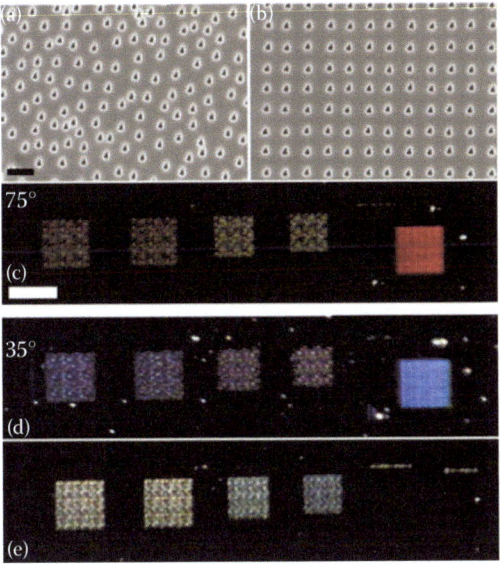

Figure 3.8 Artificial patterned photonic structures inspired by *T. rothschildi varians*. All squares except the rightmost column are squares with a random distribution of holes; the rightmost column is a square pattern of holes. (a) SEM image a 5 × (SiO/SiGe) multilayer with random hole distribution. (b) SEM image of a 5 × (SiO/SiGe) multilayer with square hole pattern. Scale bar (a, b): 1 μm. (c, d) Optical micrographs showing the same photonic structures under 75° and 35° illumination (plane of illumination is parallel with the rows of the square pattern). Scale bar: 50 μm. Right-most square is the regular hole pattern (center-to-center distance of holes: 746 nm); all others are random hole distributions (from left to right, the average hole distance is 743, 743, 609, and 580 nm). (e) Optical micrograph with illumination angle at 45° but the plane of illumination is rotated at an angle of 45° with respect to the rows of holes of the regular (square) pattern. (From Biro et al., *J. Roy. Soc., Interface/Roy.l Soc.*, 7 (47), 887–894, 2010.[30])

3.2 Three-Dimensional Structures

Three-dimensional structures, as you might expect, are periodic in all three spatial dimensions. Fabricating 3D photonic crystals should enable us to modulate light and control its reflectivity, angular dependence, etc. One of the most desirable optical effects we are attempting to create with synthetic photonic crystals is a complete, tunable bandgap. That is, a wavelength (or range of wavelengths) for which the propagation of light through the photonic structure is forbidden in every direction and polarization. The ability to have complete control over the optical properties of a material, to be able to design and fabricate these structures, has tremendous application potential in many optical and optoelectronic devices, such as sensors, displays, optical filters, lasers, optical computing, and communication.

BOX 3.1 IMPORTANCE OF LIGHT ABSORBING LAYERS

Several examples of structural color have been found that incorporate pigment (often melanin) into the biological photonic structures. These pigments, either incorporated into the structure or as a separate layer, play an important role in the production of high-purity vivid color by absorbing incoherently scattered white light.[31] Melanin has comparatively high real and imaginary parts of the refractive index. These absorbing layers and concepts have inspired several efforts to improve optical reflection in synthetic photonic structures. One-dimensional multilayers were fabricated using transparent polymer layers (poly(1,2-butadiene) (PB)) and thin (~4.5 nm) black metal layer (osmium) inspired by the multilayer structures found in Jewel beetles.[32] While the osmium layers are not optically absorbing, unlike melanin, the inclusion of the very thin layers of osmium into the multilayer created strong reflectance. Carbon is a good optical absorber, and its inclusion into colloidal photonic crystals has been shown to directly affect the production of structural color.[33]

Three-dimensional photonic crystal structures can and do appear in nature. Precious opal is formed from ordered assemblies of spheres of amorphous hydrated silicon dioxide.[34] The spheres, each approximately 250 nm in diameter, form ordered hexagonal or cubic close-packed lattices in 3D domains. The colors seen in these structures are due to the orientation and spacing of the lattice planes with respect to incident light. The most frequently studied synthetic 3D photonic crystals are the opal and inverse opal structures, which can be fabricated through colloidal techniques. Though the fabrication of these structures is challenging, the photonic band structures of these synthetic structures can be calculated.[18] In addition to colloidal self-assembly, synthetic 3D photonic crystals can be fabricated lithographically or with 3D printing. However, these techniques are expensive and not yet suitable for large-scale fabrication at high resolution and speed, which will be required for many of their potential applications.

BOX 3.2 CRYSTALLOGRAPHIC SYMMETRY AND PHOTONIC BAND STRUCTURES

In crystalline atomic materials, the specific periodic arrangement of atoms and their relation to nearest neighbor atoms are described by identifying the geometry of the unit cell of the structure and its lattice constants. The unit cell is the smallest periodically repeating arrangement of atoms in the structure, and the lattice constants are the physical dimensions of the unit cell. Identifying the specific crystallographic structure of biological photonic systems is significantly more complex than for an inorganic crystalline atomic structure.[35] The structures must be teased out of the biological systems and the 3D arrangement of the chitin reconstructed. Biological structures have been discovered with periodicity and ordering similar to that found in classical inorganic crystals such as the cubic crystal structures seen in Figure 3.9. The photonic band structure is a description of the frequency (lattice constant/wavelength) versus momentum (the direction of wave propagation).[37] The named points are related to points in the unit cell (in momentum space, the Brillouin zone corresponds to the unit cell) and are points of high symmetry.

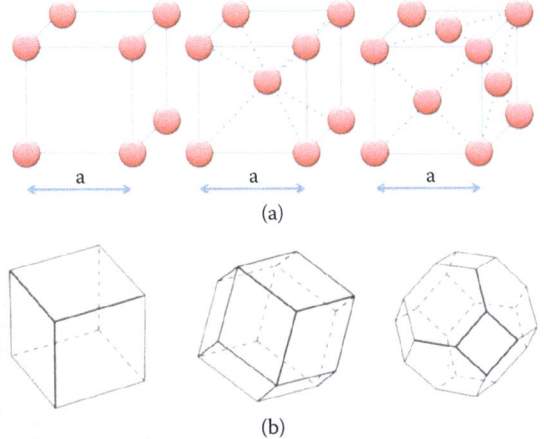

Figure 3.9 (a) Examples of cubic crystal structures: simple cubic (SC), body-centered cubic (BCC), and face-centered cubic (FCC) with lattice parameter a. (b) The first Brillouin zone for the three cubic structures SC, BCC, and FCC, respectively.

(Continued)

BOX 3.2 (*Continued*) CRYSTALLOGRAPHIC SYMMETRY AND PHOTONIC BAND STRUCTURE

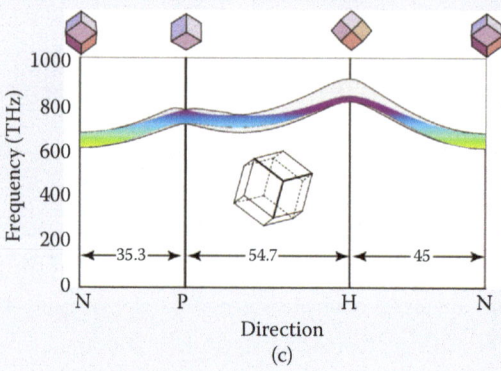

Figure 3.9 (*Continued*) (c) Calculated band structure for a BCC lattice with refractive index of 1.5 and volume fraction of 50%. (From Poladian et al., *J. Roy. Soc., Interface/Roy. Soc.*, 6 Suppl 2, S233–242, 2009.[36])

Γ is the point of highest symmetry, typically the center of the Brillouin zone. The panels in the band structure calculations represent paths between these points.

Close-packed structures are by no means limited to mineraloids. Several examples of biological 3D photonic structures have been discovered in both beetles and butterflies. The biological structures occur in one of three crystal structures: face-centered cubic (FCC), diamond, and body-centered cubic (BCC)/gyroid (at least those that have been discovered thus far). These structures are not perfectly periodic. There is significant long- and short-range disorder. Although some of these structures, such as the diamond structure, have been predicted to be able to achieve complete bandgap, none have so far been found in these biological structures. The contrast between high and low refractive indices is too small to generate omnidirectional bandgaps. However, as with many of the previous examples we have discussed, nature uses intriguing tricks to achieve similar results. One common aspect among all of the 3D biological photonic structures is the use of microdomains to achieve wide-angle reflection; they are more accurately described as polycrystals. Within a given microdomain, the photonic structures have a high degree of short-range order, but the structure has long-range disorder. Each microdomain produces vivid, saturated interference colors and combines with the reflected light from other microdomains to decrease the angle dependence.

3.2.1 Cubic Structures

Similar in structure to opaline photonic crystals, several examples of FCC structures have been found in beetles, particularly weevils and some longhorn beetles (e.g., *Prosopocera lactator* and *Pseudomyagrus waterhousei*).[22,35,38,39] Found on the Malay Peninsula and neighboring areas, the longhorn *Pseudomyagrus waterhousei* has a colorful blue-violet and black patterned carapace.[36] The blue-violet color covers most of the beetle's body, originating from teardrop-shaped scales approximately 50 μm in length. The beetle and its scales can be seen in Figure 3.10. The scales are not only blue and violet, but some also

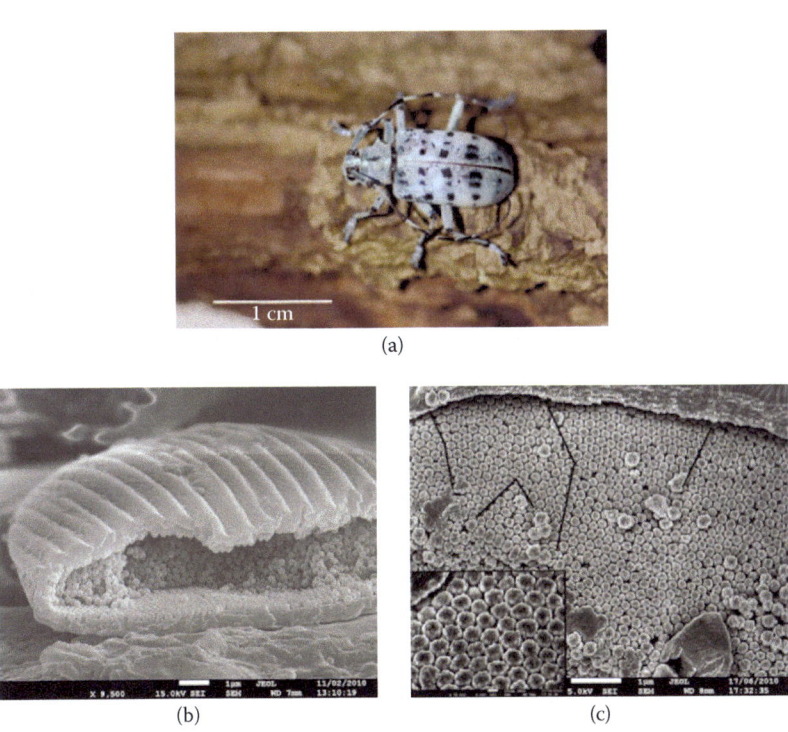

Figure 3.10 (a) Longhorn beetle *Pseudomyagrus waterhousei* showing diffuse (matte) blue coloring without iridescence. Scale bar: 1 cm. (b) SEM cross section of a scale showing the outer cortex and inner colloidal structure. Scale bar: 1 μm. (c). Chitin spherules inside the scale cortex with clearly visible domains. Scale bar: 1 μm. Inset: close-up of the spherules. (From Simonis et al., *Phys. Rev. E*, 83 (1), 2011.[38])

show spots of yellow, pink, or green. Far from the scales, color mixing of these colors averages to the diffuse, desaturated blue seen in the figure, and the reflection shows no strong angle dependence. The black color can be seen in small semirectangular patches on the beetle's carapace in regions with no scales. Spectrophotometric measurements show that the beetle reflection has a broad peak near 476 nm (blue) and a nearly flat sideband starting around 328 nm (violet). Individual scales were fractured and the cross section was examined with SEM.

The scale structure is very interesting. Down the length of the surface of the scale, long, parallel grooves separate micron-scale rib-like structures. On each rib, nanoscale flutes seem to emerge from the approximate center of each rib. The outer layer of the scale is approximately 500 nm thick and acts as a shell (cortex). Within this structure are arrays of periodically arranged, close-packed chitin colloids. In the blue-violet scales, the colloid diameter is approximately 212 nm, resulting in a peak reflectivity wavelength near 480 nm, depending on crystal orientation. As mentioned above, these periodic arrays are actually organized into polycrystalline domains, which enable angle-independent reflection when viewed from far away. Some of the polycrystalline grains are observed to reflect other colors, such as green, yellow, or pink. Within these alternately colored scales were both ordered and disordered domains containing colloids of varying diameters (ranging from 170 to 300 nm). Modeling indicated that the size of the particles in these disordered domains is primarily responsible for the alternate coloring.

Another FCC structure can be found in the bright orange scales of the weevil *Pachyrrhincus congestus pavonius* (Heller 1921).[39] Bright orange is not a common color found in insects. In the orange rings on the beetle's elytra, noniridescent scales (orange as well as some green and blue) can be found (beetle shown in Figure 3.11). As in *Pseudomyagrus waterhousei*, the scale consists of an outer cortex and an inner FCC lattice structure. The layers are made of chitin, and a regular array of voids perforates the layers. The distance between the holes is approximately 333 nm, leading to a broad reflection peak centered around 700 nm. Within the structure, there is disorder in the size of the perforations that occur within the lattice, as

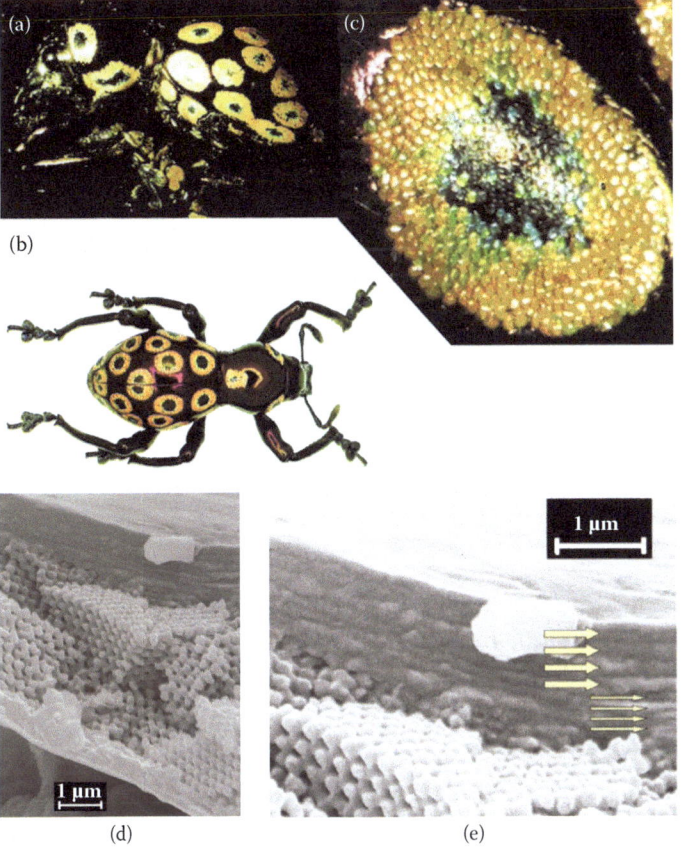

Figure 3.11 (a) Weevil *Pachyrrhynchus congestus pavonius*. (b) The specific specimen used in the study. (c) Detail of an orange spot on the exoskeleton. (d) SEM cross section showing the thick, layered cortex and the inner 3D structure (e) Detail of the layered cortex (the outer part of a scale). Two distinct multilayer structures can be seen. The upper multilayer reflects yellowish green light at a peak wavelength of 528 nm, while the lower multilayer reflects purplish blue light a peak wavelength of 447 nm. (From Welch et al., *Phys. Rev. E*, 75 (4), 2007.[39])

well as the orientation of the lattice. The top side of the cortex of a *Pachyrrhincus congestus pavonius* scale is thicker than that found in *Pseudomyagrus waterhousei*, between 1 and 2 μm. SEM shows multilayered structures in this cortex, and analysis indicates that the weevil's scales are actually a three-component system of an interior 3D polycrystal with peak wavelength around 675 nm, surrounded by a double multilayer reflector where the topmost reflects at 528 nm and bottom reflects at 447 nm.

3.2.2 Diamond Structures

The weevils *Entimus imperialis* and *Lamprocyphus augustus* achieve a similar effect via a diamond-based lattice.[40–42] Diamond lattice photonic crystal structures theoretically have the capacity to create complete bandgaps, making natural examples of great interest. *Lamprocyphus augustus* has oblong scales that create the deep green color. Each of the scales comprises an outer cortex and an inner 3D lattice. The topside of the scale cortex is thick, nearly 2 μm, as seen in the SEM image in Figure 3.12. Using very careful FIB milling, the structure was sectioned to reveal the precise scale architecture, which was then reconstructed to enable modeling of the photonic band structure. Resulting calculations showed several partial bandgaps.

3.2.3 Gyroid Structures

Gyroid structures have been found in some Papilionid and Lycaenid butterflies, such as the *Callophrys rubi* and *Paridis sesostri*.[43–45] Several species have been found so far with this type of structure, but interestingly none in beetles. *Parides sesostris*, seen in Figure 3.13, is also called the emerald-patched Cattleheart butterfly.[45] The wing has three major color areas: metallic green, deep black, and diffuse red. The reflectance spectra of the three different color areas on the butterfly wing can be seen in Figure 3.13. The green scales show a peak in reflectance at around 540 nm. The reflectance of the black portion of the wing is extremely low (<5%) throughout the visible range. The red patch shows broadband reflection above about 600 nm.

Truly stunning images of the structure of the green scales in *Parides sesostris* were achieved via helium ion microscopy (HIM), seen in Figure 3.14. HIM allows very high resolution, large depth-of-field images of these structures to be taken without a lot of the sample preparation required of SEM (such as coating the structures with metal).[46] As with many other Lepidopteran scales, ridges run longitudinally down the length of the top face of the scale. Cross-ribs connect the ridges and in this butterfly create semirectangular openings to the structures below. Microribs can be seen on the

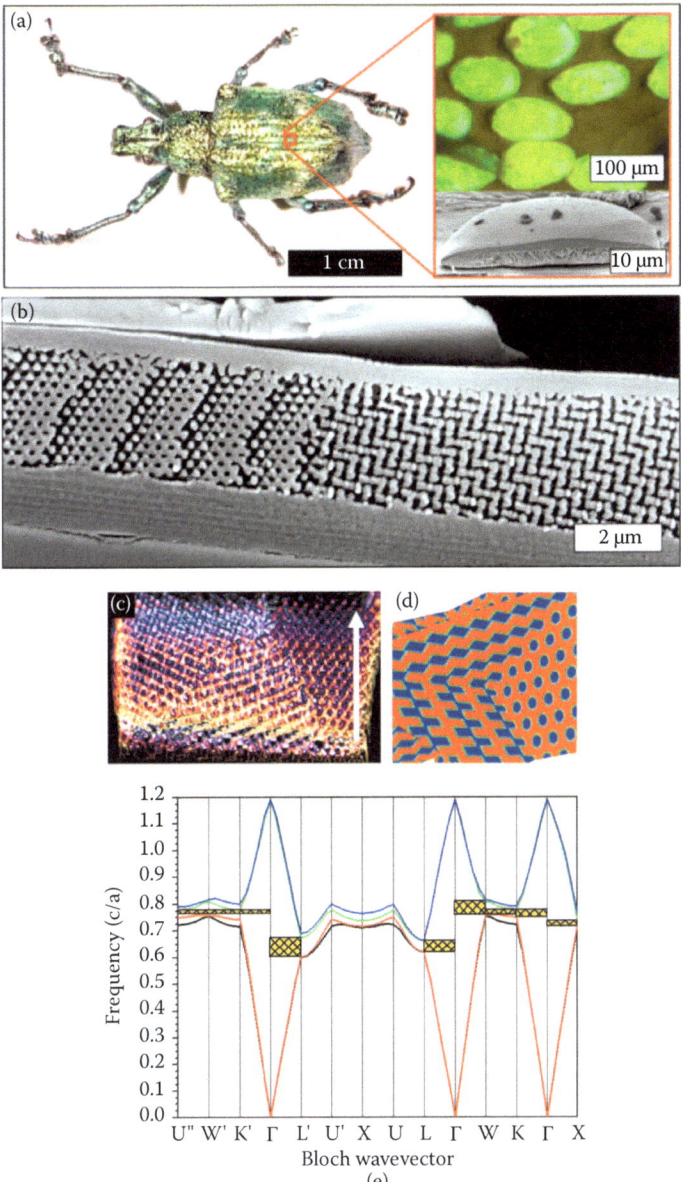

Figure 3.12 (a) The weevil *Lamprocyphus augustus*. Inset: Optical micrograph of individual scales attached to the exoskeleton under white-light illumination (top) and SEM image of cross section of a single scale. (b) SEM image of the cross section of a section of a scale. (c) 3D reconstruction (model) of the focused ion beam sectioning data. Arrow indicates direction of the milling process. (d) Dielectric function of the beetle's photonic crystal generated from 3D structure analysis showing three orthogonal planes. Air: blue; dielectric: red. (e) Photonic band diagram calculated from the beetle diamond-based structure. (Bandgaps denoted by yellow rectangles.) (From Galusha et al., *Adv. Mater.* 22 (1), 107, 2010.[42])

Figure 3.13 (a) Dorsal wings of *Parides sesostris*, Emerald-patched Cattleheart, with large, green-colored spots on the forewing and smaller, red spots on the hindwing, surrounded by black. Scale bar: 1 cm. (b) Reflectance versus wavelength of the green- and red-colored wing patches and the black frame of the upper side of the butterfly wing. Green solid line, green patches; red solid line, red patches; black solid line, black patches. (From Wilts et al., *Interface Focus*, 2 (5), 681–687, 2012.[45])

cross-ribs. A SEM of the cross section shows that this upper structure is quite thick, between 3 and 5 μm.[45] This "honeycomb" structure continues the cross-rib and rectangular air-gap structure seen from the surface. Below the honeycomb lies a 3D gyroid photonic polycrystal, which can be seen once again in HIM images.[44] An

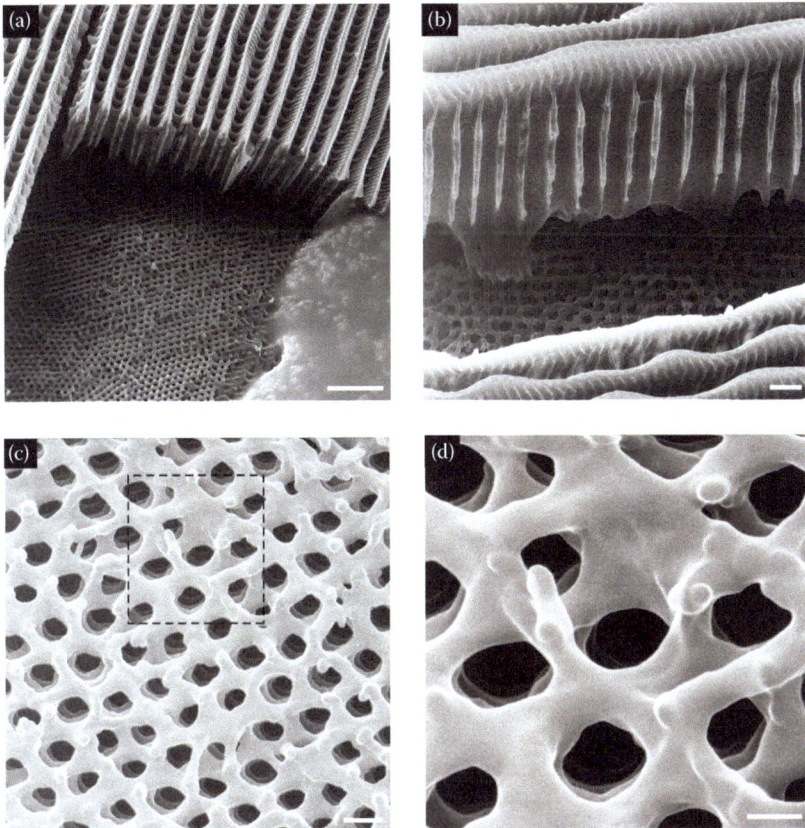

Figure 3.14 (a and b) Helium ion microscopy (HIM) images of the green scale from *Parides sesostris* showing detail of ridges and cross-ribs in cross section on the upper lamina from two different viewpoints. Scale bars and stage tilt angles: (a) 2 μm, 40° (b) 500 nm, 45°. (c) High-magnification HIM images of *Paridis sesostris* wing scale 3D photonic crystal. Box in (c) indicates the area for the image in (d). Scale bars and stage tilt angles: 200 nm, 40°. (From Boden et al., *Scanning* 34 (2), 107–120, 2012.[46])

extreme close-up image of the gyroid structure shows the ordered array of air holes in chitin. This structure can be described as a "bicontinuous, triply periodic structure with a BCC Bravais lattice symmetry."[41] This means that, for any given domain, both the air "lattice" and the chitin lattice each form a continuous 3D periodically repeating volume that interpenetrates the other, as seen in Figure 3.15. The continuous interface between these two lattices locally minimizes its surface area, meaning that the mean curvature is zero at any point. In this particular structure, the air lattice is larger than the chitin lattice. The fill-factor, along with refractive index contrast and lattice period, is an important variable influencing the reflectance properties. The gyroid structures are chiral and twist in one preferred direction or the other, resulting in preferential absorption of left- or right-circularly polarized light (circular dichroism).

Returning a bit to the optical properties of *Paridis sesostri*, the scales show an intense metallic green reflection.[36] The lattice constant of the gyroid structure has been reported at various values between approximately 220 and 300 nm with a chitin fill-fraction of about 40%.[43,45,46] The gyroid structure alone reflects at wavelengths around 540 nm but is strongly dependent on the microdomain orientation.[45] Looking at the scale from the top, the reflectance seems relatively isotropic, and angle independent. However, when looked at from below, iridescence can be seen.[36] This implies that in addition to the polycrystalline nature of the gyroid structure, the upper honeycomb structure plays a role in iridescence suppression. Further analysis and modeling show that not only is iridescence suppressed but so is the circular dichroism (polarization effects).[45] When looking at the reflection from an isolated scale through crossed polarizers the reflection seems to be relatively isotropic, but when viewing the underside through crossed polarizers polarization domains are very clearly visible. The honeycomb structure contains a pigment that is absorbent in both the UV and blue wavelength ranges. Light entering the structure has ample opportunity to interact with the absorbing pigment, due to the deep honeycomb structure. The honeycomb layer causes approximately angle-independent scattering and the pigment acts as a spectral filter, eliminating any reflecting light save for a narrow band at green wavelengths. As most of the polarization effects due to the *Parides sesostris* gyroid structure occur outside of this narrow wavelength band and are therefore filtered, the scale effectively loses the circular dichroism otherwise expected.

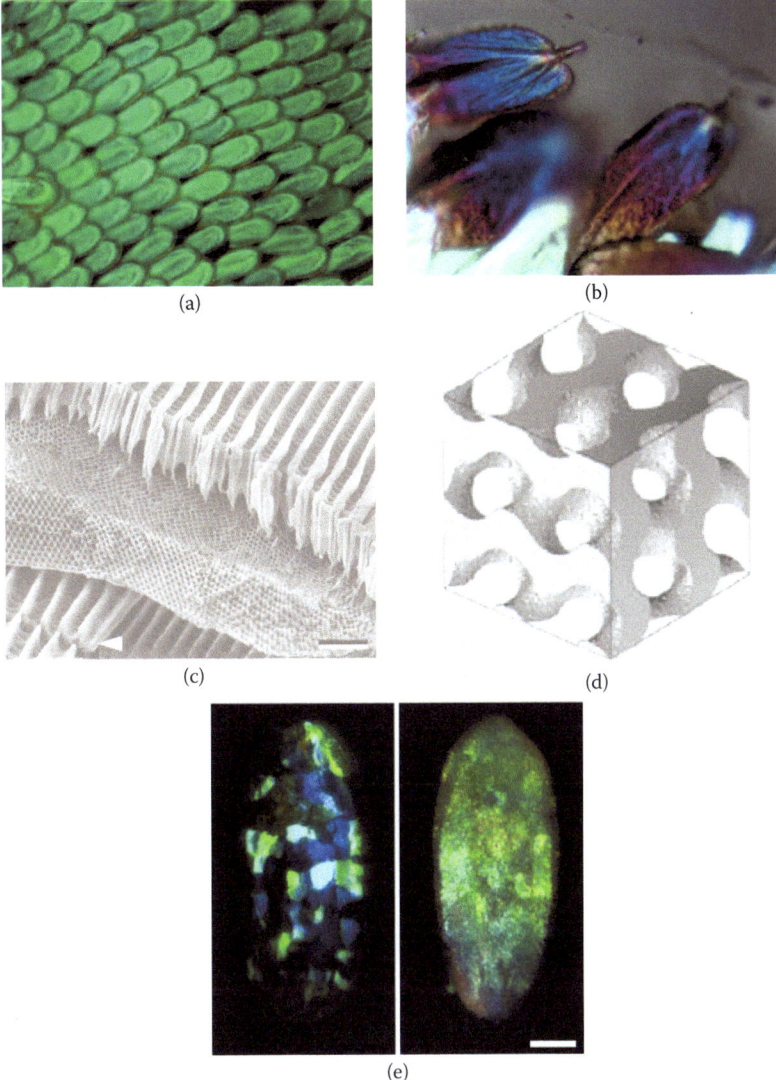

Figure 3.15 (a) Optical image of a section of the dorsal wing from *Parides sesostris* showing scales with wide-angle reflectance. (b) Optical image of the underside of individual scales. Upside-down scales show strong iridescence. (c) SEM cross section of a scale showing the ~5-μm-thick single-network gyroid photonic crystal layer below the ~5-μm honeycomb structure. Scale bar: 2 μm. (d) Calculated gyroid structure with cuticle volume fraction 0.40 (volume fraction of *Parides sesostris*.) (e) A single scale upside-down (bottom side) observed through crossed polarizers (left) and the single scale upside-up (top side) observed through crossed polarizers (right). Single domains are visible from the bottom side but are not visible in the normal scale orientation. Scale bar: 20 μm. (From Wilts et al., *Interface Focus*, 2 (5), 681–687, 2012; Poladian et al., *J. Roy. Soc. Interface*, 5 (18), 85–94, 2008; and Michielsen et al., *J. Roy. Soc. Interface* 5 (18), 85–94, 2008.[43,45,36])

BOX 3.3 FOSSILS

Fossil evidence of structural color exists from as far back as the Cambrian period, over 500 million years ago.[47] Coloration in fossils has been described in insects such as beetles and moths, mollusks, and feathers and has enabled us to attempt to determine the color in ancient insects, dinosaurs, and birds.[48-54] Some quite astonishing more recent examples of fossilized organisms and their biophotonic structures can be seen in Figure 3.16. Many of the structures found are very similar to structures we can find today. The study of color in fossils is challenging due to the fact that depositional context, geological history, age, and preservation quality, among other factors, can all affect the interpretation of data.[55] Many pigment molecules decay leaving little evidence for analysis for pigment-based color. Melanin is the only class of pigments for which we have fossil evidence and can be identified by the organelles, melanosomes, in which the pigment typically resides. Structural compression, distortion, and composition change, for example, can occur with fossilization altering the optical properties of the photonic structures responsible for color. Some fossilized organisms may show color or iridescence in the discovered specimen, while others may not. The color of the fossil does not necessarily indicate the lack or existence of photonic structures. The process of fossilization itself may induce iridescence or metallic color in a specimen. Index of refraction of the cuticle may change during fossilization. Experiments can be performed on extant, i.e., still existing, organisms that produce artificial, accelerated maturation of the photonic structures, to attempt to mimic and determine the effect of fossilization processes on the structure, such as pressure and decay.[56] The overarching question in many of these cases is whether or not the color visible in the fossils now is representative of the colors of the organisms when they were alive.

BOX 3.3 (*Continued*) FOSSILS

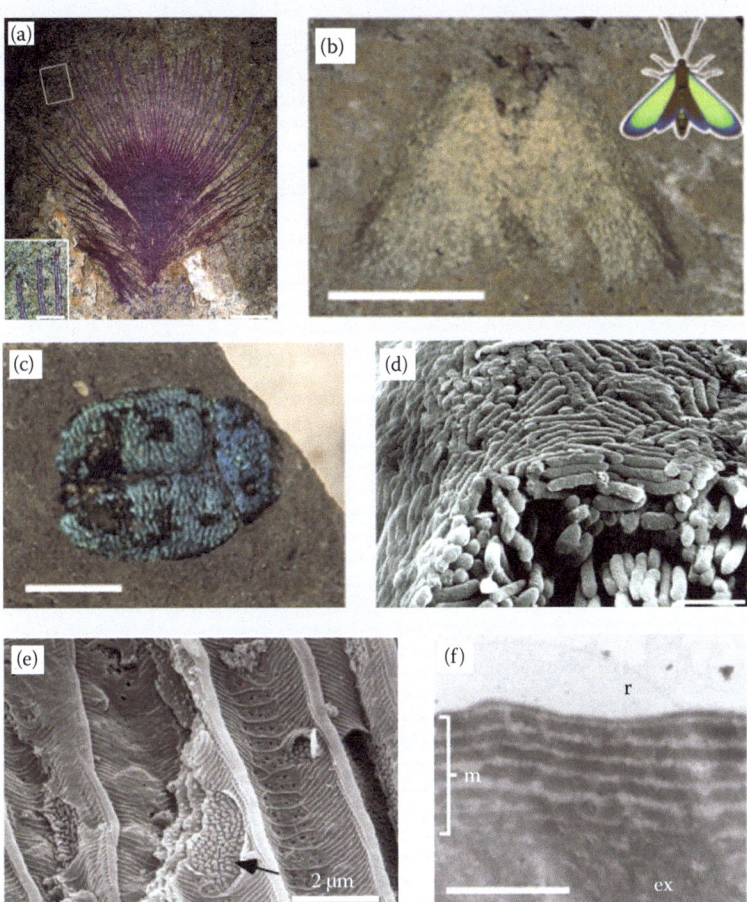

Figure 3.16 (a) Brightly colored feather fossil found at the middle Eocene Messel Oil Shale in Germany. Scale bar: 5 mm, inset 1 mm; (b) leaf beetle (Coleoptera: Chrysomelidae) from the middle Eocene of Eckfeld, Germany. Scale bar: 2 mm; (c) moth from Messel (inset, reconstruction of original wing colors). Scale bar: 5 mm. (d) SEM cross section of the feather barbule in (a) showing a thin outer layer and aligned melanosomes within. Scale bar: 1 μm. (e) SEM showing two overlapping scales of the fossilized moth in (b) with well-preserved rib and microrib structures. Arrow indicates some potentially ordered bead or rodlike structures that sit below the upper surface. Scale bar: 2 μm. (f) Transmission electron microscopy of multilayer reflector (m) in the fossilized epicuticle of a preserved unidentified beetle from the early Miocene of Clarkia, United States. Scale bar: 1 μm. (From McNamara et al., *PLoS Biol.* 9 (11), e1001200, 2011; McNamara et al., *Palaeontology* 56 (3), 557–575, 2013; Vinther et al., *Biol. Lett.* 6 (1), 128–131, 2010.[49,54,55])

3.2.4 Inspired Synthetic 3D Structures

Fabrication of these hierarchical and complex 3D architectures is a not-insignificant challenge. As with other structures, several efforts have been made to replicate 3D biophotonic structures using the biological structure itself as a template. Templating is an attractive approach to fabricating biological photonic structures, particularly these complex 3D structures, as we are not yet able to fabricate them at size scales appropriate for operation at visible wavelengths.[41] Templating allows us to explore the photonic properties of these systems, varying the constituent materials and material properties, even if we are not yet able to fully direct their architecture. Inorganic materials, silica and titania, have been used to replicate the gyroid structure found in *Callophrys rubi* (seen in Figure 3.17).[57,58] *Callophrys rubi* scales have a similar gyroid structure to that found in *Paridis sesostri* described above but lack the microns-deep honeycomb structure above. To replicate the gyroid structure, silica and titania sol–gel precursors were allowed to infiltrate the porous biological structure. Once those reactions were complete, the chitin template was removed either by acid etching in the case of a silica replica or calcination in the case of a titania replica. Because the inorganic material adopts the shape of the air lattice in the gyroid structure, these replicas are more accurately called inverse replicas. The refractive index of mesoporous silica is 1.23 versus 1.56 for chitin. While the refractive index of titania can vary with the preparation method, a value of 2.07 was used to model the optical reflection. The wavelength of the reflection is different for the inverted replicas compared to the biological structure. This is due, in part, to the fact that the inverted structure has a larger fill-fraction than the biological structure. In addition, the refractive index difference between the high and low refractive index components of the structure has changed, also shifting the reflecting wavelength. The titania replica demonstrated very high reflectance. The optical properties varied with calcination temperature. When the replica was treated at 700°C, the resulting structure had a 96% reflectivity at approximately 505 nm. Optical effects due to the chirality and broadband reflection were also observed.

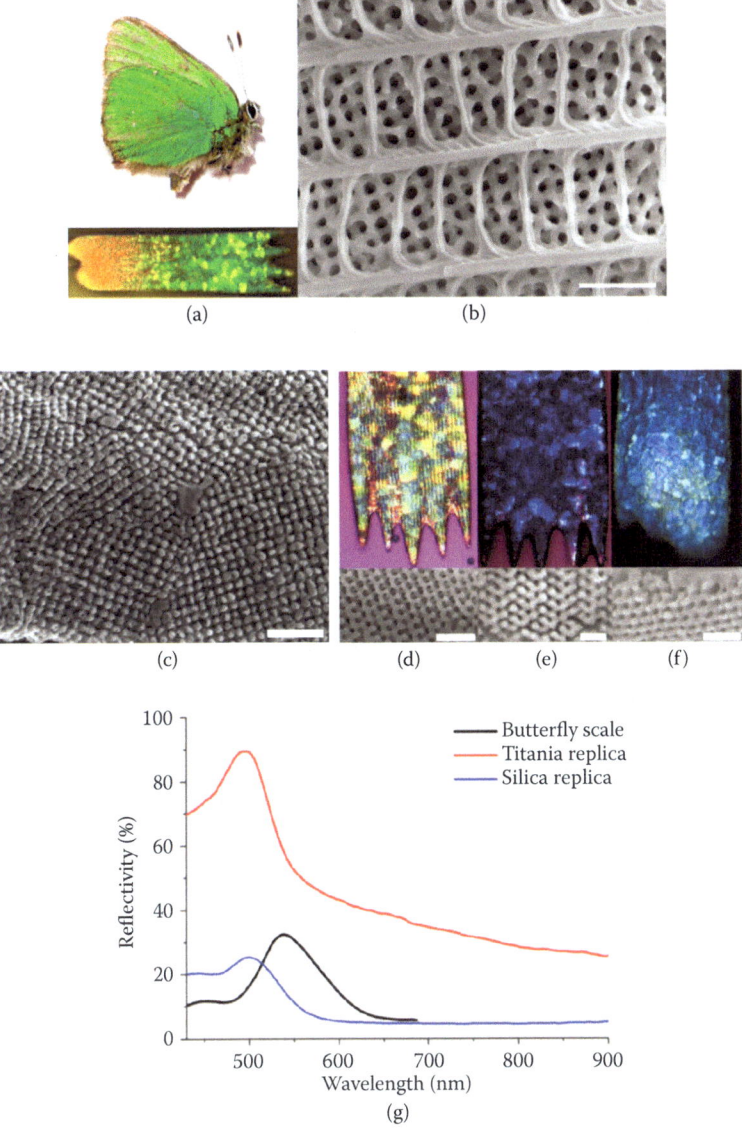

Figure 3.17 (a) Green-winged *Callophrys rubi* with optical micrograph of a single wing scale (below). Green microdomains can be seen as well as yellow and blue in the distal part of the scale. Wing scale is ~50 × 200 μm. (b) SEM of the top of the wing scale. Scale bar: 1 μm. (c) SEM image of the top surface of the titania photonic crystal replica after removal of the organic component. (d–f) Optical micrograph (top) and SEM (bottom) of the (d) butterfly wing scale; (e) silica replica; and (f) titania replica. Scale bar: 1 μm in (d) and 500 nm in (e and f). (g) Representative reflectivity spectra of *Callophrys rubi* natural scale, silica, and titania replicas. Reflection maxima centered at 540, 500, and 505 nm. (From Mille et al., *RSC Adv.*, 3 (9), 3109–3117, 2013; Mille et al., *Chem. Comm.*, 47 (35), 9873–9875, 2011.[57,58])

Titania was used to create a double inverse (or true replica) of the diamond structure found in *Lamprocyphus augustus*, seen in Figure 3.18.[42] In this case, a sacrificial sol–gel silica template was made by infiltration of the beetle lattice. The silica replica was then used as a mold and infiltrated with sol–gel titania. The silica template was then removed. By carefully controlling the crystallinity and density of the titania, a high refractive index could be achieved. Refractive indices for titania can reach 2.5, which is significantly higher than chitin. The resulting successfully replicated structures showed angle-independent reflection indicating a complete bandgap, which was confirmed by modeling. The ability to create high-quality replicas of complex biological photonic structures and tune the material and geometric properties is highly desirable.

The feature sizes of the biological photonic crystals enable the structures to interact with light at visible wavelengths. If these same structures were scaled up to larger dimensions, the structures would then interact with light at longer wavelengths such as infrared or millimeter-wave. While we cannot yet easily fabricate complex structures (like the gyroid structures) at size scales required for optical wavelengths, we can fabricate these structures at larger size scales using techniques like 3D printing and interrogate them at longer wavelengths to understand

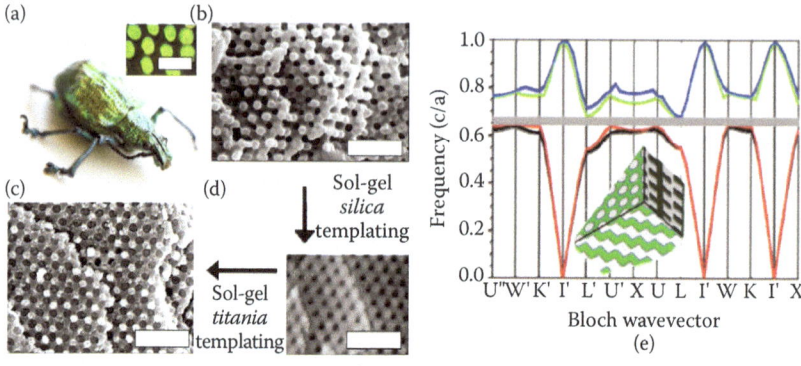

Figure 3.18 Illustration of the double-imprint sol–gel biotemplating process to create titania *Lamprocyphus augustus* replicas. (a) Photograph of the weevil and an optical image of its green-colored scales (inset). (b) SEM cross section of the original beetle biopolymer photonic structure (c) the inverse structure of the beetle photonic structure made of hybrid silica (d) and the titania replica of the beetle photonic structure templated from the hybrid silica structure. Scale bars: 200 μm (a) and 1 μm (b–d). (e) Calculated band structure diagram of a high-refractive index (RI of 2.3) *Lamprocyphus augustus*-like diamond-based lattice. Gray rectangle indicates a complete photonic band gap (PBG). Inset: Corresponding calculated dielectric function showing three orthogonal planes (air: gray; high-RI: green). (From Galusha et al., *Adv. Mater.* 22 (1), 107, 2010.[42])

better their optical properties. Researchers from the University of Exeter did exactly this to probe the photonic properties of the gyroid structure found in *Parides sesostris*.[59] Using 3D stereolithography, they fabricated a large-scale replica of the structure such that the transmission and reflection properties were optimized for the millimeter wavelengths, corresponding to frequencies from 10 to 43 GHz. Photographs of the 3D printed structure can be seen in Figure 3.19. The large structure was also designed to replicate the fill-fraction found in the biological system, meaning that the synthetic gyroid was 40% material and 60% air. This same structure was then modeled using the finite element method, which showed excellent agreement with the observed spectra. The fill-fraction of the gyroid structure provides a mechanism by which the reflected

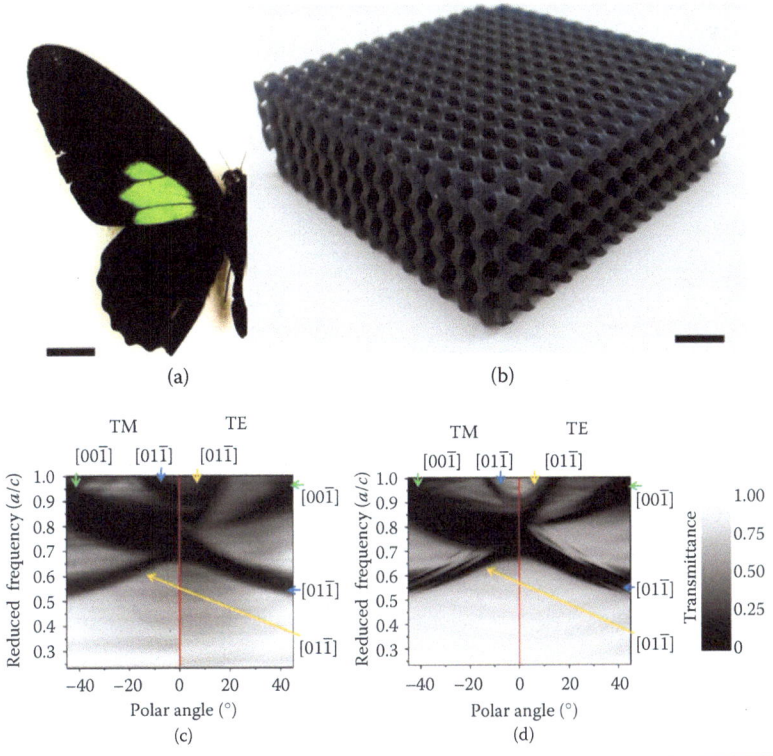

Figure 3.19 (a) *Paridessesostris*. (b) The fabricated gyroid sample with 40% material ($n = 1.656$) and 60% air. Lattice constant is 7 mm. (c) Experimental transmission data taken from the synthetically fabricated gyroid structure in the microwave regime. Polar angular data taken at a 0° azimuthal angle. (d) Theoretical transmission data from finite-element method modeling in the microwave regime. Polar angular data taken at a 0° azimuthal angle. (From Pouya et al., *Interface Focus*, 2 (5), 645–650, 2012.[59])

bandwidth can be increased. This feature combines with the microdomains, and a surface structure that diffuses reflected light, to create angle-independent color reflection. Other chiral metamaterial structures have been fabricated using an interferometric lithography technique, to fabricate 3D helical photonic crystals operating in the infrared.[60]

Opal structured photonic crystals mimic the ordered colloidal structure found in natural opals. Colloidal fabrication, where complex structures are formed from the ordered packing of nano- and microscale beads, is popular due to the low cost of the constituent components, the relative ease of fabrication of certain crystal structures, and the potential scalability to large area.[61] Synthetic opals have been commercialized since the 1970s. Opal structures are layered stacks of beads that form close-packed structures. They can be made with any colloidal material but are most often created using silica or polystyrene. With inverse opal structures, the colloidal structure is formed, and then the interstitial spaces between the beads are filled and the beads subsequently removed from the structure.

Several examples of colloidal structures are seen in Figure 3.20. Colloidal photonic crystals have been fabricated in thin or thick films and colloidal chains as well as spherical and asymmetric photonic crystal "beads."[61,62] FCC silica colloidal structural color films have been demonstrated as a potential route for security measures for currency.[63] The films are transparent and can be patterned. Multilayered structures and other structural variations can be formed to create color through color mixing, increasing or decreasing iridescence, and increasing or decreasing the reflectance. Structures have been created that combine 3D inverse opals with 2D planar structures to demonstrate fabrication approaches that combine some of the advantages of both top-down and bottom-up fabrication.[64] Spherical photonic crystals have also been encapsulated into fluid-filled spherical microcapsules.[65] An area of potential application that is particularly interesting for the colloidal bioinspired photonic beads is in paint, cosmetics, and displays where the photonic materials can replace dyes and pigments.[67] Dyes and pigments are often subject to chemical- and photobleaching. Creating approaches to photonic crystal pigments may overcome this limitation. Novel techniques enable the fabrication of photonic microcapsules—amorphous core-shell colloid assemblies encapsulated in soft membranes—that

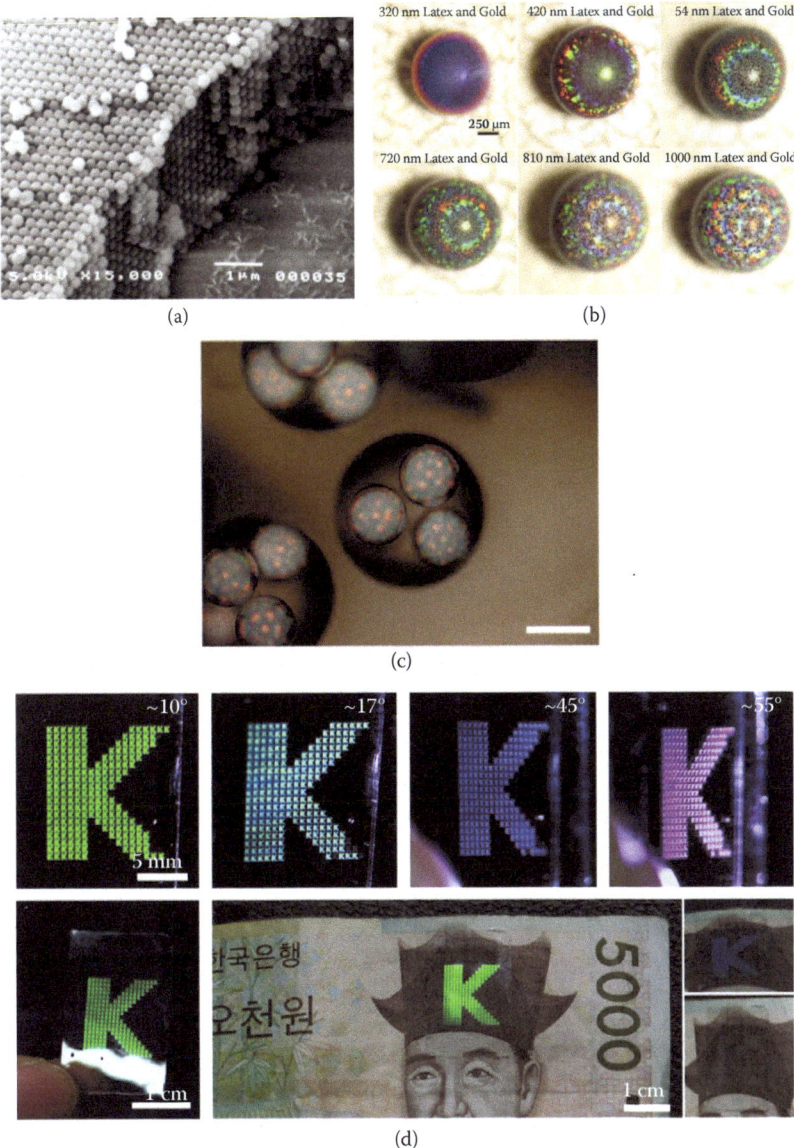

Figure 3.20 Examples of colloidal photonic crystals. (a) SEM of opal film. (b) optical microscopy images of latex and gold nanoparticle opal balls containing latex of varying sizes (c) Optical microscope images of the core-shell microparticles with triple cores. Scale bar: 200 μm. (d) Optical images of a colloidal crystal pattern forming small K's that show iridescent reflection for varying incidence angles. Optical image of a freestanding film containing a pattern of small K's. (d-lower left) Patterned photonic crystal film on a Korean bank note; the pattern exhibits bright green color under normal reflection (d-main lower image). (From Zhao et al., *Chem. Soc. Rev.* 41 (8), 3297–3317, 2012; Rastogi et al., *Adv. Mater.*, 20 (22), 4263–4268, 2008; Kim et al., *J. Am. Chem. Soc.*, 130 (18), 6040–6046, 2008; Lee et al., *Chem. Mat.*, 25 (13), 2684–2690, 2013.[62,63,65,66])

exhibit matte color that is angle independent (seen in Figure 3.21).[67] The microcapsules are ~100 μm in diameter, and the soft membrane keeps the photonic structure amorphous during bead formation, frustrating crystal formation. The photonic structure is made of core-shell colloidal particles (polystyrene as the core, and poly(N-isopropylacrylamide-co-acrylic-acid) [poly(NiPAm-AAc)] as the shell), which enables further optimization of the structure. Other

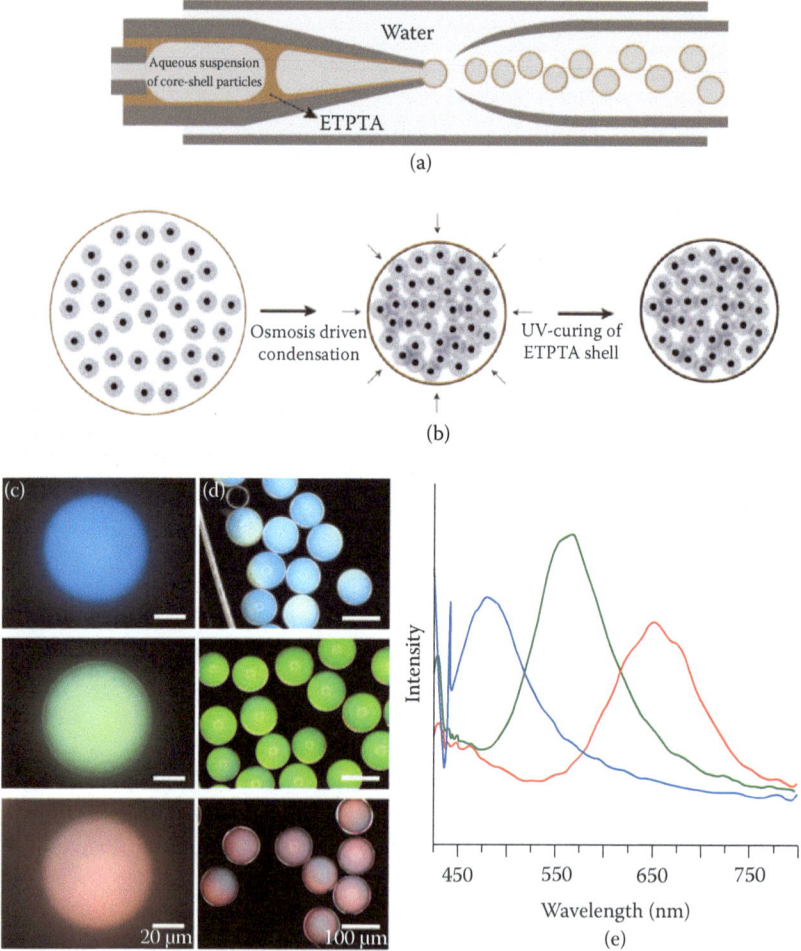

Figure 3.21 (a) Illustration of a capillary microfluidic device for the production of encapsulated amorphous colloidal photonic crystals. (b) Illustration of the amorphous structure formation by osmosis-driven condensation. (c) Bright-field (reflection) images of photonic microcapsules prepared with different shell thicknesses of core-shell particles. (d) Dark-field (reflection) images of photonic microcapsules. (e) Reflectance spectra of the microcapsules. (From Park et al., *Angewandte Chemie International Edition*, 53 (11), 2899–2903, 2014.[67])

researchers have begun to include black material (such as carbon black or magnetite) into their photonic colloidal beads to improve the color saturation.[68,69]

3.3 Nanostructures in Black and White

In addition to the examples of brilliant color described above, nature also uses structure to create or enhance its absence. Either alone on an organism or in conjunction with other structural or pigment-based colors, we find examples of both extreme black and bright white.

3.3.1 White

White reflective color is broadband reflection that covers the entire visible spectrum. Some, such as from chirped multilayers, can create a shiny, silvery, reflected color. Another type of broadband reflection results in a bright white color, which can be metallic or nonmetallic. The male Cabbage White butterflies, or *Pieris rapae*, have soft white wings

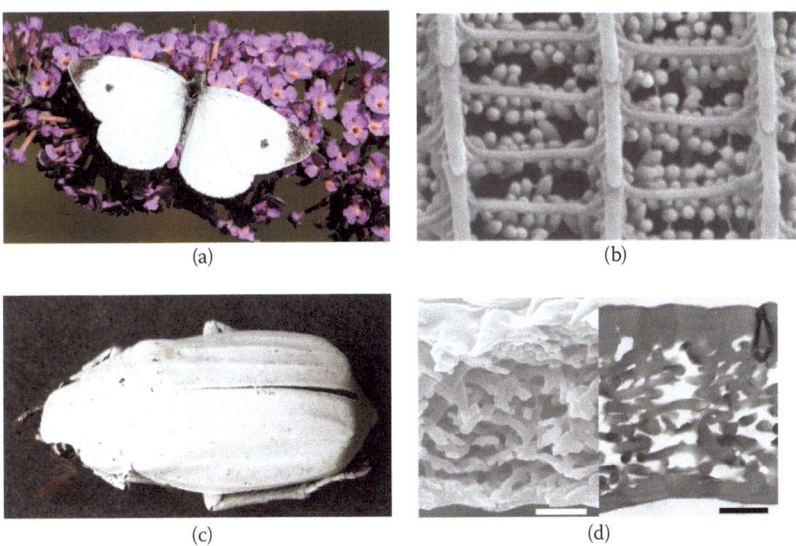

Figure 3.22 (a) *Pieris rapae*. (b) SEM images of *Pieris rapae* white wing scale, pterin pigment beads can be seen densely connected to the cross-ribs. (c) *Cyphochilus* beetle. (d) SEM (left) and transmission electron microscopy (right) images of the scale cross section showing randomly oriented interconnected filaments. (From Stavenga et al., *Proc. R. Soc. B-Biol. Sci.*, 271 (1548), 1577–1584, 2004; Vukusic et al., *Science* 315 (5810), 348, 2007; supplemental online material (SOM); and Hallam et al., *Appl. Opt.*, 48 (17), 3243–3249, 2009.[70–72])

(as seen in Figure 3.22).[4,73,74] Imaging and analysis of the wing scale ultrastructure reveal the presence of rib structures that extend the length of the long axis of the scale. The ribs are a little over approximately 2 μm apart. Between and perpendicular to the long ribs are short connecting cross-ribs spaced ~1 μm apart. Within the interstitial spaces and attached to the ribs and cross-ribs are small ellipsoidal pigment beads that are randomly distributed in the space. The beads incoherently scatter longer wavelength light while at the same time the pterin pigments in the beads absorb light in the UV range. The overlapping scales on both sides of the wing contribute to the scattering, resulting in a reflectivity in *Pieris rapae* of approximately 70% across the visible spectrum. This bright white reflectance and strong absorption (lack of reflectivity) in the UV is sexually dimorphic in other pierids, such as *Pieris rapae crucivora*. Males of *Pieris rapae crucivora* are similar to males and females of *Pieris rapae* in that their wing scale structures are similar. Both have several effective layers of highly scattering pigment beads and both reflect white light strongly, and lack UV reflectance. Females of *Pieris rapae crucivora* do not have the pterin pigment beads in their wing photonic structures, and their wing color as a result is significantly less reflective in the visible range and has a notable UV component. The bead arrays, as well as the scale structure itself, are responsible for scattering broadband light (except at the wavelengths that are absorbed). The combination of the optical scatter of visible wavelengths and the UV absorption by the pterin pigments are responsible for the white wing color.

The *Cyphochilus* spp. beetle (Figure 3.22) displays an extremely bright white color on its elytra, legs, and body scales.[70,75] The elytra of the beetle are made up of overlapping 5-μm thick scales, each approximately 250 × 100 μm. These scales contain disordered fibers made of cuticle with diameters on the order of 250 nm. These fibers fill ~70% of the scale interior, and it is this fill density, along with the diameter of the randomly aligned fibers, that enhances broadband scattering resulting in an uncommonly white appearance.

Metallic whites have been found in the South American butterfly *Argyrophorus argenteus*.[4,76] *Argyrophorus argenteus* lives at high elevations in the Andes near the border between Chile and Argentina.[76] It is one of only a few Lepidoptera that exhibit broadband metallic reflection. Unlike other Lepidoptera on which silver or gold broadband reflection may occur in small areas, *Argyrophorus argenteus*, as seen in Figure 3.23,

STRUCTURAL COLOR II: COMPLEX STRUCTURES 109

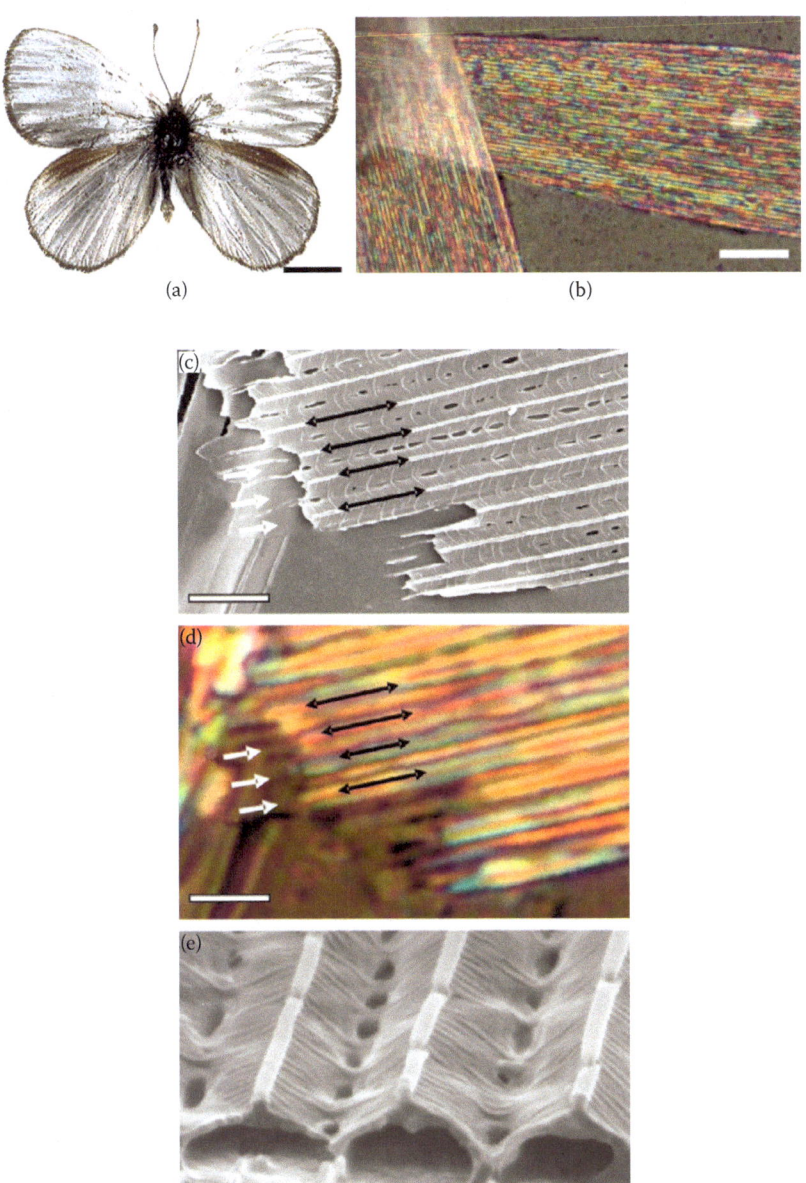

Figure 3.23 (a) Dorsal wing surfaces of *Argyrophorus argenteus*, taken in diffuse white illumination. Scale bar: 0.5 cm. (b) Optical microscope epi-mode image of two single *Argyrophorus argenteus* scales. Scale bars: 15 μm. (c) SEM of ridge structures on the scale taken from directly above the sample. Scale bar: 3 μm (d) optical micrographs of the same region on the scale. Black arrows indicate four corresponding ridge lines, white arrows show three open-ended ridge structure regions. Scale bar: 3 μm. (e) SEM taken at near grazing angle of the open-ended structures indicated by white arrows. Scale bar: 0.8 μm. (From Vukusic et al., *J. Roy. Soc. Interface*, 6 Suppl 2, S193–201, 2009.[76])

exhibits this phenomenon across both the dorsal and the ventral sides of its entire wings. The scales of the wing do not reflect uniformly, and instead light of many different wavelengths can be observed. When examined more closely, the nanostructures responsible for this reflection were very interesting. The structure used by *Argyrophorus argenteus* to create broadband metallic reflection is different than the chirped multilayer we have discussed previously. This new structure offers an interesting advantage. In the *Argyrophorus argenteus* structure seen from above, there are long, linear structures separated by irregular holes. When viewed in cross section, the structure looks like a closely packed series of hollow, pointed vault-like tubes, each roughly 1.2 μm wide, aligned parallel to the long axis of the scale. The tube apexes form the ridges that run the length of the scale when seen from above. At the interface between neighboring tubes, small (2–3 μm diameter) pores pierce the structure. The thickness of the chitin vault "ceiling" varies from 200 to 250 nm at the ridge apex to 100 nm out toward the edge, and the entire structure sits on a thin-layer base of 120 nm chitin. The air gap that exists within the tube axis therefore also changes thickness from the edge to the center of each ridge. The varying thicknesses result in a variety of colors that run along the longitudinal structures when looked at through an optical microscope. (The black arrows in all three images indicate the four same ridges/tubes.) The different colors emerge due to coherent scattering from the differing periodicities, and the silver color seen on the wing from afar results from additive color mixing.

In both the matte white beetle structures and the metallic white butterfly structure, broadband reflection was achieved in a form factor that has significant implications for applications requiring white pigments or enhancing pigmentary brightness.[4,70,76,71] A large potential market for structures capable of bright white may be the paper and textile industry.[71] Whiteness in paper is often created with mineral inclusions, mineral coatings, and dyes. The *Cyphochilus* beetle structures achieve a comparable or better whiteness with only 5 μm thickness, 25× thinner than conventional uncoated wood-free papers.[70] Inspired by the beetle structures, attempts have been made to optimize the size, density, and pore spacing in mineral coatings to improve scattering.[71] Also inspired by the beetle structures, nanofiber webs of polyurethane, polymethyl methacrylate, and poly(ethylene terephthalate) were electrospun onto a substrate. The bioinspired webs were optimized for broadband brightness by varying the polymer concentration, the fiber structure, and fiber size, and the resulting best reflectance approached 85% between 400 and 700 nm. The white brightness of this bioinspired nanofiber web exceeds that of the natural beetle.

The metallic white structure offers an interesting advantage and potential inspiration because in a normal natural multilayer achieving mirrorlike, broadband reflection typically requires multilayers 50 μm thick or more.[76] This is due to the comparatively small difference in refractive index between chitin and air. The total thickness of the butterfly scale is significantly less than that, 1–5 μm, and yet also achieves broadband reflection. The reflection was not specular, however, and was rather diffuse, which may serve a cryptic purpose for the butterfly. This structure may have implications for ultrathin synthetic reflectors (which have the potential for higher refractive index difference between material components). The authors also created a large-scale model using a 3D printing technique, which they were able to interrogate using microwave wavelengths. The ultrathin synthetic broadband reflector approach may be applicable at other wavelengths beyond the visible range. In addition, the ability to achieve significant broadband reflection in a structure that is permeable, flexible, and fault tolerant in an ultrathin form factor in both the matte and metallic structures could be very useful as a way to enhance the reflection of optical emitters, such as LEDs and OLEDs.[70]

3.3.2 Black

Structural or structurally assisted black is very much of interest to the photonics community.[77–80] Ultralow reflectance (or Ultra absorptive) materials, called super black materials, are useful for applications where stray light at specific wavelengths may impede the operation of optical components. One potential application is in low-reflectance coatings for sensitive optical equipment such as star trackers, which would improve the quality of high-resolution imagers. They are also potentially useful in applications where high optical absorption may improve performance such as improving the capture efficiency of solar panels.

Many blacks and browns are due to the presence of pigments, such as melanin, which absorb light across a large range of visible wavelengths. These colors can make an animal less conspicuous by making its edges less distinct; it can enhance the brightness of a more colorful area by providing contrast; and it can perform a needed biological function, such as enhancing the organism's ability to regulate body temperature. The "blackness" of the color can be enhanced by nanostructures, which reduce specular (mirrorlike) reflection and direct light toward light-absorbing pigments. *Bitis gabonica rhinoceros* (seen in Figure 3.24)

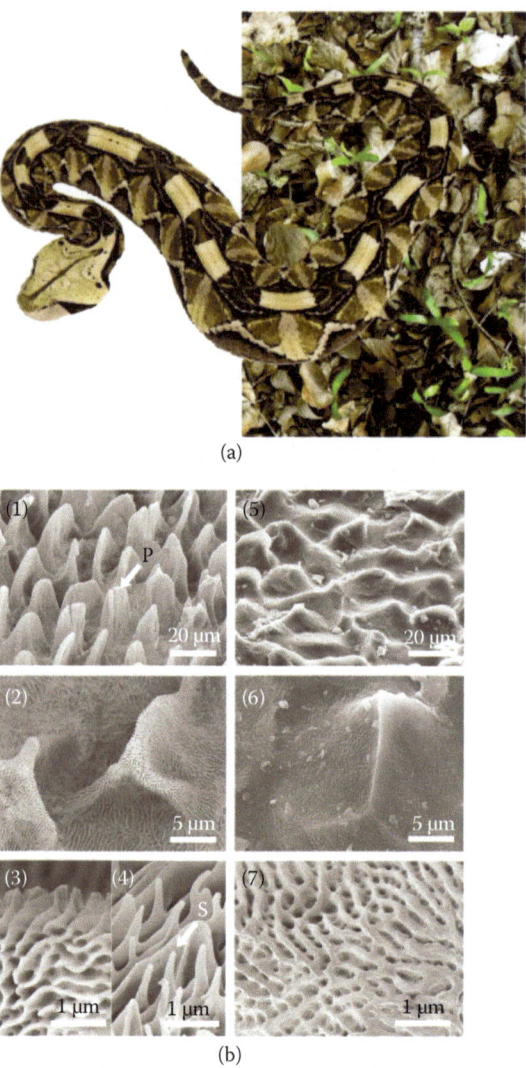

Figure 3.24 (a) West African Gaboon viper *Bitis rhinoceros* partly on white and leafy background. (b) SEM images of dorsal scale microornamentation. (1–7) Leaf-like structures at the surface. (3 and 4) Branched ridges on the leaf-like structures on a black dorsal scale. (5) Ridges with spinules (S) between the leaf-like structures (black dorsal scale). (6 and 7) Structures on a pale dorsal scale. (From Spinner et al., *Sci Rep*, 3, 8, 2013.[81])

is a large, poisonous viper from West Africa.[81] The snake's overall skin color is relatively pale with a pattern of brown, white, and very deep black spots. The black spots are described as having an exceptional spatial depth, and a "black velvet" appearance with subtle gloss appearing only at certain angles. The brown, white, and black pattern renders the snake very effectively camouflaged in its environment. The scales of snakes often have what is called microornamentation, which serves some nonphotonic purposes such as to aid in friction reduction and dirt shedding, as well as resulting in interesting optical properties. By examining the shed skin, some of the details of color display production in these snakes can be determined. On the black areas of the scales, small, leaf-like microstructures, on average, approximately 30 µm high, can be found. These tall structures are fairly densely packed, with average density around 1900/mm². The surface of these microstructures is covered in small, branching ridges (~600 nm in height) each only 60 nm wide. The ridges are connected to each other by small cross-ribs, and between the ridges and cross-ribs are pits. The black portions of the snakeskin are the only parts that contain pigment. The nanostructure reinforces and enhances the scale blackness. The surface structures and the pigmentary absorption combine to reduce the reflectance of the scale, resulting in reflectance of less than ~11% in these spots across visible wavelengths.

Butterflies provide another example of structurally enhanced black in nature. Several examples of black structural color in butterflies are known to be linked to the butterfly's need for thermal regulation such as with *Troides plateni* and *Pachliopta aristolochiae*, which will be discussed further in Chapter 5. The male butterfly *Papilio ulysses* (seen in Figure 3.25) has two regions of blackness on its dorsal wings, one matte and one with some sheen, both with even more impressive optical blackness quality than the viper described above.[82] The reflectivity of these black spots on *Papilio ulysses* is extremely low with a scale reflectivity of less than 2% and a calculated optical absorption of 90%–95% of normally incident light across all visible wavelengths. The structure of the scales of *Papilio ulysses* is known as the inverse V-type where the longitudinal ridges found on the scales are triangular in shape, as seen in HIM images (Figure 3.25).[46] In the 2- to 3-µm space between ridges, a random network of walls and cross-ribs forms a honeycomb-like lattice structure extending to the scale base

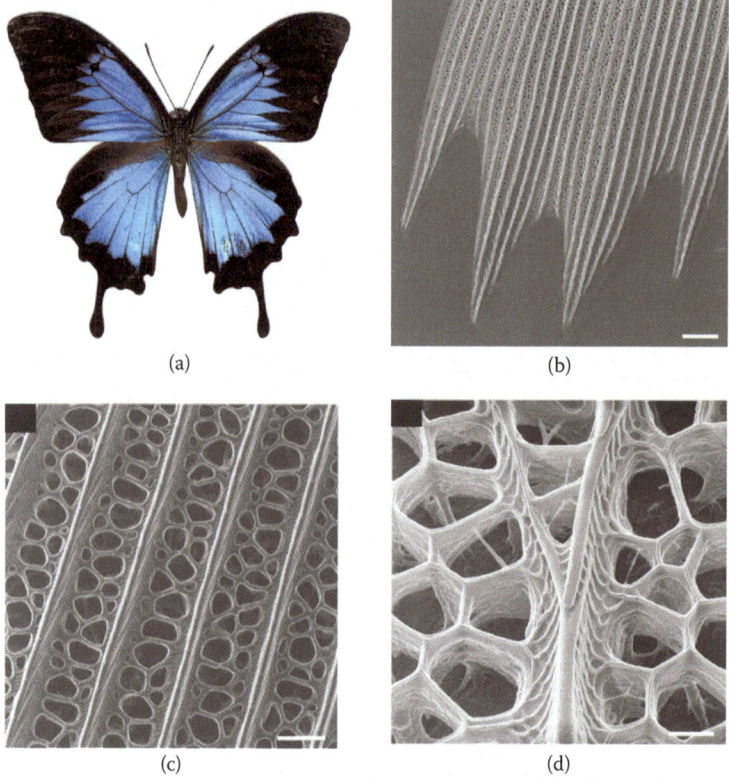

Figure 3.25 (a) *Papilio ulysses* butterfly (b) HIM images of *Papilio ulysses* black ground scale. Scale bar: 10 μm, 0° tilt; (c) scale bar: 2 μm, 0° tilt; (d) scale bar: 400 nm, 21° tilt. (From Boden et al.)

below. On the walls of the triangular ridges, pointing perpendicularly to the ridge, are nanoscale structures (on the order of 50–100 nm). Light that hits the scale is very efficiently scattered toward the absorptive pigment. The tapering of the ridge and the nanostructures on the ridge walls act as a graded index material, which reduces the magnitude of reflection at the surface.

The micro- and nanostructures on the scales of *Ornithoptera goliath*, seen in Figure 3.26, are also in the form of an inverse V-type, and *O. goliath* has been used as a model to create super black antireflective structures.[78,83] In this approach, the chitin in butterfly wing itself is converted into an amorphous carbon structure by vacuum sintering. The measured spectra, as well as a finite difference time domain

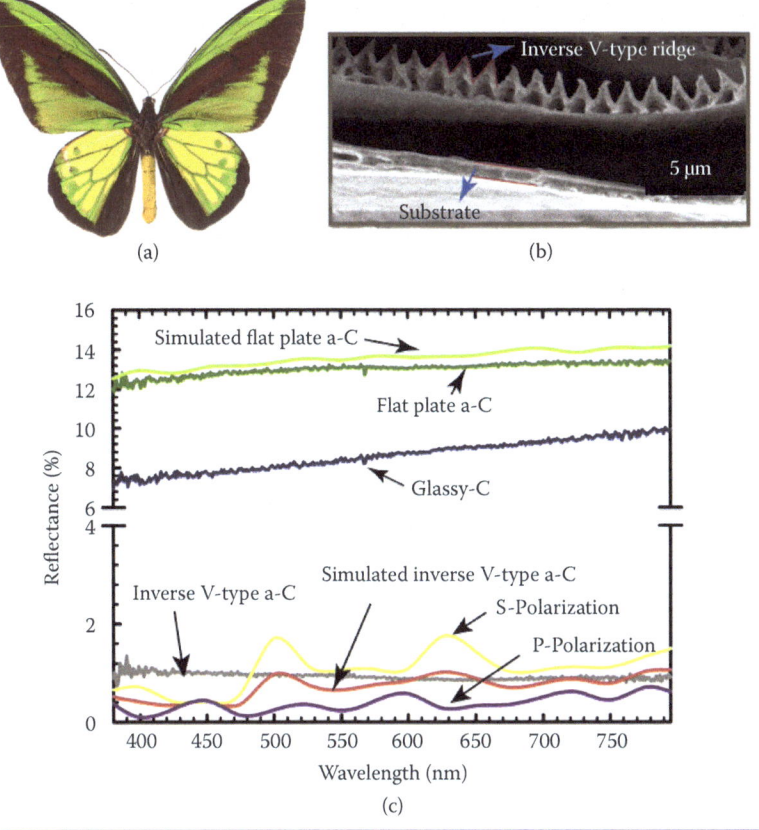

Figure 3.26 (a) *Ornithoptera goliath*. (b) Field emission scanning electronic microscope cross section of the black wing scales. (c) Measured and calculated reflectance spectra of the inverse V-type structured amorphous carbon film and the flat-plate amorphous carbon film. (From Zhao et al., *Carbon*, 49 (3), 877–883, 2011.[78])

calculation for the reflectance of the amorphous carbon structure, can be seen in Figure 3.26. The biological structure-based film has a reflectance of only 1%, and absorption of 99%, over the entire visible wavelength range, which is comparable to other super black materials. Other examples of super black materials are surface coatings, such as carbon nanotube forests and etched nickel phosphorous coatings, and research and development of super black materials are still an active area of interest.[77,79,80,84] Bio-based super black materials are of interest because, at 5 μm, the bio-super black films are significantly thinner than (only 15% of the size of) the other structures.[78] The amorphous carbon butterfly material is, for example, stable, corrosion resistant, and flexible and is potentially useful for sensors, optical instruments, thermal detectors, and solar cells.

References

1. P. Vukusic, J. R. Sambles, C. R. Lawrence, and R. J. Wootton, *Proc. Roy. Soc. Lon. Ser. B-Biol. Sci.* **269** (1486), 7–14 (2002).
2. A. L. Ingram, V. Lousse, A. R. Parker, and J. P. Vigneron, *J. Roy. Soc. Interface* **5** (28), 1387–1390 (2008).
3. J. P. Vigneron, P. Simonis, A. Aiello, A. Bay, D. M. Windsor, J. F. Colomer, and M. Rassart, *Phys. Rev.* E **82** (2), 8 (2010).
4. K. L. Yu, T. X. Fan, S. Lou, and D. Zhang, *Prog. Mater. Sci.* **58** (6), 825–873 (2013).
5. E. Van Hooijdonk, S. Berthier, and J. P. Vigneron, *J. Appl. Phys.* **112** (7), 8 (2012).
6. A. Herman, C. Vandenbem, O. Deparis, P. Simonis, and J. P. Vigneron, in *Nanophotonic Materials Viii*, edited by S. Cabrini, and T. Mokari (Spie-Int Soc Optical Engineering, Bellingham, 2011), Vol. 8094.
7. E. Van Hooijdonk, C. Barthou, J. P. Vigneron, and S. Berthier, *J. Opt. Soc. Am. B-Opt. Phys.* **29** (5), 1104–1111 (2012).
8. J. P. Vigneron, K. Kertesz, Z. Vertesy, M. Rassart, V. Lousse, Z. Balint, and L. P. Biro, *Phys. Rev.* E **78** (2), 9 (2008).
9. L. C. Zhang, H. Lin, C. S. Jin, H. J. Zhou, and T. L. Hu, *Opt. Lett.* **34** (6), 818–820 (2009).
10. D. L. Voronov, M. Ahn, E. H. Anderson, R. Cambie, C. H. Chang, L. I. Goray, E. M. Gullikson, R. K. Heilmann, F. Salmassi, M. L. Schattenburg, T. Warwick, V. V. Yashchuk, and H. A. Padmore, *Adv. X-Ray/Euv Optics Components V* **7802** (2010).
11. D. L. Voronov, E. H. Anderson, R. Cambie, S. Dhuey, E. M. Gullikson, F. Salmassi, T. Warwick, V. V. Yashchuk, and H. A. Padmore, *Nucl. Instr. Methods Phys. Res. A* **649** (1), 156–159 (2011).
12. R. Potyrailo and R. R. Naik, in *Annual Review of Materials Research, Vol 43*, edited by D. R. Clarke (Annual Reviews, Palo Alto, 2013), Vol. 43, pp. 307–334.
13. J. Y. Huang, X. D. Wang, and Z. L. Wang, *Nano lett.* **6** (10), 2325–2331 (2006).
14. S. H. Kang, T. Y. Tai, and T. H. Fang, *Curr. Appl. Phys.* **10** (2), 625–630 (2010).
15. M. R. Weatherspoon, Y. Cai, M. Crne, M. Srinivasarao, and K. H. Sandhage, *Angewandte Chemie-International Edition* **47** (41), 7921–7923 (2008).
16. A. Saito, Y. Miyamura, Y. Ishikawa, J. Murase, M. Akai-Kasaya, and Y. Kuwahara, *Adv. Fabrication Technol. Micro/Nano Optics Photon. Ii* **7205**, 9 (2009).
17. A. Saito, *Sci. Technol. Adv. Mater.* **13** (2), 1 (2012).

18. L. P. Biro and J. P. Vigneron, *Laser Photon. Rev.* **5** (1), 27–51 (2011).
19. W. L. Wang, W. Zhang, J. J. Gu, Q. L. Liu, T. Deng, D. Zhang, and H. Q. Lin, *Sci Rep* **3**, 9 (2013).
20. C. Bamann, G. Nagel and E. Bamberg, *Curr. Opin. Neurobiol.* **20** (5), 610–616 (2010).
21. S. Vignolini, P. J. Rudall, A. V. Rowland, A. Reed, E. Moyroud, R. B. Faden, J. J. Baumberg, B. J. Glover, and U. Steiner, *Proc. Natl Acad. Sci. U S A* **109** (39), 15712–15715 (2012).
22. A. E. Seago, P. Brady, J. P. Vigneron, and T. D. Schultz, *J. Ro,y. Soc., Interface/Roy. Soc.* **6 Suppl 2**, S165–184 (2009).
23. A. R. Parker and N. Martini, *Optics Laser Technol.* **38** (4–6), 315–322 (2006).
24. http://www.kentdisplays.com.
25. http://www.chiralphotonics.com.
26. *Eco-friendly Optics: Spider Silk's Hidden Talents Brought to Light for Applications in Biosensors, Lasers, Microchips.* http://www.osa.org/en-us/about_osa/newsroom/newsreleases/2012/eco-friendly_optics_spider_silks_hidden/. (Angela Stark, The Optical Society, 2012).
27. J. Hwang, M. H. Song, B. Park, S. Nishimura, T. Toyooka, J. W. Wu, Y. Takanishi, K. Ishikawa, and H. Takezoe, *Nat. Mater.* **4** (5), 383–387 (2005).
28. M. H. Song, B. C. Park, K. C. Shin, T. Ohta, Y. Tsunoda, H. Hoshi, Y. Takanishi, K. Ishikawa, J. Watanabe, S. Nishimura, T. Toyooka, Z. G. Zhu, T. M. Swager, and H. Takezoe, *Adv. Mater.* **16** (9–10), 779 (2004).
29. A. Matranga, S. Baig, J. Boland, C. Newton, T. Taphouse, G. Wells, and S. Kitson, *Adv. Mater.* **25** (4), 520–523 (2013).
30. L. P. Biro, K. Kertesz, E. Horvath, G. I. Mark, G. Molnar, Z. Vertesy, J. F. Tsai, A. Kun, Z. Balint, and J. P. Vigneron, *J. Roy. Soc., Interface/Roy.l Soc.* **7** (47), 887–894 (2010).
31. M. D. Shawkey and G. E. Hill, *J. Exp. Biol.* **209** (7), 1245–1250 (2006).
32. H. Yabu, T. Nakanishi, Y. Hirai, and M. Shimomura, *J. Mater. Chem.* **21** (39), 15154–15156 (2011).
33. D. P. Josephson, E. J. Popczun, and A. Stein, *J. Phys. Chem. C* **117** (26), 13585–13592 (2013).
34. R. Withnall, J. Silver, T. G. Ireland, S. Zhang, and G. R. Fern, *Opt. Laser Technol.* **43** (2), 401–409 (2011).
35. J.-F. Colomer, P. Simonis, A. Bay, P. Cloetens, H. Suhonen, M. Rassart, C. Vandenbem, and J. P. Vigneron, *Phys. Rev. E* **85** (1) (2012).
36. L. Poladian, S. Wickham, K. Lee, and M. C. Large, *J. Roy. Soc., Interface/Roy. Soc.* **6 Suppl 2**, S233–242 (2009).
37. C. Lopez, *J. Opt. A-Pure Appl. Opt.* **8** (5), R1–R14 (2006).
38. P. Simonis and J. P. Vigneron, *Phys. Rev. E* **83** (1) (2011).
39. V. Welch, V. Lousse, O. Deparis, A. Parker, and J. P. Vigneron, *Phys. Rev. E* **75** (4) (2007).
40. B. D. Wilts, K. Michielsen, J. Kuipers, H. De Raedt, and D. G. Stavenga, *Proc. R. Soc. B-Biol. Sci.* **279** (1738), 2524–2530 (2012).
41. M. R. Jorgensen and M. H. Bartl, *J. Mater. Chem.* **21** (29), 10583–10591 (2011).

42. J. W. Galusha, M. R. Jorgensen, and M. H. Bartl, *Adv. Mater.* **22** (1), 107–110 (2010).
43. K. Michielsen and D. G. Stavenga, *J. Roy. Soc. Interface* **5** (18), 85–94 (2008).
44. G. E. Schroder-Turk, S. Wickham, H. Averdunk, F. Brink, J. D. F. Gerald, L. Poladian, M. C. J. Large, and S. T. Hyde, *J. Struct. Biol.* **174** (2), 290–295 (2011).
45. B. D. Wilts, K. Michielsen, H. De Raedt, and D. G. Stavenga, *Interface Focus* **2** (5), 681–687 (2012).
46. S. A. Boden, A. Asadollahbaik, H. N. Rutt, and D. M. Bagnall, *Scanning* **34** (2), 107–120 (2012).
47. A. R. Parker, *J. Opt. A-Pure and Applied Optics* (2000).
48. M. E. McNamara, D. E. Briggs, P. J. Orr, H. Noh, and H. Cao, *Proc. Biol. Sci./Roy. Soc.* (2011).
49. M. E. McNamara, D. E. Briggs, P. J. Orr, S. Wedmann, H. Noh, and H. Cao, *PLoS Biol.* **9** (11), e1001200 (2011).
50. Q. G. Li, K. Q. Gao, J. Vinther, M. D. Shawkey, J. A. Clarke, L. D' Alba, Q. J. Meng, D. E. G. Briggs, and R. O. Prum, *Science* **327** (5971), 1369–1372 (2010).
51. R. C. McKellar, B. D. Chatterton, A. P. Wolfe, and P. J. Currie, *Science* **333** (6049), 1619–1622 (2011).
52. M. A. Norell, *Science* **333** (6049), 1590–1591 (2011).
53. F. Zhang, S. L. Kearns, P. J. Orr, M. J. Benton, Z. Zhou, D. Johnson, X. Xu, and X. Wang, *Nature* **463** (7284), 1075–1078 (2010).
54. J. Vinther, D. E. Briggs, J. Clarke, G. Mayr, and R. O. Prum, *Biol. Lett.* **6** (1), 128–131 (2010).
55. M. E. McNamara, *Palaeontology* **56** (3), 557–575 (2013).
56. M. E. McNamara, D. E. G. Briggs, P. J. Orr, N. S. Gupta, E. R. Locatelli, L. Qiu, H. Yang, Z. R. Wang, H. Noh, and H. Cao, *Geology* **41** (4), 487–490 (2013).
57. C. Mille, E. C. Tyrode, and R. W. Corkery, *RSC Adv.* **3** (9), 3109–3117 (2013).
58. C. Mille, E. C. Tyrode, and R. W. Corkery, *Chem. Comm.* **47** (35), 9873–9875 (2011).
59. C. Pouya and P. Vukusic, *Interface Focus* **2** (5), 645–650 (2012).
60. A. K. Raub and S. R. J. Brueck, *J. Vac. Sci. Technol. B* **29** (6) (2011).
61. J. Y. Wang and J. T. Zhu, *Eur. Polym. J.* **49** (11), 3420–3433 (2013).
62. Y. Zhao, Z. Xie, H. Gu, C. Zhu, and Z. Gu, *Chem. Soc. Rev.* **41** (8), 3297–3317 (2012).
63. H. S. Lee, T. S. Shim, H. Hwang, S. M. Yang, and S. H. Kim, *Chem. Mater.* **25** (13), 2684–2690 (2013).
64. I. B. Burgess, J. Aizenberg, and M. Loncar, *Bioinspir. Biomim.* **8** (4) (2013).
65. S. H. Kim, S. J. Jeon, and S. M. Yang, *J. Am. Chem. Soc.* **130** (18), 6040–6046 (2008).
66. V. Rastogi, S. Melle, O. G. Calderon, A. A. Garcia, M. Marquez, and O. D. Velev, *Adv. Mater.* **20** (22), 4263–4268 (2008).

67. J.-G. Park, S.-H. Kim, S. Magkiriadou, T. M. Choi, Y.-S. Kim, and V. N. Manoharan, *Angewandte Chemie International Edition* **53** (11), 2899–2903 (2014).
68. Y. Takeoka, S. Yoshioka, A. Takano, S. Arai, K. Nueangnoraj, H. Nishihara, M. Teshima, Y. Ohtsuka, and T. Seki, *Angewandte Chemie-International Edition* **52** (28), 7261–7265 (2013).
69. Y. Takeoka, S. Yoshioka, M. Teshima, A. Takano, M. Harun-Ur-Rashid, and T. Seki, *Sci Rep* **3** (2013).
70. P. Vukusic, B. Hallam, and J. Noyes, *Science* **315** (5810), 348 (2007).
71. B. T. Hallam, A. G. Hiorns, and P. Vukusic, *Appl. Opt.* **48** (17), 3243–3249 (2009).
72. D. G. Stavenga, S. Stowe, K. Siebke, J. Zeil, and K. Arikawa, *Proc. R. Soc. B-Biol. Sci.* **271** (1548), 1577–1584 (2004).
73. S. M. Luke, P. Vukusic, and B. Hallam, *Opt. Express* **17** (17), 14729–14743 (2009).
74. N. I. Morehouse, P. Vukusic, and R. Rutowski, *Proc. R. Soc. B-Biol. Sci.* **274** (1608), 359–366 (2007).
75. P. Vukusic, *Ophth. Physiol. Opt.* **30** (5), 435–445 (2010).
76. P. Vukusic, R. Kelly, and I. Hooper, *J. Roy. Soc. Interface* **6 Suppl 2**, S193–201 (2009).
77. R. Beckhusen, *Army Goes Goth With 'Super-Black' Materials* http://www.wired.com/dangerroom/2012/12/army-goth/.
78. Q. B. Zhao, T. X. Fan, J. A. Ding, D. Zhang, Q. X. Guo, and M. Kamada, *Carbon* **49** (3), 877–883 (2011).
79. M. Hammer, *Mini craters key to 'blackest ever black'* http://www.newscientist.com/article/dn3356-mini-craters-key-to-blackest-ever-black.html-.UvWN13mEzwI.
80. D. Quick, NASA's new super-black nanotube-based material is good news for star-gazers http://www.gizmag.com/super-black-material/20434/.
81. M. Spinner, A. Kovalev, S. N. Gorb, and G. Westhoff, *Sci Rep* **3**, 8 (2013).
82. P. Vukusic, J. R. Sambles, and C. R. Lawrence, *Proc. Roy. Soc. Lo. Ser. B-Biol. Sci.* **271**, S237–S239 (2004).
83. Q. B. Zhao, X. M. Guo, T. X. Fan, J. Ding, D. Zhang, and Q. X. Guo, *Soft Matter* **7** (24), 11433–11439 (2011).
84. NASA, Milestone in quest to advance emerging super-black nanotechnology http://www.sciencedaily.com/releases/2013/07/130717183927.htm.

4

Dynamic, Adaptive Color

4.1 Color Changing Organisms as Inspiration

Perhaps more intriguing than the static color mechanisms found in nature are the numerous examples of organisms that can change color dynamically. Many organisms change color with season, in response to their environment, or for communication. These biological photonic systems are responsive to stimuli, either internal stimuli such as the release of hormones or neurotransmitters, or external stimuli such as humidity or environmental toxins. The ability, in some form, is surprisingly widespread; examples abound in land and marine animals and plants. Among these active biological photonic systems are a select few examples that are even more fascinating than the others from an applications perspective. Their ability to control color change is so profound, and the structures that do so are sufficiently complex that we are yet unable to replicate the structures or their complete performance. Within these organisms are photonic systems that offer potential inspiration for a wide range of new applications, and specific advantages over existing technologies. Many of these applications are similar to those we have already discussed for the static structures, such as sensors, components for optical circuits, fashion, and security. One of the biggest areas of potential impact and interest is in displays, particularly in less conventional technologies such as flexible displays. Bioinspired dynamic photonic structures may enable completely new paradigms such as biodegradable and biocompatible displays.

4.2 The Expanding Display Industry

Modern display technology is a vast area of research and development. A tremendous number of applications exist for technologies that can change optical response on demand. Established display technologies such as liquid-crystal display (LCD), for example, are currently used in most televisions, computer monitors, tablets, and cellphones. But as the display industry expands beyond conventional display applications and market, there is a définite demand for innovation as new applications are developed. Smartphones and tablets have been a big application driver, but now also electronic paper, digital signage, wearable displays, smartcards, and more may potentially be enabled. Alternate display spaces such as walls, windows, clothing, and other materials can become possible. Foldable displays, for example, that are thin and rollable (such as the imagined devices in Figure 4.1) have been described as "the future of mainstream display market."[1] Transparent displays and lighting panels are another area of potential interest.[2] Transparent displays could enable heads up data navigation or dashboard information through the windshield of a car or plane, helmet visor, or glasses, and project video onto windows or eyeglasses or for storefront displays. The total global display market is expected to reach nearly $165 billion by 2017.[3] This value includes both conventional display technology as well as the predicted markets for these alternate technologies like transparent and flexible displays.

New potential applications are calling for a wider array of device properties. For many of these applications (for example, again, flexible and transparent displays) we need to turn to less conventional

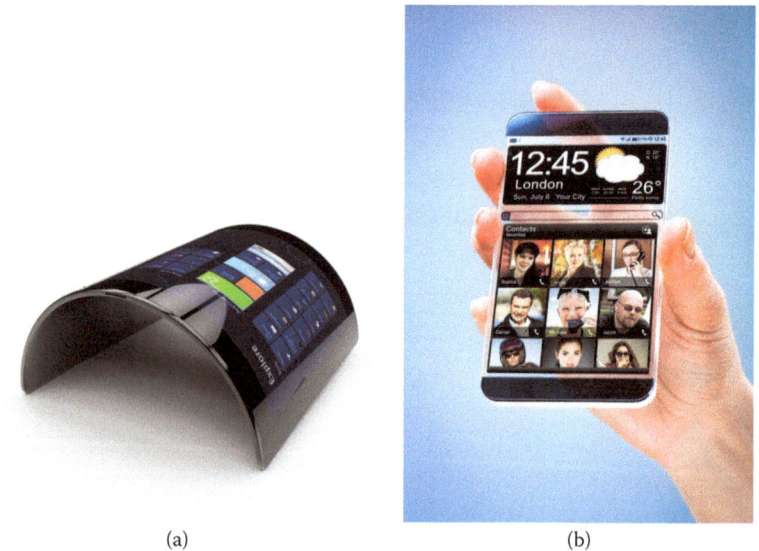

(a)   (b)

Figure 4.1 Artists' renderings of flexible and transparent displays.

approaches. For flexible display technologies the device must have mechanical performance, flexibility, and robustness, that do not interfere with its optical performance. For high-performance devices, many requirements, such as an adaptive reflective surface, multicolor contrast, polarization, reflectance, diffusivity, and texture need to be controlled simultaneously without optical losses.[4] Creating active full color is a challenge for many of these alternative displays. Reflective displays are highly desirable because they offer a benefit in low power consumption compared to typical displays that require backlighting. They also offer the benefit of improved viewability under bright light. However, under low light conditions their weakness becomes apparent. In addition, today's reflective displays, such as those used in the kindle e-book reader, are monochromatic, and slow to refresh, compared to the fast colorful LCD or LED displays, for example.

Another growing area of interest is in the arena of biodegradable and biocompatible displays. Modern technology changes at a breakneck pace, and last year's models find their way to landfills and into the environment. Devices that degrade benignly at the end of the product lifetime are of great interest to the consumer, as long as they do not sacrifice performance.[5,6] New possibilities in clinical healthcare applications may exist for diagnostics and monitoring of patients, such as smart bandages, or diagnostic implants, both in and out of the clinic.[7] Finally, personal fashion applications, such as video tattoos may be of interest.[8] Biological photonic structures may offer intriguing designs for flexible, conformal, robust, and organic photonic structures, which may enable these applications.

Some of the dynamic photonics approaches rely on actuating static biological photonic structures; others are investigating the examples of organisms capable of controllable active color change. Nature once again provides a host of intriguing design approaches for study, evolved millions of years in the past and still used by organisms today. Modern displays from small, high-resolution cellphones, TVs, and computer screens to extralarge advertising billboards all have certain common needs: high resolution, high contrast, wide viewing angle, long lifetime, fast switching speed and very low power consumption, simplicity of manufacturing, scalability, low cost, bright full color, and video rate capability, among others. Understanding how biological color change occurs can offer new ways to approach these spaces. Researchers in universities and industry alike are searching for ways to combine the best of these two worlds.

4.3 Nature's "Unconventional" Display Technologies

Many organisms change color over the course of their lifetime. In some organisms, these changes are gradual and shift with health or season, maturity, or stage of life. The process is slow, occurring over days, weeks, or seasons.[9,10] We humans, for example, look forward to getting our summer tan. For us, exposure to ultraviolet (UV) radiation triggers increased melanin pigment fabrication in specialized cells in our skin. The pigment is transferred to surrounding cells, leading to large areas of brown pigmentation.[11] Even more fascinating than these gradual, seasonal changes are the active and dynamic color changes of which some animals are capable. More rapid (physiological) color change tends to be for crypsis and communication. This often dramatic color change is most interesting to scientists for display technologies. The timescale of transformation can range from a few hours to within milliseconds. It is controlled either through the animal's endocrine system or its nervous system, and it is the latter that results in near instantaneous transformation.[12] These animals interest us because we can see in their behavior direct analogy to display technologies, but at the same time they are completely unlike the conventional displays we use daily. Scientific investigation of these systems has reached a point where we can begin to understand the interconnected functionality of the complex photonic structures within these creatures, and draw inspiration from their architecture.

There are many examples of animals, such as those in Figure 4.2, that are able to change color for communication, or camouflage. Lizards, cephalopods, insects such as beetles, some spiders, and fish including teleosts and seahorses can all be found that demonstrate relatively rapid dynamic color change.[9,13–15] One example is the crab spider, *Misumena vatia*, which changes color between white and yellow by the metabolic creation and elimination of pigment types within its chromatophores.[16] The reversible color change is relatively slow; it happens over several days and occurs to match the color of the spider to the flowers upon which the spider sits. The male

Figure 4.2 Examples of animals that can change color fairly rapidly. (a) The crab spider, *Misumena vatia*, (b) lizard *Urosaurus ornatus*, (c) the veiled chameleon, and (d) seahorse. (Adapted from Insausti, T.C., and Casas, J., *J. Exp. Biol.* **211** (5), 780–789, 2008; *U. ornatus* image courtesy of Brad and Lynn Weinert.)[16]

common tree lizard *Urosaurus ornatus* has brightly colored belly and throat patches.[17] The belly patches change color from dull amber/light green to a vibrant blue with increase in temperature in the lizard's environment. It is possible that this temperature-dependent change to a high-visibility color is a sexual signal, indicating that he is available to the ladies when the temperature is right for that type of activity. When the temperature is less than perfect, the lizard reverts to the duller coloration and is better hidden from predators.

Arguably, the most well-known example of rapid color change is the chameleon.[18] In English, the word has come both to imply the animal and to describe a person who has the ability to rapidly and completely change or disguise themselves. Chameleons change their coloring not for camouflage (crypsis) but for mood, or thermal control. However, it is the cephalopods (octopus, squid, and cuttlefish) that truly shine when it comes to rapid and dynamic color change.

4.4 Cephalopods

By far, the most complex, and perhaps most studied, of all color-changing animals is the cephalopod. Coleoid cephalopods (squid, octopus, and cuttlefish [see Figure 4.3]) are shell-less mollusks. There are over 700 known extant species, and many of them demonstrate the ability to change color dynamically.[12,19] Cephalopods exist in all oceans, from polar to tropic, and at most depths. They are able to change color over the entire spectrum from visible to infrared (IR) wavelengths. In addition to color, they can also display (and see) polarized light. Despite being color blind, they are experts at disruptive patterning and mimicry. How they achieve this amazing capability is the subject of a significant amount of study. Cephalopods use dynamic color change for either communication or camouflage, to stand out from or blend into their environment, as seen in Figure 4.4.[12,20] Their displays serve as an interspecific (between different species) and intraspecific (within same species)

Figure 4.3 Cephalopods: (a) unknown squid, (b) reef octopus, and (c) the common cuttlefish *Sepia officinalis*.

DYNAMIC, ADAPTIVE COLOR 127

Figure 4.4 Examples of color and textural changes in cephalopods: (a) spots and stripes on the Pharao cuttlefish *Sepia prashadi*, (b) countershading in the Broadclub cuttlefish *Sepia latimanus*, (c) highly defined stripes on a pajama squid, (d) splotches on an octopus, (e) localized spots to match with the coral on a reef octopus *Octopus cyaneus*, and (f) an example of textural variation in the Papuan Cuttlefish *Sepia papuensis*.

signaling mechanism to communicate aggressive (agonistic) or sexual messages. Cephalopod skin can change its color, brightness/contrast, and texture into one of a wide variety of pre-set patterns. An individual cephalopod is able to display as many as 15–50 different body patterns in its repertoire. Their patterns incorporate changes in the chromatic and textural aspects of the skin on their head, arms, and mantle (the body wall behind the head), as well as postural, and motion aspects. The chromatic (optical) properties of their skin are dependent on specialized pigment and structural color cells in the dermis. Three-dimensional (3D) textural control comes from muscles under the skin that can push the skin outward; these skin extensions, called papillae, can reach to more than a centimeter in height.[12,21] Their use of color, pattern, physical shape change, and mimicry achieve astounding feats of disguise and makes them very effective at hiding against rocks, plants, or the ocean floor. An example can be seen in Figure 4.5 where an octopus, *Octopus vulgaris*, very effectively hides on an available plant, revealing itself when the diver gets too close.[22] Cephalopods can even display multiple signals at once, simultaneously communicating separate messages to multiple observers.[20] The entire display process operates under neuromuscular control and is almost instantaneous. Some can change multiple times per minute for hours.[19,23]

Cuttlefish live in diverse aquatic environments. The visual cues that they use for camouflage come from sand, gravel, rocks, algae, and corals.[20] For camouflage, the cephalopods, particularly octopus and cuttlefish, mimic the contrast and overall brightness of background of the area both underneath and beside the animals, as well as the general pattern of bright/dark elements of background on which they are hiding.[20] There are three common camouflage patterns (as seen in Figure 4.6): uniform—a relatively even coloring for environments such as sand, or very small pebbles, where there is minimal contrast variation (or none); mottle—where the cephalopod presents skin coloration that looks like small, broken patches or spots of color for environments such as coral, or pebbles, where the small light and dark patches splotches throughout skin when the background has moderate

Second:frame 0:00   0:08 (270 msec)   2:02 (2,070 msec)

Figure 4.5 An *Octopus vulgaris* being startled. Initially colored and textured to match the aquatic plant on which it rests, the octopus changes color and texture completely in approximately 2 seconds. (From Hanlon, R., *Curr. Biol.*, 17 (11), R400–R404, 2007.)[22]

Figure 4.6 Examples of the three primary patterns for crypsis employed by the cuttlefish. Uniform, mottle, and disruptive. (From Hanlon, R., *Curr. Biol.*, 17 (11), R400–R404, 2007.)[22]

contrast; and disruptive—irregular patches of different sizes, shapes, color, scale, and contrast. These patterns also effectively break up the outline of the animal. The animal can estimate degree of graininess as well as take into account its own size when determining which pattern type to use. Besides background matching, some cephalopods also display deceptive resemblance to other aquatic animals as well as countershading, where the dorsal portion of the animal (the top-side) is a darker color than the ventral side (the bottom side of the animal.)[12,24,25] Countershading helps them match the lighter background when seen from below, and the darker background when seen from above.

Displays in cephalopods can communicate, for example, fighting intent (agonistic) and threatening (deimatic) displays or sexual and gender displays.[12] With the Intense Zebra display (seen in the common cuttlefish *Sepia officinalis* in Figure 4.7) the animal exhibits very high-contrast dark and light stripes as an honest signal of fighting intent. An example of a deimatic display can also be seen in Figure 4.7 where the animal spreads and flattens itself. The color in the center of its body pales, whereas the periphery darkens, and two dark spots appear on the mantle perhaps changing the appearance of size. Cephalopod control of their body patterning is so rapid that they are capable of producing dynamic, almost video-like body patterns. In the "passing cloud" or "passing wave" display, the cephalopod creates dark bands that move down the body from the mantle tip to arm tip mimicking movement of clouds, or waves on ocean floor.[26] This display can be flickered very rapidly, effectively mesmerizing or distracting prey.[27] While difficult to capture in still photographs, there are many fascinating videos of both slow and rapid "Passing Cloud" cephalopod

Figure 4.7 (a) Deimatic and (b) agonistic displays in *S. officinalis*. (From King A.J., and Adamo, S.A., *J. Exp. Biol.* 209 (6), 1101–1111, 2006.)[71]

displays that can be found on the Internet, and it is well worth taking the time to view them. Signaling types are highly conserved, meaning many cephalopods demonstrate similar signals.[28]

4.5 Architectures of Dynamic Biological Photonics

Dynamic photonic systems of the cephalopods, fish, and lizards share some common features. Natural color is produced through pigment, structural color, or (most often) a combination thereof, and this is no different for dynamic systems. The photonic systems in these creatures employ a variety and combination of three primary components: chromatophores, iridophores, and leucophores.[4,9,12,19] In organisms with dynamic photonic structures, active chromatophores, iridophores, and in at least one example, active leucophores drive the color change.

Chromatophores (meaning "color bearers" in Greek) are cells in which biological pigment molecules are contained, synthesized, or translocated.[9] The pigments in these organs absorb and reflect specific wavelengths of light depending on the molecules. Thus far, four different types of pigment cells have been identified: melanophores, which produce black to brown pigments; red/orange erythrophores; yellow xanthophores; and blue cyanophores. The latter is a comparatively recent discovery, identified in some tropical fish.[29] By expanding or retracting distinct groups of chromatophores, cephalopods can darken their entire body surface, or produce array of patterns, bands, stripes, and spots as seen in Figure 4.4.[28]

The remaining two photonic components, iridophores and leucophores, are structural color in nature. Iridophores are cells that contain multilayer reflectors made of protein plates interspaced by cytoplasm (the viscous goo that surrounds the cells' interior structures). The multilayer stack produces iridescent color via constructive interference, and the resulting optical response is based on the refractive indices of the components, the dimensions of the plates and stack, and the angles of incident and observed light. These structures often produce reflection at the smaller wavelengths of visible light, blue, cyan, or green-colored iridescence, supplementing the longer wavelength pigmentary chromatophore colors. Leucophores are long, flat cells that are covered in thousands of small, knobby colloids (leucosomes). The leucosomes are colorless themselves, but provide sites for incoherent scattering. The result is diffuse reflection of ambient light, described under white light illumination as creamy, or chalky white.[30] Leucophores provide a solid, opaque background and enhance contrast of the wavelength-specific structures.

Figure 4.8 shows examples of these components and the optical effects they create in cephalopod skin.[32] These photonic structures lie in separate layers in the dermis of the animals. Many cephalopods have transparent muscle and tissue; the tissue has evolved to reduce scattered light.[28] Their dermis and epidermis allow light to penetrate

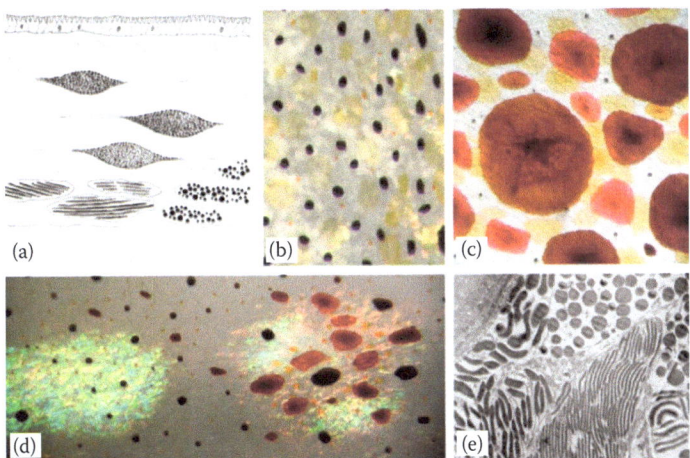

Figure 4.8 (a) Illustration of cross section of cephalopod skin showing the location of chromatophores, iridophores, and leucophores (b) Close-up of cuttlefish *Sepia officinalis* skin showing chromatophores (yellow, expanded; dark brown, partially retracted; orange, retracted) and white leucophores. Scale bar: 1 mm. (c) Close-up of squid *Loligo pealeii* skin showing brown, red and yellow chromatophores. Scale bar: 1 mm. (d) Combination of chromatophores and iridophores to illustrate the range of colors. Scale bar: 1 mm. (e) Electron micrograph showing iridophore plates and spherical leucophores of S. *officinalis* skin. Scale bar: 1 mm (From Mathger et al., *J. Roy. Soc. Interface*, 6, S149–S163, 2009.)[72]

into the underlying photonic structure and return. When all of the chromatophores are retracted, some cephalopods become translucent or even fully transparent. The octopus *Japetella heathi* and squid *Onychoteuthis banksii* can switch between transparency and pigmentation. In general, the active structures are layered within the transparent dermis with the chromatophores lying above the iridophores and leucophores. Figure 4.9 shows an electron microscopy image of a vertical section of *Octopus vulgaris* skin showing the layering of the chromatophore above iridophores and leucophores.[33] The type and arrangement of chromatophores within the epidermis varies with species, and within different regions of the animal's body.

While exceptions exist (see Box 4.1) most dynamic color change in fauna is driven by a combination of this common set of components. Some sources describe the pigment cells alone as chromatophores, and describe their color as "true color," to differentiate them from the cells that produce structural color. Others use the word chromatophore to describe any cell type that produces color, including structural color cells. Here, we use chromatophore to describe only the pigment cells, and refer to iridophores and leucophores separately.

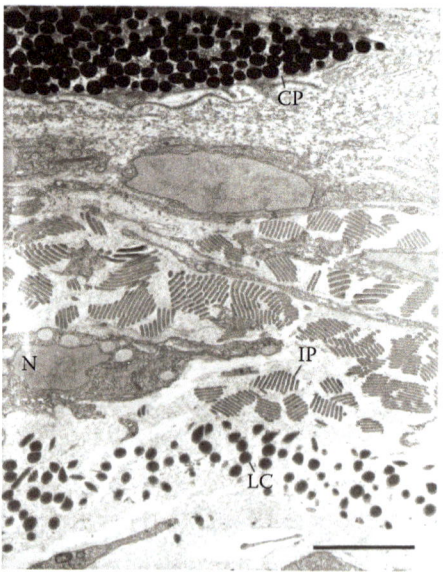

Figure 4.9 Low-power electron micrograph of a vertical section of *Octopus vulgaris* skin showing the layering of the chromatophore above the iridophores and leucophores. CP, chromatophore; IP, iridosomal platelets; N, nucleus; LC, leucophores clubs. Scale bar: 5 μm. (From Froesch, D. and Messenger, J.B., *J. Zool.*, 186, 163–173, 1978.)[73]

DYNAMIC, ADAPTIVE COLOR 133

BOX 4.1 CASE OF THE GOLDEN TORTOISE BEETLE

The Panamanian tortoise beetle is quite lovely; its elytra, as seen in Figure 4.10, is shiny and golden. *Charidotella egregia* (Bohemian, 1855) is a relatively small beetle, typically less than 1 cm in length.[31] It is covered by a strong transparent armor.

Figure 4.10 The Panamanian tortoise beetle, *Charidotella egregia*. (a) Beetle switching from gold (top) to red (bottom). (b) Cross section of the elytron of the beetle. Red pigment can be seen beneath the outer surface of the elytron (the multilayer reflector cannot be seen at this scale). (c) Transmission electron microscopy (TEM) of a *C. egregria* cuticle. Irregularities can be seen in the material density parallel to the layers. Porous patches are distributed irregularly in each layer of the stack. (d) Scanning electron microscopy (SEM) of a damaged portion of cuticle. Small channels can be seen (indicated by yellow arrows) which may play a role in fluid transport. (From Vigneron et al., *Phys. Rev. E*, 76 (3), (2007.)[31]

(*Continued*)

BOX 4.1 (*Continued*) CASE OF THE GOLDEN TORTOISE BEETLE

The beetle's dorsal cuticle, the outer covering on the beetle's back, is a bright metallic, specular gold. When *C. egrigia* is disturbed, the beautiful gold color changes rapidly to a wide-angle diffusive matte red. The color change begins at the periphery of the wing case and external and central parts of prothorax (the front-most section of thorax) moving inward. Scanning electron microscopy (SEM), TEM, and optical microscopy of the cuticle combined with spectrophotometry and modeling reveal something quite interesting. Unlike the other systems described in this chapter, the tortoise beetle does not rely on adaptive chromatophores to change its color. Instead, the entire chitinous cuticle consists of a porous chirped multilayer stack that sits on top of a layer of red pigment. The chirped multilayer is a wide-band metallic reflector with the layers varying in thickness with depth into the structure, as discussed in Chapter 2. When in its disturbed state, the reflective multilayer becomes transparent, revealing the pigment layer underneath. How does it do this? The pores in the chirped multilayer seem to be the key to the transformation. Evidence suggests that at rest, the beetle maintains a level of hydration such that the pores are filled with liquid. The layer then has a homogeneous refractive index (RI) that differs from the RI of the material between the plates. But when disturbed, the beetle expels that liquid. The dry pores act as scattering centers, destroying the Bragg coherence necessary for constructive interference, and the cuticle becomes translucent revealing the pigment layer underneath. It is such a simple trick, with such spectacular results.

4.6 Chromatophores

The cephalopod chromatophores are organs that expand and contract under direct neural control of the animal.[12] Within the chromatophore membrane is a sac containing pigment granules, called the cytoelastic sacculus (see Figure 4.11). Attached radially around the sac are 15–25 enervated muscle fibers. These radial muscles are electrically coupled

DYNAMIC, ADAPTIVE COLOR

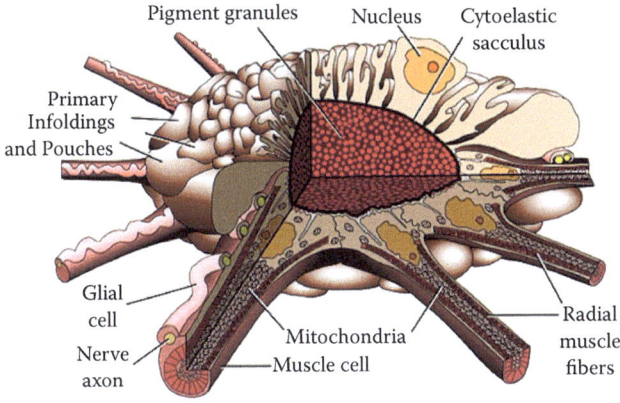

Figure 4.11 Illustration of a chromatophore. (From Cloney, R. A., and Brocco, S. L., *Am. Zool.*, 23 (3), 581–592, 1983; Florey and the Tree of Life web project.)[30,74]

to their neighbors, with the nerves that follow a winding course down the muscle. (This meandering course helps accommodate stretch. This same strategy, although not perhaps inspired by the cephalopod, has been demonstrated in stretchable interconnects to enable flexible integrated circuits.[32]) At rest, the cytoelastic sac is contracted and very small.[12] When the muscles are triggered, they pull the sac in all directions, stretching it thin. Because the expansion is under the animal's direct neural control, the activity is as fast as a few hundred milliseconds. The cell membrane that surrounds it is highly folded in the resting state, but unfolds flat over the sac as it stretches. When in its resting, contracted state, the chromatophore can be as small as micrometers in diameter; during expansion the chromatophore can expand out to millimeter diameter. The surface area can change as much as 500%.[4,33] Cephalopods with larger and more complex repertoire of patterns have more chromatophores at a higher density.[20] The squid *Loligo plei* has large chromatophores, approximately 100 μm in the contracted state, extending to 1.5 mm in the expanded state. But *L. plei* has a relatively low density of chromatophores, with only about 8/mm². *Octopus vulgaris*, in contrast, has much smaller chromatophores. When expanded, they reach only 300 μm diameter. However, *Octopus vulgaris* has a much higher density of chromatophores on their mantle, approximately 230/mm².[12] This makes the animal capable of highly detailed patterns. Some species have three separate color chromatophores, while others may have fewer.[28] *Octopus vulgaris*, for example, has pale yellow chromatophores deep in the skin with red and then black closest to the surface, blue/green iridophores lie underneath across the entire area, and white leucophores occur in clumps below the chromatophores. The squid *L. opalescens* has yellow, red, and brown chromatophores, and iridophores. However, squid generally have no leucophores. Cephalopods are able to regenerate chromatophores that are lost to damage.[26]

The pigment granules that reside within the sac are tethered to each other and the sac by microfilaments a few tens of nanometers in diameter. Figure 4.12 shows an image of a partially contracted chromatophore. The muscle fibers that surround and actuate the pigment sac can be seen in the image. Scanning electron micrographs of the pigment sac shows the pigment granules tethered to each other. When the sac is expanded, the pigments move along with the sac and form a thin layer. The filament tethers enable the pigments to distribute uniformly within the chromatophore during stretch.[12,19] Recent work out

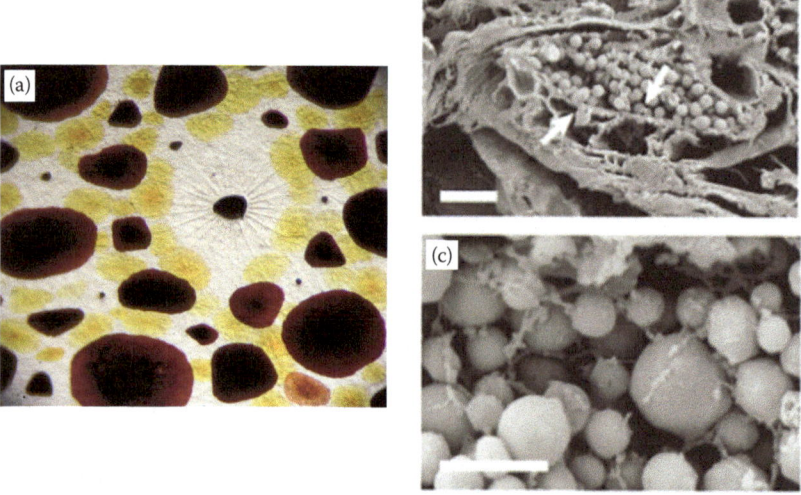

Figure 4.12 (a) Yellow, red, and brown chromatophores of the squid *L. pealeii* (transmitted light). Muscle fibers attached to pigment sac can be seen in the semicontracted brown chromatophore (slightly above center). (b) SEM of a chromatophore cross section prepared from wax section. CS, cytoelastic sacculus. Scale bar: 2 μm. (c) Tethered pigment granules within the chromatophore. Scale bar: 1 μm. (From Mathger, L.M. and Hanlon R.T., *Pigment Cell Melanoma Res.*, 25 (3), 295–296, 2012; Deravi et al., *J. Roy. Soc. Interface*, 11 (93), 2014.)[28,33]

of Harvard University and the Marine Biological Laboratory indicates that, in at least one example, the pigment granules themselves are not simple ommochrome pigments and are more complex than previously thought.[33] In their extended state, the layers of pigment granules in the cuttlefish, *Sepia officinalis*, chromatophore may be as few as three granules thick, yet still maintain their strong color reflectance and contrast. The pigment granules show two interesting features, which may indicate that the chromatophore color is at least partially dependent on structure. First is the presence of reflectin and crystallin proteins within the granules. Reflectin and crystallin are high refractive index proteins (RI as high as ~1.59 for reflectin) that are also found in many biological structural color reflectors and lenses, such as the iridophores and leucophores in cephalopods. Second, the granules, when isolated, fluoresce at wavelengths between 650 and 720 nm. When aggregated, however, the fluorescence decreases, likely due to increased scattering and absorption from proximity to other granules. Exactly how these interesting properties influence the color production is still unknown, but it is indicative of potential structural dependence in even this pigmentary component of the cephalopod photonic system.

Not all chromatophores operate under the same principles as the cephalopod chromatophores. Related in structure are the chromatophores found (for example) in chameleons and zebrafish.[4,34,35] In neither of these animals is the chromatophore under direct neural control as it is in cephalopods. Instead, they are under hormonal control, and as such color change is a much slower process than in the cephalopods. Lizards, such as the chameleon, have chromatophores and iridophores (like the gecko described in Chapter 2). In chameleons, the melanophore resides deep in the dermis. Their transparent dermis has multiple layers of chromatophores as well as a layer of iridophores, and the chromatosomes migrate within the cells and spread out their color within the dermal layer.[18,31] Zebrafish also have melanophores. In both of these cases, the melanophore acts like a reservoir. Activation causes transportation of the pigment up to spread nearer to the skin surface. In zebrafish, the pigment is a fluid and transportation is accomplished by muscle-actuated pressure, where in chameleons pigment granules travel up small channels. The size and number of chromatophores, analogous to resolution in display technology, is highly species dependent.

4.7 Chromatophore-Inspired Structures

Examples of approaches that are inspired by the chromatophore concept primarily focus on electrically triggered movement of pigment. Several approaches investigate concepts for reflective full-color display applications specifically, and focus on conventional microfabrication techniques.[4,36] Others are inspired by the compliant nature of the chromatophore system and its potential for "soft active surfaces" for applications that could require flexible, compliant optically active systems, such as camouflage, thermal control, and active photovoltaics.[35] Researchers from the University of Bristol, for example, have explored the latter using dielectric elastomers—electroactive deformable polymers—inspired by both the expanding and contracting cephalopod chromatophore, and the pigment transport of the zebrafish chromatophore. Electroactive polymers are used extensively as compliant polymer artificial muscles. When a dielectric elastomeric film is coated on each side with a compliant electrode, and a voltage is applied across the film, the film compresses transversely causing planar expansion. The film releases to its original state when the voltage is no longer applied. This property of the film can then be used to couple the electrically driven strain to an optical effect, much like in natural chromatophores.

The expanding and contracting functionality found in the cephalopods inspired one fairly simple but effective approach to synthetic chromatophores. In this case, the electrode painted on an elastomeric film acts as both the trigger and the pigment. When voltage was applied, the central electrode-coated area expanded 235%. However, the process required kilovolts to do so. Several variations on this theme were fabricated: an inverse of the above artificial chromatophore, where the electrode is painted on to surround a transparent circle in the center, which can then be filled with a pigmented gel, as well as a grouping of multiple artificial chromatophores to mimic the clustering and interconnection seen in chromatophores in cephalopod skin. In addition to the artificial actuated film model, a system to hydrostatically move pigment in and out of an artificial cell mimicking the zebrafish model of chromatophore was also developed. The device consisted of a 13-mm wide central chamber cut from a silicone layer sandwiched between two glass slides. This central chamber is connected to two elastomeric pumps. The system is pre-loaded with two fluids, the opaque fluid (watered down inkjet printer ink) and a transparent fluid (white spirit). When the fluids are injected the pumps bulge, loading the system with elastic energy. The fluids have roughly the same viscosity, and are immiscible, so they do not mix when loaded; they form a discrete transition between one fluid and the other. When neither pump is activated, the system is at equilibrium. When one of the pumps is activated, by applying a voltage, the elastomer relaxes. To bring the pressure in the system back to equilibrium, the opposite pump contracts. The contraction causes the fluid to move toward the active side. So, by activating one pump or the other, the researchers could change the color of the cell from black to transparent and back by pushing one fluid or the other into the central chamber. The solvent clings to the inner surfaces of the entire system. When the opaque fluid is pushed in, a thin layer of solvent remains, preventing the opaque ink from clinging to the cell. The opaque ink can then move in and out of the cell without leaving residual ink behind. In tests, one or the other elastomeric pump was actuated, and the resulting fluid transition captured on video. The University of Bristol was able to show that, in just about 10 seconds, the cell could change from fully transparent to fully opaque, and approximately the same time to return to transparency, when the pump was activated. Because of the type of elastomer, which was chosen for its ability to accept high strain, full actuation required very high voltage (kV) application.

On a scale much closer to what would be required for most applications, and in a scalable microfabrication approach, there are several examples of devices targeting electronic paper applications. The first, an electrokinetic approach, creates artificial chromatophores inspired by the spread and contraction of pigment driven by an electric field.[4,37] In this approach, small pits are fabricated on a base layer (substrate) with electrodes. The pigments, in this case, are electrically charged particles. When a voltage is applied, the electrically charged pigment particles are attracted to the small pits effectively contracting the chromatic element into the pit area. The field-induced reaction is fast, hundreds of milliseconds. A white backing substrate beneath the now "clear" area provides white reflection. When the voltage is off, the pigment is released from the pit and spreads over the surface of the pixel. Much like the layered chromatophore structures in the cephalopod, pixels of multiple colors can be layered, potentially enabling full color images.

Another impressive approach uses surface tension to drive movement of pigmented fluid from a reservoir to the pixel and back.[36] An illustration of the electrofluidic device is shown in Figure 4.13. The device consists of a top and bottom substrate between which is

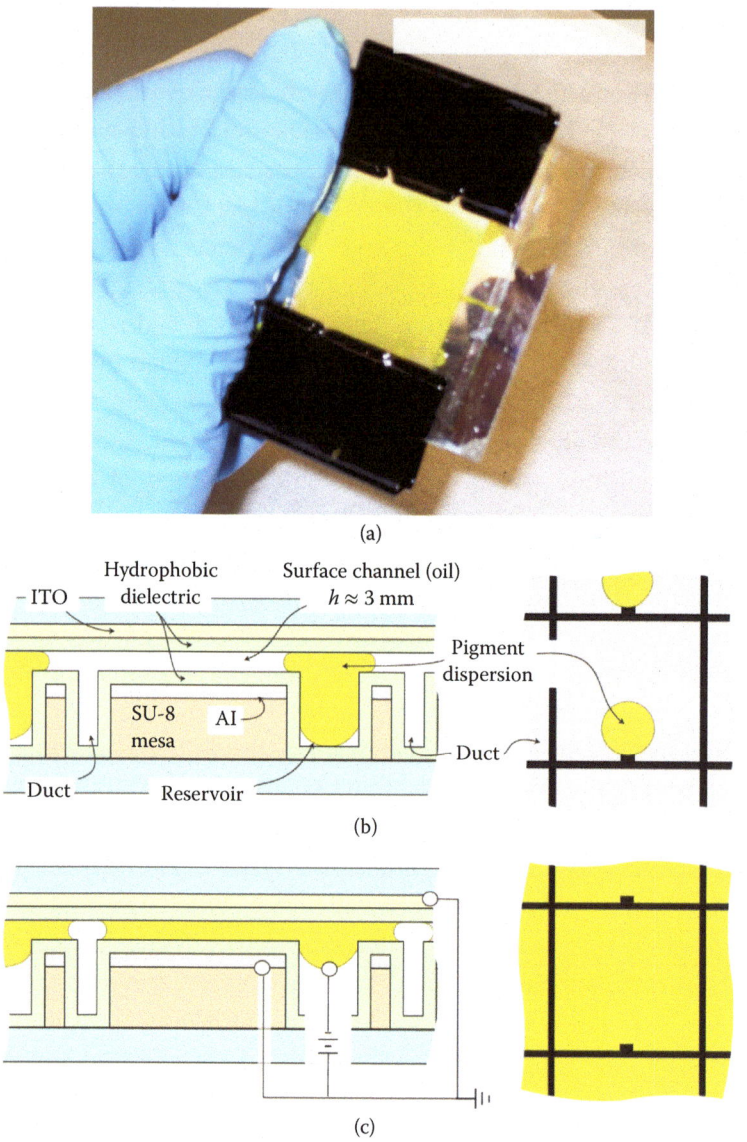

Figure 4.13 (a) Bright-field image of a 170 DPI direct-drive demonstrator with ~30,000 pixels. (b) Cross section (left) and top view (right) of the pixels with no voltage. (c) Cross section (left) and top view (right) of the pixels with applied voltage causing pigment dispersion to fill the surface channel. (From Heikenfeld et al., *Nat. Photon.*, 3 (5), 292–296, 2009.)[36]

sandwiched the electrofluidic pixel: a reflector-backed shallow channel connected to a fluid-filled reservoir. When no voltage is applied, the pigment fluid is driven by surface tension into the reservoir, and the reflective layer is revealed. When a voltage is applied, the fluid is released, spreading out into the larger area channel creating a highly reflective colored pixel. The reservoir can be as small as only 5%–10% of the total channel area, providing a high contrast between its ON and OFF states. While switching into the ON state does require a fairly high voltage (~12 V), the process is very fast: t_{on} = ~50 ms, and t_{off} = ~25 ms for a 150-μm pixel. This device also has the advantage that it can be fabricated using conventional photolithography, making it compatible with other device manufacturing techniques. There are many potential areas of improvement for the already impressive demonstrated device. Alternate pigment fluids could increase the saturation and hue of the color, as well as the fluid dynamics of the electrofluidic device, alternate reflector materials (wide-band nanostructured reflectors as seen in lizards and cephalopods, for example) could increase the brightness of the "white" screen, as well as improving the pigment color performance, and highly absorbing layers could be added to increase contrast. The systems can be improved for electronic paper applications in designs where the color pixels can be stacked to create multicolor devices, they can be bi-stable (where power is required only to switch between two states), and increase contrast and improve fabrication.[38,39]

Most recently, an exciting collaboration of researchers have demonstrated an adaptive optoelectronic camouflage system inspired not only by the color-changing ability of the cephalopod but also by their dynamic light-sensing capability.[40] Seen in Figure 4.14, the resulting device is a flexible 16 × 16 multiplexed array of thermochromic and photosensitive pixels. The total array size is about 1.4 × 1.4 cm. The entire array was fabricated on a flexible substrate and is extremely thin, so the array can bend while retaining complete functionality. The photosensitive pixels can detect light from either above or below the array. Photodetection triggers localized heating, which changes the color of the pixel above from opaque (in this case black) to clear. In the figure, light shining through a patterned filter below the array drives the pixel transformation in about 1–2 seconds. The details of the pixels and array construction can be seen in Figure 4.15. The pixel architecture is specifically described in analogy to the cephalopod inspiration.

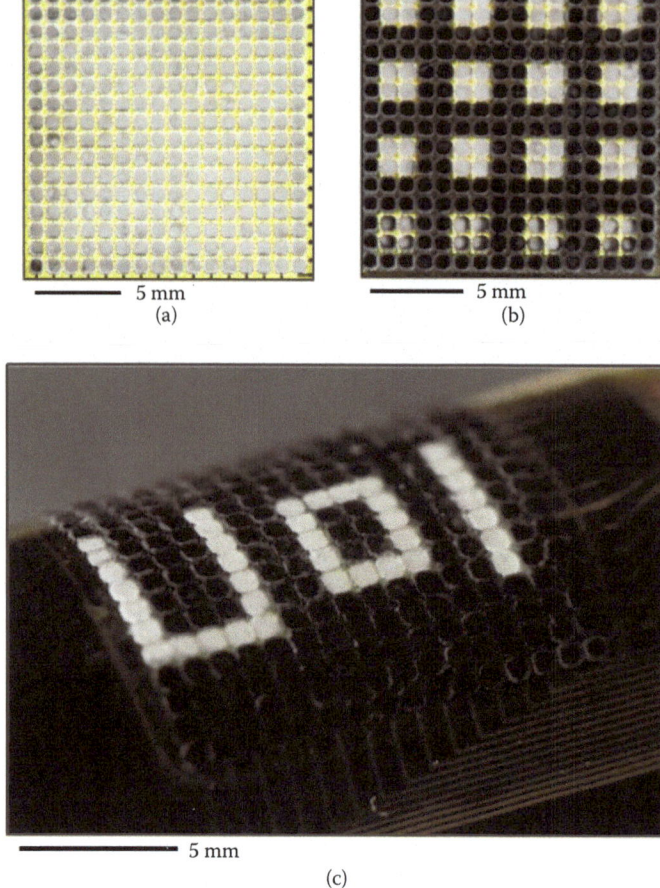

Figure 4.14 Illustration of metachrosis for several different static patterns. Top view image of the device in a uniform geometry (a), and large (b) squares. (c) Image of a device in operation while bent, while showing the text pattern "U o I". (From Yu et al., *Proc. Natl Acad. Sci. U S A*, 111 (36), 12998–13003, 2014.[40])

Each pixel consists of a chromatophore analog, a bright reflective background (leucophore) layer, an actuation mechanism (analogous to chromatophore muscles), and a photodetector on a flexible substrate. The photodetector architecture is similar to the molecules (opsins) found in cephalopod skin that are postulated to serve as distributed light sensors. The color-changing unit is composed of a patterned dye and polymer composite layer, a white reflective thin silver layer, and a thin silicon diode. The dye is thermosensitive and reversibly changes color above a critical temperature from its normal opaque color to clear; the dye used in this study had a transition temperature around

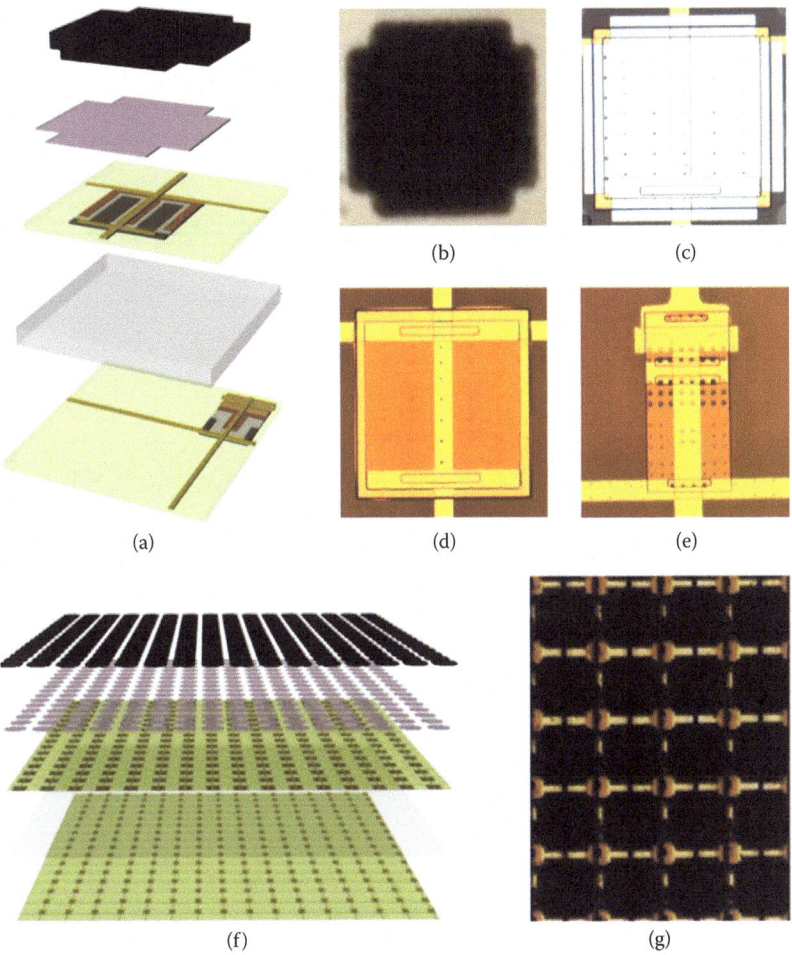

Figure 4.15 Illustrations and images of adaptive camouflage systems that incorporate essential design features found in the skins of cephalopods. (a) Exploded-view illustration of a single unit cell that highlights the different components and the multilayer architecture. The first and second layers from the top correspond to the leucodye composite (artificial chromatophore; ac) and the Ag white reflective background (artificial leucodye; al). The third layer supports an ultrathin silicon diode for actuation, with a role analogous to that of the muscle fibers that modulate the cephalopod's chromatophore (artificial muscle; am). The bottom layer, separated from the third layer by PDMS, provides distributed, multiplexed photodetection, similar to a postulated function of opsin proteins found throughout the cephalopod skin (artificial opsin; ao). (b) Optical image of a thermochromic equivalent to a chromatophore. (c) Optical micrograph of the Ag layer and the silicon diode. (d) Optical image of the diode. (e) Optical image of a photodiode and an associated blocking diode for multiplexing. (f) Exploded view of illustration of a 16 × 16 array of interconnected unit cells in a full, adaptive camouflage skin. (g) Top-view image of such a device. (From Yu et al., *Proc. Natl Acad. Sci. U S A*, 111 (36), 12998–13003, 2014.)[40]

47°C. The silver layer becomes visible when the dye transitions from opaque to clear. The silicon diode serves as the heating element. The photosensitive element consists of a photodiode and a multiplexing switch (blocking diode). The color unit is shaped like a square with notched corners, and the photosensor elements sit in the exposed corner notches. The resulting array is an angle-independent bi-chromic display that offers interesting possibilities for future devices.

4.8 Dynamic Structural Color: Iridophores and Leucophores

4.8.1 Iridophores

The iridophores are an active component of the potent cephalopod communication tool. Iridophores are typically less than 1 mm in width.[4] In the iridophores are stacks of up to 30 thin plates of reflectin protein ($n \sim 1.59$) or guanine ($n \sim 1.86$), and cytoplasm ($n \sim 1.36$) (see Figure 4.16), which act as a multilayer reflector.[41–45] The thicknesses

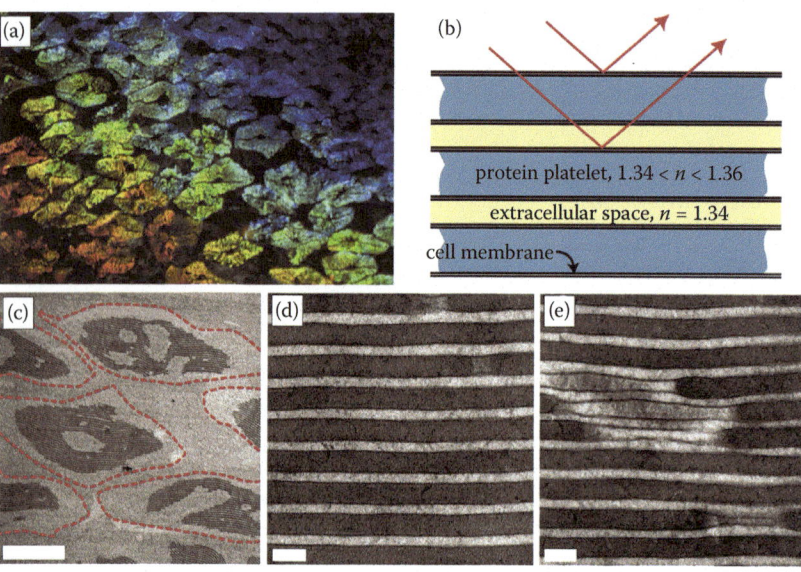

Figure 4.16 (a) Darkfield microscopy image of excised skin tissue from *Loligo pealeii*. 40× magnification under white light illumination. Size of iridophores ranging between 20 and 40 μm in width and between 2 and 5 μm in thickness. The dark area in the center of each cell is the cell nucleus. (b) Illustration of an intracellular Bragg reflector. (c–e) TEM images of ultrathin cross sections of green iridophore cells fixed with uranyl acetate. (c) Scale bar: 5 μm. (d and e) Scale bar: 200 nm. Dark areas are electron-dense, protein-filled lamellae. Cell boundaries are outlined in red. (From Tao et al., *Biomaterials*, 31 (5), 793–801, 2010.)[75]

of each reflecting layer and the spacing between the layers is typically a fraction of the reflecting wavelength. The constructive interference that occurs produces highly angle-dependent reflection. The reflected wavelength is dependent on the size, spacing, and orientation of the platelets and the light and viewing angle. These iridophores occur in patches, splotches, or stripes on the skin. The interaction between the iridophores and chromatophores may take on many roles in the color production of the system. Iridophores provide the blue- and green-color capability lacking from the pigment-based chromatophores, as well as reinforcing the color and contrast of the chromatophores (Figure 4.17).

The reflecting properties of the iridophores are dynamically tunable in some species. The reflectance can change as much as 80%, and shift the peak reflectance over ~100 nm.[4] The squid, *L. pealeii*, for example, change their iridescence during agonistic (fighting) displays, altering the reflected wavelength by as much as 100 nm. Some, such as pelagic squid, use their iridophores to direct light inward to the body. The mantle (the main body wall) of the squid and other cephalopods is mostly transparent, and these iridophores channel light downward. When viewed from below, the light comes from within the mantle, perhaps mitigating shadows and allowing the squid to match the intensity and wavelength of the light above.[19] Iridophores

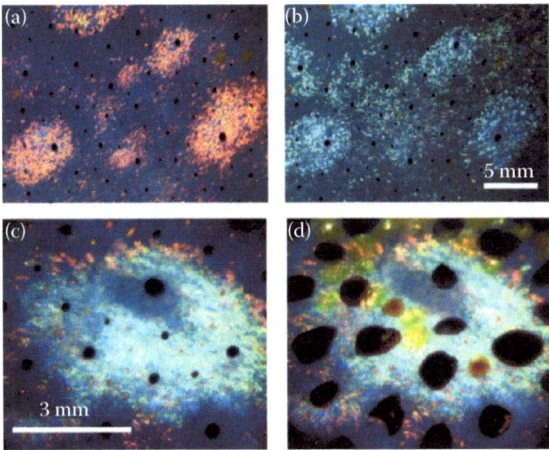

Figure 4.17 (a and b) Iridophore "splotches" viewed at near normal (red reflectance) (a) and 45° incidence (blue–green reflectance) (b) under white light illumination. Identical iridophores shown in both images; images illustrate the spectral shift seen with increasing viewing angle. (c and d) Iridophores retracted (c) and expanded (d) chromatophores. (From Mathger L.M., and Hanlon R.T., *Biol. Lett.*, 2 (4), 494–496, 2006.)[46]

are also responsible for the reflection of polarized light seen on the arms and head of some species (particularly females).[45,46] The release of acetylcholine changes the arrangement of the reflectin proteins in the iridophores through a reversible phosphorylation process, altering both the RI and spacing of iridophore plates.[4] Somewhat slower than muscle-controlled chromatophore, shifting chromatophores up to 1 minute, some *L. pealeii* also have IR ($\lambda \sim 800$ nm) reflecting collars, though it is not known why as IR, light does not travel well in water. Nor is it known how they achieve it.[44]

There are other examples of marine animals that also have dynamic iridophores, as seen in Figure 4.18. The Paradise whiptail has a red stripe on its body and head.[47] It can change color from blue to red in less than 1 second (~0.25 seconds for 465–650 nm.). In this fish, the active iridophores are controlled by the sympathetic nervous system.[48] The damselfish can also change the distance between the iridophore guanine platelets. Under stressful conditions, this fish changes reflected color from blue to yellow by also changing the spacing of the reflecting plates. The blue-ringed octopus, *Hapalochlaena lunulata* "flashes" its rings to signal aggression.[49] In this case, the actual multilayer is static, but muscles surrounding the iridophores enable the octopus to enhance or hide the bright blue rings. The neon tetra operates under a slightly different model. The neon tetras, *Paracheirodon innesi*, exhibit structural color that produces cyan.[50] This fish is green throughout the day, but changes to violet-blue at night. In this case, the iridophores operate under a "venetian blind" mechanism, where the platelets tilt, rather than changing shape.[48]

4.8.2 Leucophores

Leucophores, as in Figure 4.19, are long, flat cells that are covered in thousands of small, knobby bits (leucosomes).[31,56] The leucosomes are colorless themselves, but provide sites for incoherent scattering. Leucophores can contain thousands of the micrometer-sized particles; ~12,000 particles were found in a single leucophore of the common cuttlefish *Sepia officinalis*. The leucosome particles are made of sulfated glycoproteins and reflectin and range in size from 200 to 2000 nm (average 759 ± 203 nm). As the leucosomes scatter all wavelengths of light (UV through near-infrared [NIR], 300–1000 nm) in all directions,

Figure 4.18 Other marine animals capable of dynamic iridophore control. (a) the Paradise whiptail, (b) the blue-ringed octopus, and (c) Neon Tetra *Paracheirodon innessi*.

the resulting reflection is a broad-band, creamy opaque white when exposed, as well as providing contrast and background to the reflections produced by the other photonic color components. Reflectance can range from ~35% to 75%. As with the other photonic components, the leucophores provide an optical response that is not altered by any shape change of the animal. The cells are flexible, stretchable, and compliant. The leucophores reflectance properties might be useful for high-quality

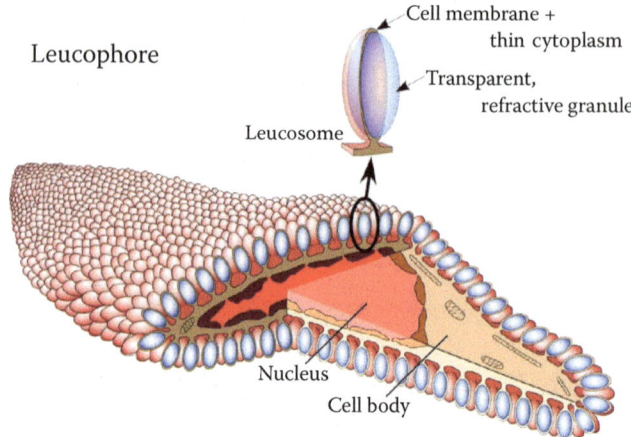

Figure 4.19 Illustration of the leucophores of *Octopus dofleini*. (From Cloney, R.A., and Brocco, S.L., *Am. Zool.*, 23 (3), 581–592, 1983; the Tree of Life web project http://tolweb.org/notes/?note_id=2646.)[30,76]

Figure 4.20 (a) Dorsal aspect of juxtaposed male and female *Doryteuthis opalescens* squid showing sexually dimorphic structural color. Male shows pronounced testis within its mantle. Female shows correspondingly positioned white stripe in the dermis as well as the iridescent stripes in the dermis on the dorsal–lateral surface of the mantle just beneath the fin attachment. (b) *D. opalescens* mantles laid out flat after being cut longitudinally on the ventral side and removal of all the internal organs. Males are on left, females are on right. The prominent iridescent stripes (iridocytes) and white stripe (leucophores), are found only in the female squid. Also seen in the dermis of both genders is the comparatively fainter iridescence of the typical dorsal iridocytes. (From Demartini et al., *J. Exp. Biol.*, 216 (19), 3733–3741, 2013.)[43]

wide-band back-reflectors, uniform backlighting for LEDs, and color mixing for white LEDs. Most leucophores are completely passive. But there is one example of active adaptive leucophores found in the female Loligonid squid *Doryteuthis opalescens*.[43] The females (seen in Figure 4.20) have a stripe on their mantle that can change from transparent to white reflection in response to acetylcholine. It is theorized that the white shape mimics the appearance of the white testis structure found in the males. The mechanism by which *D. opalescens* accomplishes this interesting effect is not yet fully determined.

4.9 Actuating Structural Color

Moving away from the naturally dynamic photonic systems we've discussed to this point, we now look again at structural color. Structural color in nature is, for the most part, static. An organism's coloration is set during the growth of those particular structures at specific points in its lifecycle. While the color may change slightly with age or health, on the whole, most of these structures are fairly fixed. For applications, structural color offers some interesting advantages over traditional pigmentary color production, such as the ability to produce iridescent or metallic color as well as resistance to photobleaching. The colors that we see (due to structural color) emerge from the delicate interplay of the nanoscale structures with incident light, where the physical parameters of the photonic crystals: size, shape, order, periodicity, etc., drive the spectral response. Because the spectra are exquisitely sensitive to the physical properties of the structure, small perturbations may create large changes in reflected color. This sensitivity to perturbation is another advantage of structural color particularly when it comes to creating active, dynamic color systems. Many have been inspired by this very useful relationship between structure and performance to create not only synthetic static structures but also dynamically tunable bioinspired photonic systems. As with the tremendous work going on to develop photonic crystals, a lot of research is occurring to create photonic crystal structures that are responsive, with properties suitable for a variation of applications. Concurrent work between the biological photonics community and the strictly synthetic photonics community means there is significant potential for cross-pollination of creative ideas. Some approaches are explicitly inspired by a biological example, while others are described as bioinspired because they are actuating a structural color system.

To create a dynamically tunable photonic crystal, one must incorporate materials into the structure that are responsive to stimuli. This responsive material gives us the ability to controllably change one or more of the physical parameters of the system (e.g., RI, lattice spacing,

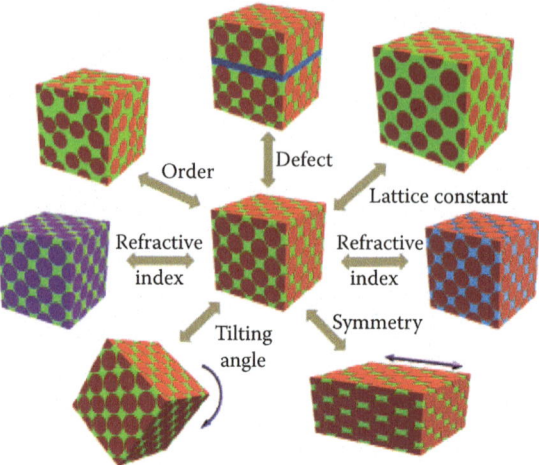

Figure 4.21 Illustration of how the 3D responsive photonic crystal structure can be tuned. (From Ge and Yin, *Angew. Chem. Int. Ed.* 50 (7), 1492–1522, 2011)[52]

viewing angle, etc., as seen in Figure 4.21) and therefore change the reflected wavelength. There are a host of potential mechanisms by which we might actuate a photonic crystal structure: mechanical, electrical, magnetic, and chemical approaches, as well as temperature variation, light, and others have all been attempted.[47,50,52] For many applications, such as displays, one would like to create a single structure, which is capable of being tuned across the entire range of desired wavelengths (e.g., across all of the visible spectrum). Some of the challenges faced by structural color approaches may be limited tunability of the reflectance (e.g., may only cover a narrow wavelength range), the change may not be completely reversible or may have a short lifetime of tunability, integration into existing devices may be a challenge, or their response times may be slow. Bioinspired approaches may offer novel designs and insights in actuating structural color to help overcome these challenges.

As in creating static photonic crystals, approaches can be described as top–down or bottom–up. Both approaches have strengths and weaknesses when it comes to photonic crystal fabrication and incorporation of responsive materials. Top–down approaches, such as lithographic approaches, offer control at the expense of cost and time. Bottom–up approaches, such as colloidal self-assembly, are rapid and inexpensive, but often have defects and are more limited in potential structural design variation. Many of the dynamic structural color approaches thus far center on multilayers or colloids, which relate to the multilayer and opaline structures found in nature. Colloidal fabrication is well controlled for many of the material systems most popularly used, such as polystyrene and silica. The structure, size, and composition of the particle, (or particle core and shells, for a core–shell system) can all be specifically designed and fabricated. These particles can be functionalized to control the behavior of the colloids in the subsequently fabricated photonic structure, for example, the interactions between colloids or the dispersion in a particular medium or suspension. Multilayers are relatively easy to fabricate, and can be fabricated from or incorporate block co-polymers, liquid crystals, elastomers, and other functional materials. Dynamically tunable defect layers may be introduced to disrupt the color response. Structural color systems constructed with a responsive material, therefore, couple the crystal structure to the mechanism of stimulation.

4.9.1 Refractive Index Modulation

One mechanism of change is the dynamic introduction of materials that alter the RI of the system. Most structural color systems in nature combine a structural material (such as a biopolymer) and air gaps in a complex nanoarchitecture, and many synthetic structures are similarly constructed. Colloidal structures such as opal and inverse opal, for example, have interstitial spaces that can be infiltrated by any number of materials. By infiltrating the open spaces in a photonic structure with a second material of sufficiently different RI, the RI of the system is changed affecting the reflectance of the structure. This could be, for example, a solvent, liquid crystal, or biomolecules such as antibodies or DNA. RI changes by filling the interstitial spaces with solvents, for example, causes a color change in the optical response of the structure, as seen in the color change on the butterfly wing when it is dipped in ethanol, shown in Figure 2.7. Infiltration of a liquid into the empty spaces depends on the ability of that liquid to flow and maintain contact with the surface of the structure you are infiltrating (its wettability). The surfaces can be functionalized, i.e., specific molecules can be used to coat the surfaces, to improve or inhibit the wettability on a surface, such as the inner surface of an inverse opal film.

Researchers at Harvard University were able to pattern the functionalization of the inner walls of an inverse opal film to essentially localize the ability of various fluids to infiltrate the structure.[53] The patterning can be done sequentially, and with more than one molecule (as seen in Figure 4.22), so that a single film was "encoded" for

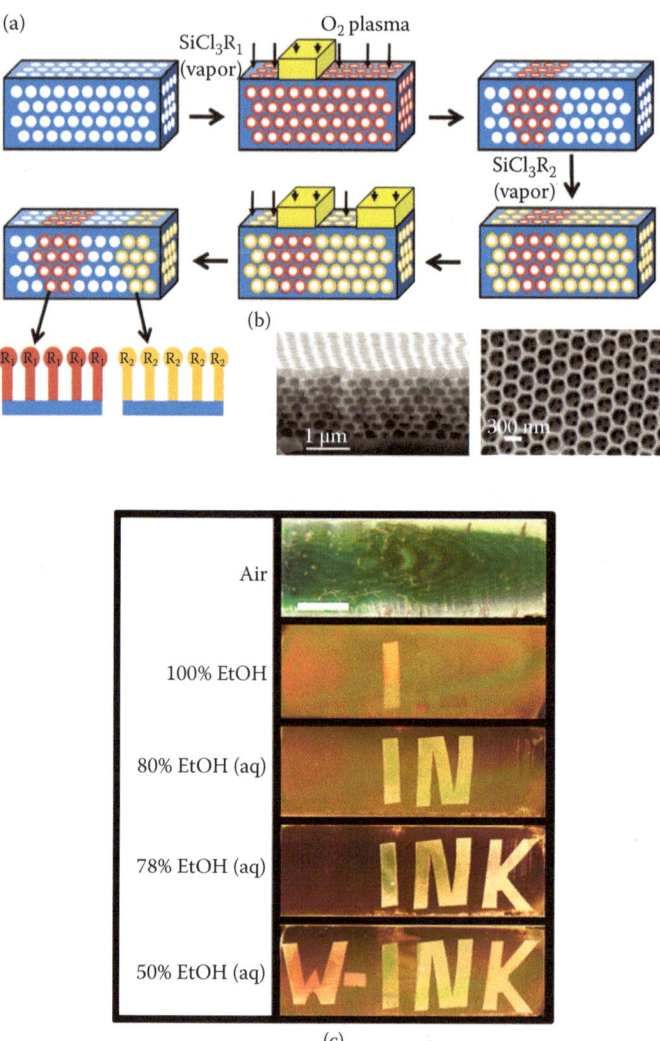

Figure 4.22 (a) Illustration of the procedure to serially functionalize the inner surfaces of inverse opal films (IOFs). Procedure repeats iterations of exposing the entire film to alkylchlorosilane vapors, followed by selective exposure to oxygen plasma, masked through slabs of PDMS, sealed on top of the film. (b) SEM images of IOFs, left: cross section; right: top view. (c) Optical images of an IOF in which the word "W-INK" is sequentially encoded. In different water–ethanol mixtures, different letters appear. (From Burgess et al., *J. Am. Chem. Soc.*, 134 (2), 1374–1374, 2012.)[53]

multiple different surface functionalities enabling specific infiltration by only certain fluids. Dubbed "watermark-ink" ("W-INK"), this concept was demonstrated by patterning the word "W-INK" into a film. Prior to submersion in fluid, the text is not visible on the film. However, subsequent submersion in four different water–ethanol mixtures revealed a color change as those liquids infiltrated the inverse opal photonic structure. Other demonstrations have shown to differentiate between the different solvents—water, acetone, and isopropanol. Using liquids with more highly varying refractive indexes, wider color variation in the subsequently filled structure could be obtained. Applications for these types of approaches may be in security (anticounterfeiting), colorimetric sensing, and writable displays. Robust, visible color change is seen upon infiltration. However, infiltration is a slow process compared to approaches that depend on other mechanisms (e.g., electric fields). This is particularly true when erasing or rewriting, where elimination of the infiltrating fluid may need to depend on evaporation, potentially limiting their reversibility. In addition, because many of these liquids do evaporate, long-term stability is also a challenge.

4.9.2 Mechanical Deformation

Structures can be actuated dynamically by pressure, stretching or swelling, and deswelling.[52] These mechanical mechanisms of actuation are not only for specialized applications such as humidity or pressure sensors but also for more conventional applications such as displays and electronic writing applications. Elastomeric inverse opal photonic crystal structures created from a soft polymer, such as those from Opalux Inc., can deform under compressive (mechanical) stress, as seen in Figure 4.23.[54,55] The stress changes the shape and spacing of these voids in the inverse opal structures. A 15 kPa compression causes a 100-nm change in void diameter, from a sphere to an ellipsoid, as well as reducing the spacing between the voids from ~276 to 188 nm. This change in structure causes a shift in the peak wavelength of ~60 nm toward shorter wavelengths (blue shifted) When the elastomer is compressed, small variations in the compression can be captured and small features down to 5 microns can be distinguished. The developers of this technology envision a potential application as

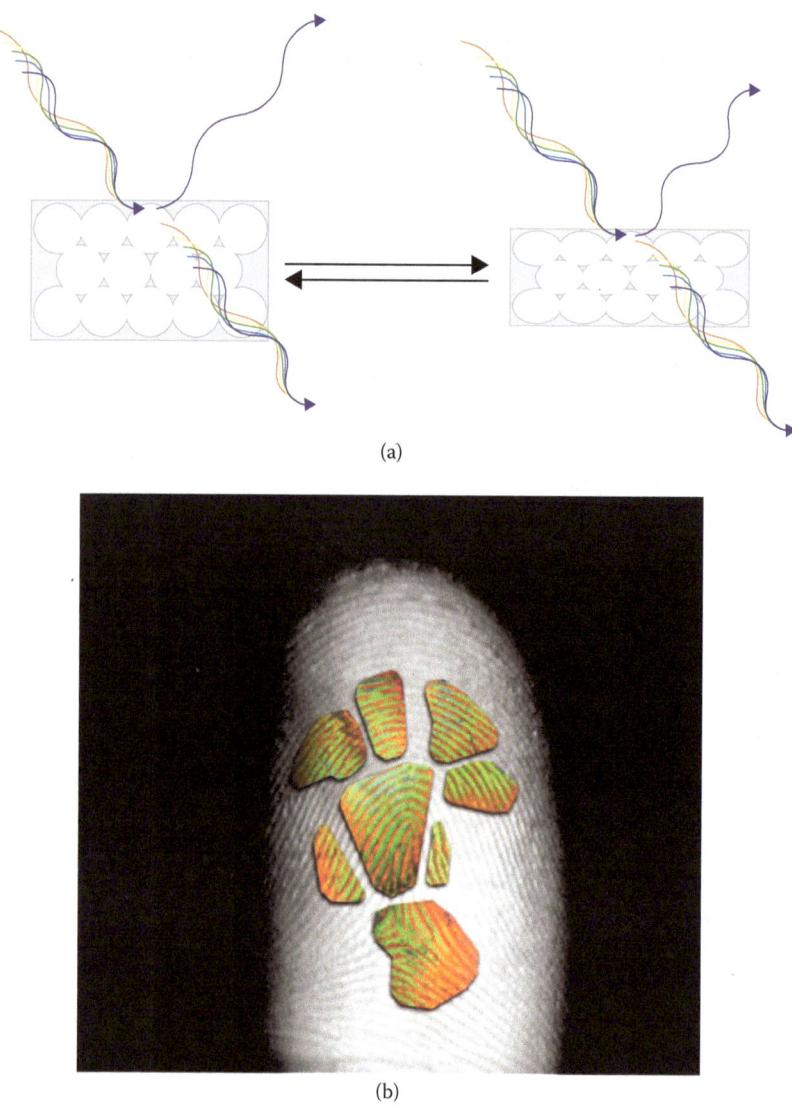

Figure 4.23 (a) Elast-Ink is an elastomer inverse opal. Compression reduces the lattice constant and blue-shifts the color. (b) A full-color fingerprint visualized using an elastic photonic crystal, overlaid onto a grayscale image of an index finger. (From Ozin, G.A., and Arsenault. A.C., *Mater. Today*, 11 (7–8), 44–51, 2008.)[54]

fingerprint reader for biometrics and identification. The high resolution visualization of the small features means that it might be possible to capture information about fingerprints, such as the ridge line structures and pressure distribution (pressure and time dependence). These structural changes are reversible and robust, allowing repeated use.

Researchers at Cornell University use a different approach and created a technique for printing erasable, high-resolution, multicolor images for photonic paper and ink applications.[56] To create the photonic structure, uniform polystyrene nanoparticles approximately 200 nm in diameter are allowed to self-assemble into a (~100 layer, 40 μm thick) face-centered cubic (FCC) structure on a slide. The structure is then filled in with PDMS (polydimethylsiloxane)-infused elastomer and crystalized. To "write" on the substrate, the group deposits (transparent) silicone oils of varying molecular weights. The oil diffuses into the elastomeric crystal and causes a localized swelling in the PDMS structure, shifting the wavelength of the reflectance. Figure 4.24 shows an image of the colloidal photonic

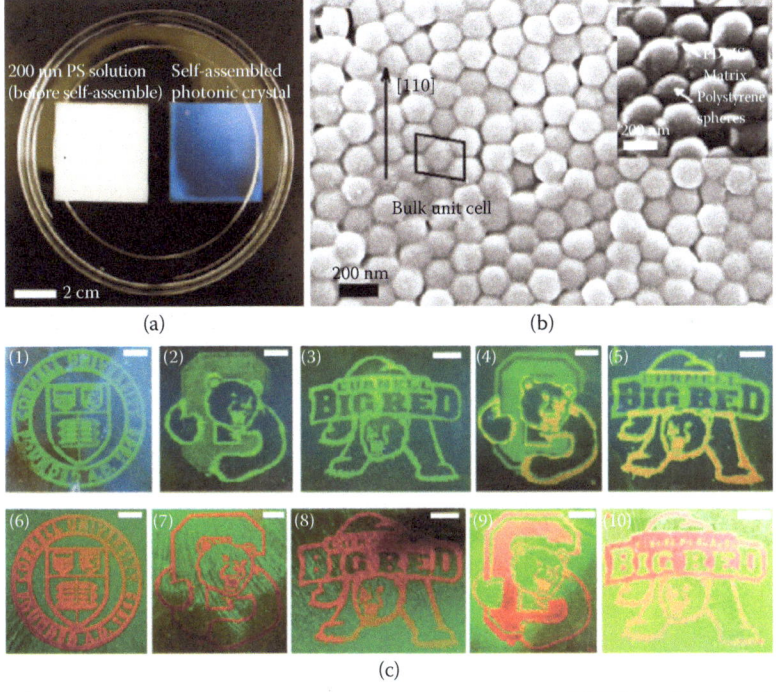

Figure 4.24 (a) Colloidal photonic crystal self-assembled on glass slides: initial amorphous state (left) and the postcrystallization FCC structure with photonic band gap of 489 nm. (b) SEM image of the self-assembled crystal structure. The inset image was taken at a 30° tilted angle. (c) High-resolution multicolor images printed on blue and green substrates: (1–3) Series of images represented with a single printed color, green (T11), on a blue CPC substrate. (4, 5) Multicolor green and orange images printed using T11 and T03 oil, respectively. (6–8) Same images shown in (1–3) printed in red (T03) on a green CPC substrate. (9, 10) Multicolor red (T03) and orange (T11) images on a green substrate. All scale bars: 1 mm. (From Kang et al., *Langmuir*, 27 (16), 9676–9680, 2011.)[56]

crystal structure, as well as examples of pictures "written" using the silicone oils. Lighter oils penetrate deeper into the structure, causing the reflectance to move more toward longer wavelengths (toward red). The images are "erased" by applying a very low-molecular weight oil, which mixes and dilutes the existing "ink" and then rapidly evaporates. Using a modified inkjet printer, the group was then able to demonstrate printing multicolor images with a resolution as high as 200 μm. The timescale of the color-change process, however, took 5–7 minutes to form; the process depends on diffusion, which requires no external power but occurs comparatively slowly. Erasing occurred on the order of seconds. The stability of the resulting images depended strongly on the molecular weight of the ink used. High-molecular weight oils could last for days, whereas the lightest oils lasted for only a few minutes.

4.9.3 Field-Induced Modulation

Magnetic and electric fields can be used as an actuating mechanism, inducing rotation, or expansion and contraction of a photonic crystal structure. Materials such as Janus particles (particles that have multiple domains with separate specific surface properties) and liquid crystals have demonstrated field-inducible color change.[50,57] Magnetic field-based approaches use the field direction to alter the orientation of a magnetically coupled photonic crystal structure. An interesting example of the magnetic field modulation-based approaches can be seen in Figure 4.25. Tunable structural color created by changing the orientation of magnetochromatic microspheres, which use angle-dependent reflection to their advantage.[52,58] The microspheres are nanoparticle-based photonic crystals that are formed into micronscale colloids. The colloids are composed of ordered superparamagnetic Fe_3O_4@SiO_2 core-shell particles that form into nanochains. The nanochains are then embedded in a polymer matrix to form larger colloids called a microdroplet. The droplet is then UV cured to polymerize and fix the alignment. The magnetic nanochains remain ordered within the microsphere and the ordered nanochains align to the field direction of an applied magnetic field. Because the nanoparticles embedded within the microspheres are ordered, diffraction occurs and a color-specific reflection is produced. When an external

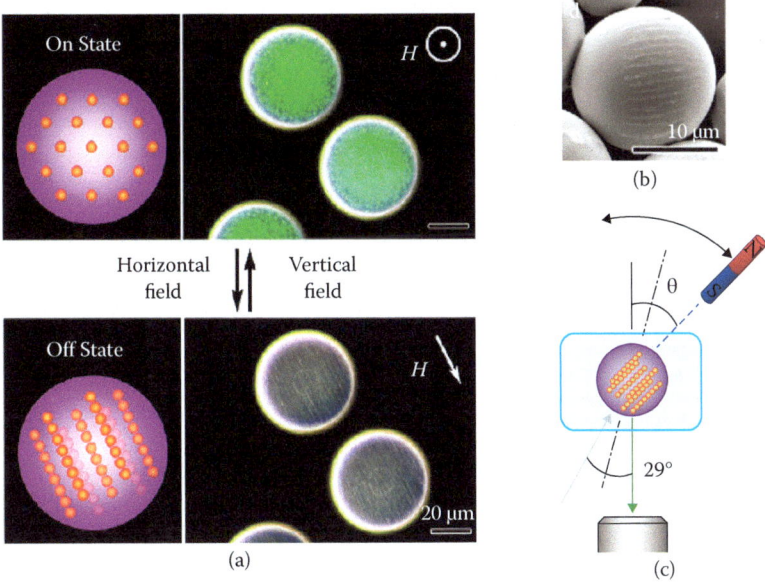

Figure 4.25 (a) Illustration and optical microscopy images illustrating the change in optical properties caused by rotating the chain-like photonic structures in magnetic fields. (b) Side-view SEM images of a microsphere. Some $Fe_3O_4@SiO_2$ particle chains are visible, aligned on the surface along the longitudinal direction. Many particle chains are embedded inside the microspheres (c) Illustrations of the experimental setup for studying the angular dependence of the diffraction property of the magnetochromatic microspheres. (From Ge et al., *J. Am. Chem. Soc.*, 131 (43), 15687–15694, 2009.)[58]

magnetic field is applied, the nanochain microspheres also want to rotate, changing orientation to align with the field direction. The diffraction of the photonic crystal structure, therefore, also changes. In a vertical magnetic field, the long axes of the nanochain microspheres rotate so that the nanochains stand vertically turning the diffraction "on" and producing a reflected color. When the magnetic field is horizontal, the nanochain microspheres rotate 90° and the nanochains lay horizontally eliminating the diffraction. The microspheres then look brown due to the native brown of the iron oxide in the droplets. The diffraction can be switched between on and off states, as well as all intermediate states, reversibly with application and removal of the magnetic field. The direction of the magnetic field and, therefore, the direction of the microspheres can be kept at specific intermediate tilting angles. Colors produced by the structures are determined by the nanoparticle and microsphere size and variation of field strength and orientation. These structures can be cast into elastic polymer films.

The resulting film would be flexible, foldable, and moldable into various shapes while still displaying magnetic tunable colors. One can create predesigned color patterns or letters with color contrast. This approach has potential use in displays as well as forgery protection and security. The approach has several advantages: stable color with a fast on and off time, and the ability to apply the field without direct contact to the microspheres. However, there are some inherent weaknesses as well: the reflectivity is relatively week due to the low number density within the polymer matrix, as well as potential long-term stability concerns due to slow evaporation through the polymer network.

Opalux, Inc. uses electric field modulation to drive a display technology based on a nanocomposite colloidal photonic crystal that they call P-Ink.[59-64] The ink is made of a cross-linked silica microsphere/metallopolymer composite that self-assembles into an opal photonic crystal structure. The opal structure is embedded in a matrix of redox-active polyferrocenylsilane gel. Sandwiching the material between electrodes, along with an electrolyte, creates an electrochemical pixel. When an oxidative or reductive potential applied, the system reversibly shrinks or swells accordingly changing the spacing of the photonic crystal structure, and therefore the color of reflected light (see Figure 4.26). Inverse opal structures create a nanopore array with significantly more area for the electrolyte to infiltrate the structure.[62] Removing the silica nanoparticles from the structure (creating the inverse opal structure) allowed the electrolyte to flow into the (now empty) spaces, enhancing electron and ion transport, improving switching speed and voltage requirements. The wavelength is continuously and reversibly tunable from the UV all the way out to the NIR. This means that displays require only one addressable pixel type, rather than the typically more complicated arrangements of red, blue, and green that are built now. The structures give them relatively narrow spectral peaks, 75–150 nm, and a reflectivity of approximately 60%, which combine to give good color saturation and reflection. Switching between any spectral color takes less than 0.2 seconds. This technology is also bistable for 2 hours after removing the electrical field, and has impressively low power requirements (switching voltage <1.5 V DC, and current below 100 mA). In addition, the devices are quite thin, only 0.3–0.5 mm thick. Manufacturing of these devices can be done via roll-to-roll processing, and are manufactured on standard ITO (indium

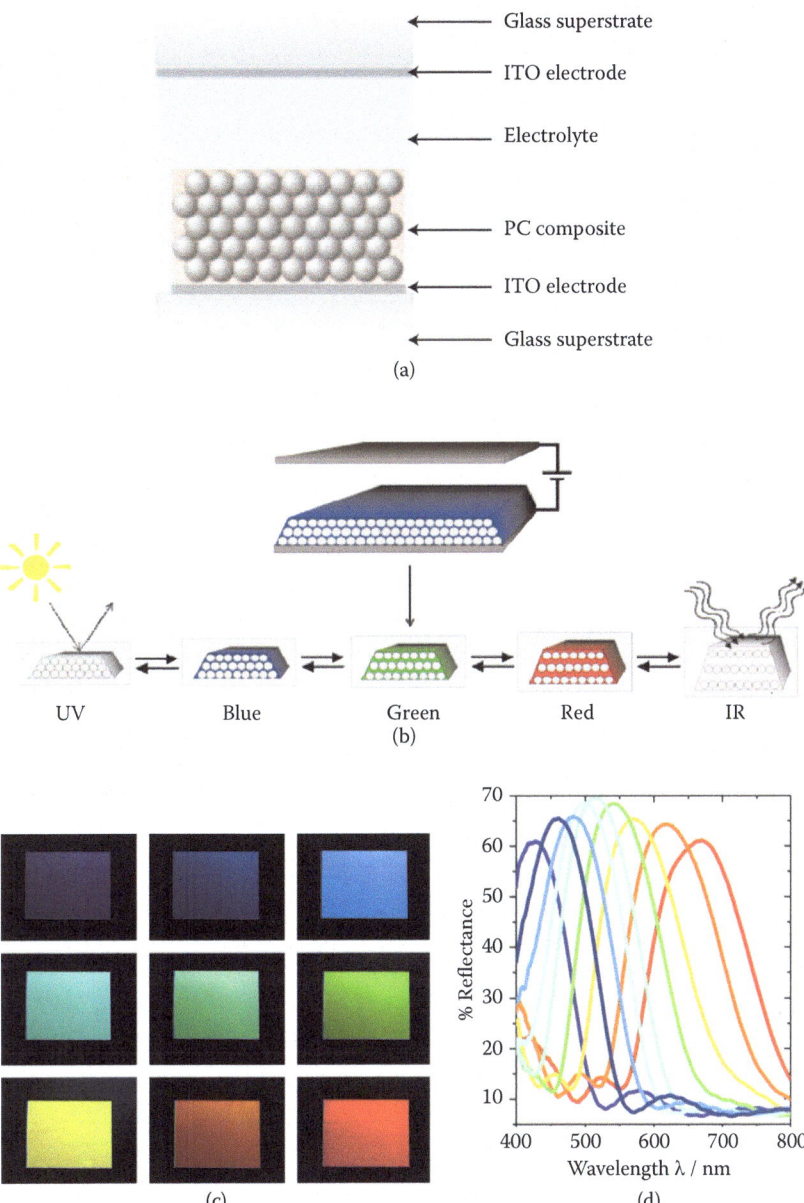

Figure 4.26 Illustration of a two-electrode photonic crystal electrochemical cell marked with relevant components. The P-Ink film is approximately 3–5 mm in thickness; cell gap can range from tens to hundreds of micrometers (b) The reflected color can span UV through visible to infrared via expansion and contraction of the spacing between spherical cavities. Pixels use 1.5 V DC or less. (c) Images of the display window of a P-Ink device showing progressively red-shifted color states from violet to red (d) in the corresponding reflectivity curves. (From Arsenault et al., *Nature Photon.*, 1 (8), 468–472, 2007; Arsenault et al., *Proc. SPIE*, 8613, 7, 2013; and Wang et al., *Information Display Magazine*, Society for Information Display, New York, NY, 2011.)[61,63,64]

tin oxide)–PET substrates, which means that the fabrication process is quite scalable. These devices are scalable, robust, cost-competitive, and interesting for a host of potential applications from mobile electronics, wearables (apparel and articles), smart packaging, architectural glass, reflective e-paper devices, and more. They can be fabricated on curved surfaces or integrated into flexible devices. Challenge remains, however, particularly with the angular dependence of the reflection affecting viewing angle, as well as achieving video rate switching and color control hue, lightness, and saturation of each pixel.

Another example is the system inspired by the dynamically tunable spacing of cephalopod and teleost iridophores as seen in Figure 4.27.[65] Researchers at MIT (Massachusets Institute of Technology) are making lamellar (layered) structures out of self-assembled block co-polymer (BCP) polystyrene-*b*-poly(2-vinylpyridine) (PS-P2VP). BCPs are polymers that phase separate into multiple periodic domains. The structure of the domains is controllable by the choice of monomers that make up the copolymer as well as the polymerization conditions. The active cell was the BCP film and an electrolyte sandwiched between two transparent electrode-coated glass slides. Application of a voltage caused the film to contract, which changes the wavelength of reflection. By

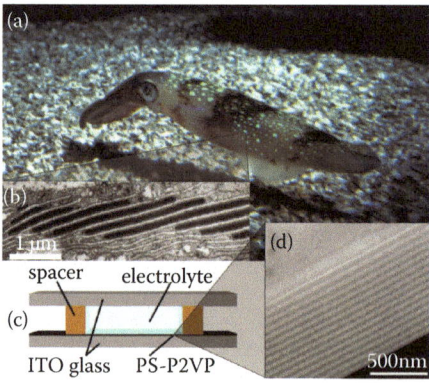

Figure 4.27 (a) Squid, *Loligo pealeii*. (b) Cross-sectional TEM image of squid iridophore. Scale bar: 1 μm. (c) Illustration of a simple electrochemical cell: 1st ITO-coated glass slide, gel-forming photonic block copolymer multilayer, spacers electrolyte, 2nd ITO-coated glass slide (d) TEM of lamellar block copolymer multilayer system. Scale bar: 500 nm. The block copolymer reflectors contained approximately 15 periods, compared to ~5–50 periods in cephalopods (varies with species and bodily location). (a) Image courtesy of Dr. R. Hanlon, Marine Biological Laboratory, Woods Hole, MA 02543; (b) adapted from Cooper, K. M., et al., *Cell Tissue Res.* 259 (15), 1990. [2], with permission. Copyright 1990, Springer. (From Walish, J.J. et al., *Adv. Mater.*, **21** (30), 3078–3081, 2009)[65]

DYNAMIC, ADAPTIVE COLOR

changing the salt concentration in the solvent, the wavelengths could be tuned from the UV to NIR (350–1600 nm).[66] The group demonstrated a proof of concept electrochromic cell, which comprised a slide with a transparent electrode layer (ITO), followed by a spin-coated PS-P2VP film.[65] The cell was filled with 2,2,2-trifluoroethanol (TFE) electrolyte, with a second ITO layer on the top slide. In its initial state, after the electrolyte had been added to the film, the reflection was red (675 nm). When 5 V was applied to the cell, the color changed to green (563 nm). Maximum reflectivity achieved was 40%. Applying a larger voltage, 10 V, did reduce the wavelength further, into the 450–500 nm range, but the color was no longer uniform. This was most likely due to small imperfections in the cell spacing, or film thickness, to make inexpensive, robust, high-contrast, low-energy displays.

A final example of tunable color technology based on structural color is the Qualcomm's Mirasol™ reflective displays inspired by the photonic structures on butterfly wings.[67,68] The active element consists of a deformable reflective membrane on the bottom, separated by an air gap from a thin-film interference stack above, creating an optical resonance cavity (see Figure 4.28). Light travels through the thin-film stack to the reflective membrane and back. As described in Chapter 2, constructive and destructive interference of the reflected beams leads

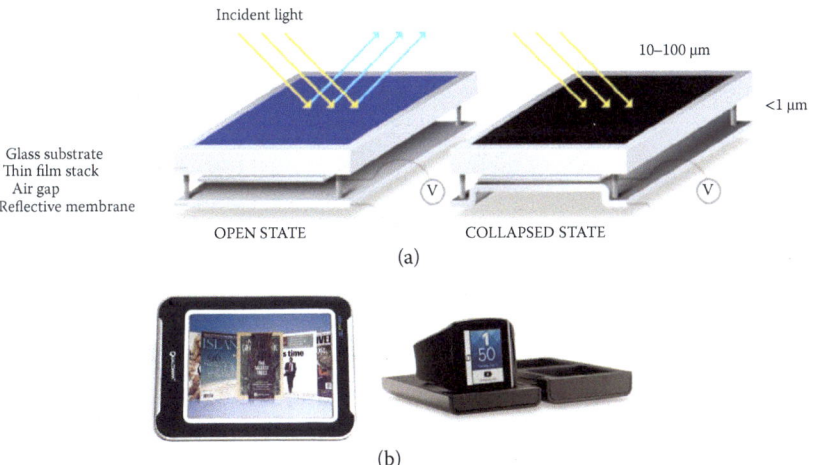

Figure 4.28 (a) Basic structure of the Mirasol IMOD Pixel. The active element consists of a deformable reflective membrane on the bottom, separated by an air gap from a thin-film interference stack above, creating an optical resonance cavity. (b) Image of a Mirasol display and Toq watch. Courtesy of Qualcomm. (From http://www.qualcomm.com/products/mirasol.)[77]

to the amplification of a narrow range of wavelengths, and elimination/reduction in others. The wavelength of the element can be tuned by the size of the air gap, allowing green, blue, and red elements to be fabricated out of the same basic structure. When in its open state, the reflector and thin-film stack are separated, and the element reflects its designed color. When a voltage is applied above a certain threshold value, electrostatic forces cause the reflective membrane to deform and be pulled up toward the stack, eliminating the air gap. This causes the reflected wavelength to shift to UV, which humans cannot see, and the pixel appears black. To return the pixel to its visible color, the voltage is reduced below a different threshold value and the membrane returns to its open state. In between is a hold-state voltage value at which the open or closed state can be held, and at which the device can be "pulsed" to change state rapidly. This device is described as being bi-stable—it can hold either of its two potential states with a minimal cost to the system.

Qualcomm has been developing this system for several years and claim that the structure's design enables very fast switching speeds and wide viewing angles (compared to conventional display technologies), as well as high contrast and reflectivity (compared to paper).[68,69] The technology has won several awards over the last few years. Several e-readers in Korea, China, and Taiwan have used the Mirasol technology.[67] Recently, Qualcomm has released the Toq™ smartwatch, which uses a Mirasol touchscreen display.[70] Novel color reflective displays are taking a long time to get commercially ready. Especially given the advances in battery life and reduced power consumption enabling the already champion LCD technology to continue to succeed.

4.9.4 Other Approaches

In addition to the examples above, here are other mechanisms that are also being explored. Thermally switchable phase change materials (both organic and inorganic) alter the RI with changing temperature. LCDs, of course, are the technology to beat for displays, where the electrical field changes the orientation of the liquid crystal molecules, acting as a light filter. Liquid crystal materials can also be incorporated into other structures, such as inverse opal films (for thermal or electric field-induced changes). There are also very interesting sensor approaches, which can be pH responsive, where infiltration of material by the target molecules (organic vapors, liquid solvents, water vapor [humidity], etc.) change the refractive index and spacing. These systems are also useful as biosensors, where a binding event to DNA or antibody functionalization changes the optical response. These approaches can be useful for detection of biomarkers, biological molecules, toxins, or otherwise dangerous materials.

References

1. *Markets and Markets Blog:Foldable Display Market Drivers and Future Challenges—An Overview* http://www.marketsandmarketsblog.com/foldable-display-market-drivers-and-future-challenges-an-overview.html.
2. *Seeing things: A new transparent display system could provide heads-up data, January 21, 2014,* http://newsoffice.mit.edu/2014/seeing-things-a-new-transparent-display-system-could-provide-heads-up-data-0121.
3. Global Display Market worth $164.24 Billion by 2017, http://www.marketsandmarkets.com/PressReleases/display.asp.
4. E. Kreit, L. M. Mathger, R. T. Hanlon, P. B. Dennis, R. R. Naik, E. Forsythe, and J. Heikenfeld, *J. Roy. Soc. Interface* **10** (78) (2013).
5. Interview of Dr. John Rogers by Melissa Blocks. 2012, http://www.npr.org/2012/09/27/161909578/biodegradable-electronics-could-end-toxic-trash.
6. Priya Ganapati, 2009, http://www.wired.com/2009/08/samsung-cornphone/.
7. C. vaughan, *ASU Magazine* (2013).
8. http://www.huffingtonpost.com/2009/11/20/led-tattoos-could-turn-sk_n_365376.html.
9. E. Leclercq, J. F. Taylor, and H. Migaud, Fish. *Fish* **11** (2), 159–193 (2010).
10. H. M. Fox and G. Vevers, *The Nature of Animal Colors.* (Sidewick and Jackson, London, 1960.)
11. M. Wallin, *Nature's Palette: How Animals, Including Humans, Produce Colors.* Bioscience Explained, Vol 1, No 2, http://www.bioscience-explained.org.
12. J. B. Messenger, *Biol. Rev.* **76** (4), 473–528 (2001).
13. J. C. Murphy, H. K. Voris, and M. Auliya, *Raffles Bull. Zool.* **53** (2), 271–275 (2005).
14. K. D. L. Umbers, *J. Insect Physiol.* **57** (9), 1198–1204 (2011).
15. A. S. Rundus, D. H. Owings, S. S. Joshi, E. Chinn, and N. Giannini, *Proc. Natl Acad. Sci. U S A* **104** (36), 14372–14376 (2007).
16. T. C. Insausti and J. Casas, *J. Exp. Biol.* **211** (5), 780–789 (2008).
17. R. L. Morrison, W. C. Sherbrooke, and S. K. FrostMason, *Copeia* (4), 804–812 (1996).
18. A. E. Best, *Ann. Sci.* **24**, 147–167 (1968).

19. L. M. Mathger, E. J. Denton, N. J. Marshall, and R. T. Hanlon, *J. Roy. Soc., Interface/Roy. Soc.* **6 Suppl 2**, S149–163 (2009).
20. A. Barbosa, L. Litman, and R. T. Hanlon, *J. Comp. Physiol. A Neuroethol. Sens. Neural Behav. Physiol.* **194** (4), 405–413 (2008).
21. J. J. Allen, L. M. Mathger, A. Barbosa, and R. T. Hanlon, *J. Comp. Physiol. A Neuroethol. Sens. Neural Behav. Physiol.* **195** (6), 547–555 (2009).
22. R. Hanlon, *Curr. Biol.* **17** (11), R400–R404 (2007).
23. J. B. Messenger, *Biol. Rev.* **76** (4), 473–528 (2001).
24. N. MD and H. FG, *Molluscan Res.* **25**, 57–70 (2006).
25. C. L. Huffard, N. Saarman, H. Hamilton, and W. B. Simison, *Biol. J. Linnean Soc.* **101** (1), 68–77 (2010).
26. J. Yacob, A. C. Lewis, A. Gosling, D. H. J. St Hilaire, L. Tesar, M. McRae, and N. J. Tublitz, *J. Exp Biol.* **214** (20), 3423–3432 (2011).
27. S. A. Adamo, K. Ehgoetz, C. Sangster, and I. Whitehorne, *Biol. Bull.* **210** (3), 192–200 (2006).
28. L. M. Mathger and R. T. Hanlon, *Pigment Cell Melanoma Res.* **25** (3), 295–296 (2012).
29. M. Goda and R. Fujii, *Zool. Sci.* **12** (6), 811–813 (1995).
30. R. A. Cloney and S. L. Brocco, *Am. Zool.* **23** (3), 581–592 (1983).
31. J. P. Vigneron, J. M. Pasteels, D. M. Windsor, Z. Vertesy, M. Rassart, T. Seldrum, J. Dumont, O. Deparis, V. Lousse, L. P. Biro, D. Ertz, and V. Welch, *Phys. Rev. E* **76** (3) (2007).
32. D. H. Kim, N. S. Lu, Y. G. Huang, and J. A. Rogers, *MRS Bull.* **37** (3), 226–235 (2012).
33. L. F. Deravi, A. P. Magyar, S. P. Sheehy, G. R. R. Bell, L. M. Mathger, S. L. Senft, T. J. Wardill, W. S. Lane, A. M. Kuzirian, R. T. Hanlon, E. L. Hu, and K. K. Parker, *J. Roy. Soc. Interface* **11** (93) (2014).
34. D. W. Logan, S. F. Burn, and I. J. Jackson, *Pigm. Cell. Res.* **19** (3), 206–213 (2006).
35. J. Rossiter, B. Yap, and A. Conn, *Bioinspir. Biomim.* **7** (3) (2012).
36. J. Heikenfeld, K. Zhou, E. Kreit, B. Raj, S. Yang, B. Sun, A. Milarcik, L. Clapp, and R. Schwartz, *Nature Photonics* **3** (5), 292–296 (2009).
37. J.-S. Yeo, T. Emery, G. Combs, V. Korthuis, J. Mabeck, R. Hoffman, T. Koch, Z.-L. Zhou, and D. Henze, SID *Symp. Dig. Tech. Pap.* **41** (1), 1041–1044 (2010).
38. S. Yang, E. Kreit, J. Heikenfeld, K. C. Zhou, and Ieee, in *2010 18th Biennial University/Government/Industry Micro-Nano Symposium* (2010).
39. M. Hagedon, S. Yang, A. Russell, and J. Heikenfeld, *Nature Commun.* **3** (2012).
40. C. Yu, Y. Li, X. Zhang, X. Huang, V. Malyarchuk, S. Wang, Y. Shi, L. Gao, Y. Su, Y. Zhang, H. Xu, R. T. Hanlon, Y. Huang, and J. A. Rogers, *Proc. Natl Acad/Sci. U S A* **111** (36), 12998–13003 (2014).
41. M. Izumi, A. M. Sweeney, D. DeMartini, J. C. Weaver, M. L. Powers, A. Tao, T. V. Silvas, R. M. Kramer, W. J. Crookes-Goodson, L. M. Mathger, R. R. Naik, R. T. Hanlon, and D. E. Morse, *J. Roy. Soc. Interface* **7** (44), 549–560 (2010).

42. D. DeMartini, M. Izumi, A. Tao, and D. Morse, *J. Shellfish Res.* **30** (3), 1000–1001 (2011).
43. D. G. DeMartini, A. Ghoshal, E. Pandolfi, A. T. Weaver, M. Baum, and D. E. Morse, *J. Exp. Biol.* **216** (19), 3733–3741 (2013).
44. L. M. Mathger and R. T. Hanlon, Cell Tissue Res. **329** (1), 179–186 (2007).
45. J. G. Boal, N. Shashar, M. M. Grable, K. H. Vaughan, E. R. Loew, and R. T. Hanlon, *Behaviour* **141**, 837–861 (2004).
46. L. M. Mathger and R. T. Hanlon, *Biol. Lett.* **2** (4), 494–496 (2006).
47. K. L. Yu, T. X. Fan, S. Lou, and D. Zhang, *Prog. Mater. Sci.* **58** (6), 825–873 (2013).
48. H. Fudouzi, *Sci. Technol. Adv. Mater.* **12** (6) (2011).
49. L. M. Mathger, G. R. R. Bell, A. M. Kuzirian, J. J. Allen, and R. T. Hanlon, *J. Exp. Biol.* **215** (21), 3752–3757 (2012).
50. Y. Zhao, Z. Xie, H. Gu, C. Zhu, and Z. Gu, *Chem. Soc. Rev.* **41** (8), 3297–3317 (2012).
51. L. M. Mathger, S. L. Senft, M. Gao, S. Karaveli, G. R. R. Bell, R. Zia, A. M. Kuzirian, P. B. Dennis, W. J. Crookes-Goodson, R. R. Naik, G. W. Kattawar, and R. T. Hanlon, *Adv. Funct. Mater.* **23** (32), 3980–3989 (2013).
52. J. Ge, and Y. Yin, *Angew. Chem. Int. Ed.* 50 (7), 1492–1522 (2011).
53. I. B. Burgess, L. Mishchenko, B. D. Hatton, M. Kolle, M. Loncar, and J. Aizenberg, *J. Am. Chem. Soc.* **134** (2), 1374–1374 (2012).
54. G. A. Ozin and A. C. Arsenault, *Mater. Today* **11** (7–8), 44–51 (2008).
55. A. C. Arsenault, T. J. Clark, G. Von Freymann, L. Cademartiri, R. Sapienza, J. Bertolotti, E. Vekris, S. Wong, V. Kitaev, I. Manners, R. Z. Wang, S. John, D. Wiersma, and G. A. Ozin, *Nature Mater.* **5** (3), 179–184 (2006).
56. P. G. Kang, S. O. Ogunbo, and D. Erickson, *Langmuir* **27** (16), 9676–9680 (2011).
57. S. H. Kim, S. J. Jeon, W. C. Jeong, H. S. Park, and S. M. Yang, *Adv. Mater.* **20** (21), 4129–4134 (2008).
58. J. P. Ge, H. Lee, L. He, J. Kim, Z. D. Lu, H. Kim, J. Goebl, S. Kwon, and Y. D. Yin, *J. Am. Chem. Soc.* **131** (43), 15687–15694 (2009).
59. *Opalux Inc.* http://www.opalux.com.
60. A. C. Arsenault, D. P. Puzzo, A. Ghoussoub, I. Manners, and G. A. Ozin, *J. Soc. Inf. Disp.* **15** (12), 1095–1098 (2007).
61. A. C. Arsenault, D. P. Puzzo, I. Manners, and G. A. Ozin, *Nature Photon.* **1** (8), 468–472 (2007).
62. D. P. Puzzo, A. C. Arsenault, I. Manners, and G. A. Ozin, *Angew. Chem. Int. Ed.* **48**, (5), 943–947 (2009).
63. A. C. Arsenault, H. Wang, E. Henderson, F. Kerins, U. Kamp, L. D. Bonifacio, P. H. Law, and G. A. Ozin, *Proc. SPIE* **8613**, 7 (2013).
64. H. Wang, F. Kerins, U. Kamp, L. Bonifacio, A. C. Arsenault, and G. A. Ozin, in *Information Display Magazine* (Society for Information Display, New York, NY, 2011), Vol. 27, pp. 26–29.

65. J. J. Walish, Y. Kang, R. A. Mickiewicz, and E. L. Thomas, *Adv. Mater.* **21** (30), 3078–3081 (2009).
66. Y. Kang, J. J. Walish, T. Gorishnyy, and E. L. Thomas, *Nature Mater.* **6** (12), 957–960 (2007).
67. http://www.mirasoldisplays.com.
68. *Qualcomm: Operating Principles of the IMOD Drive.* (July 6, 2011).
69. *Qualcomm: Fast Facts.* (February 14, 2012).
70. https://toq.qualcomm.com.
71. A. J. King and S. A. Adamo, *J. Exp. Biol.* **209** (6), 1101–1111 (2006).
72. L. M. Mathger, E. J. Denton, N. J. Marshall, and R. T. Hanlon, *J. Roy. Soc. Interface* **6**, S149–S163 (2009).
73. D. Froesch and J. B. Messenger, *J. Zool.* **186** (OCT), 163–173 (1978).
74. R. Young, M. Vecchione, K. Mangold, http://tolweb.org/accessory/Cephalopod_Chromatophore?acc_id=2038.
75. A. R. Tao, D. G. DeMartini, M. Izumi, A. M. Sweeney, A. L. Holt, and D. E. Morse, *Biomaterials* **31** (5), 793–801 (2010).
76. R. Young, M. Vecchione, http://tolweb.org/notes/?note_id=2646.
77. http://www.qualcomm.com/products/mirasol.

5

Vision Systems

5.1 Inspiring Vision

Vision plays an enormous role in the evolution and survival of many animal species. Roughly two-thirds of all fauna have some form of photonic receptor system.[1] This is not unique to animals; plants also have a complex organization of photoreceptor and signaling systems.[2,3] However, when most of us think of eyes and vision we tend to think of not only the organs that are capable of sensing light but also specifically those that are capable of forming images.[1] Approximately 30% of all fauna have what we would define as image-forming eyes. Even among image-forming eyes, there is a great deal of variety in the performance of biological vision systems. Some, like the primitive eye of the nautilus, are capable only of forming crude images; they are simple collections of photoreceptors and a pinhole opening with only the most basic comparison of light intensity. Our own eyes, in comparison, are extremely complex, sophisticated optical systems capable of forming high-resolution detailed images. Figure 5.1 shows examples of the variety of animal eyes. Vision is extremely integrated into animal behavior. Image-forming vision has played a significant role in the evolution of color displays. It influences both predator and prey behavior and is used for locating food, navigation, and identification of mating suitability. Currently, biological image-capturing photonic systems are attracting a great deal of interest among scientists and technologists due to their sophisticated structure and functionality under severe size and power restrictions. Biological eyes have shown remarkable structural complexity integrating heterogeneous components across many length scales, nanometer through centimeter; the functionality comes from the organization of functional molecules, cells, and larger scale biomaterial-based optical components. They are

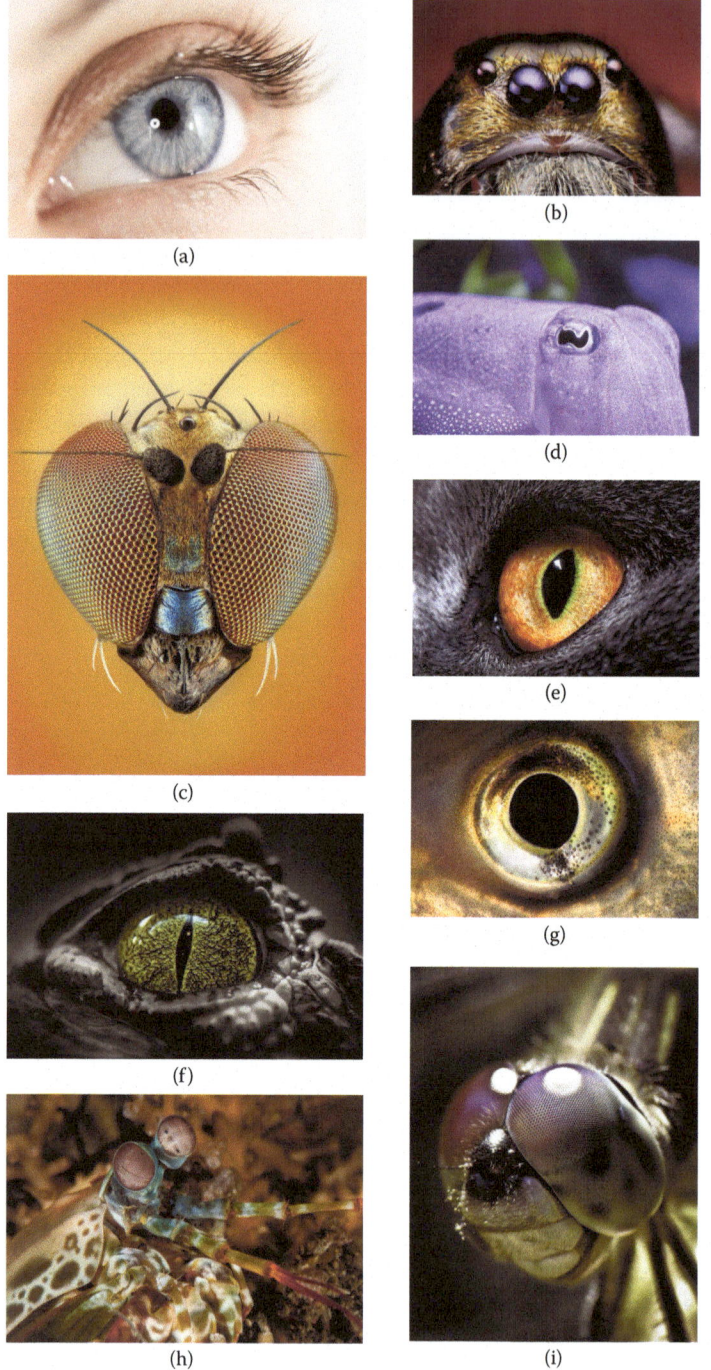

Figure 5.1 Examples of animal eyes. (a) Human being, (b) jumping spider, (c) long-legged fly, (d) cuttlefish, (e) cat, (f) crocodile, (g) fish, (h) mantis shrimp, and (i) dragonfly.

optical powerhouses. The diversity of vision systems found in nature provides a variety of potential synthetic design options with desirable operational properties, particularly compared to conventional designs.[4–7] We can find examples of visual systems that offer high visual acuity, some of which are specific to objects in motion (motion hyperacuity) or improved photosensitivity in low-light environments, wide fields of view (FOVs), polarization vision, and improved aberration correction and depth of field. Investigations of biological vision systems have even identified new states of matter.[8] Often, these systems have more integrated optical components and come in a significantly more compact package compared to conventional imaging systems. In addition, biological vision systems are composed of a variety of organic materials, offering the potential for extremely compact synthetic systems that ensure high performance, are capable of biological signal amplification, and are biodegradable and biocompatible. There are many potential applications for technologies that emerge from these inspiring biological systems. Extremely compact cameras are of interest not only for cell phones but also for use in small autonomous and aerospace vehicles, integration into other devices, and prosthetics. These bioinspired designs may be used to enable new approaches to navigation and sensing and in small systems such as unmanned aerial vehicles (UAVs).[5,7,9,10] Biocompatible optics have potential applications in medical systems, such as endoscopes and implantables.[11] They may be used in solar applications such as photovoltaics and thermophotovoltaics to reduce losses and increase efficiency.[12] Some other potential applications are in three-dimensional imaging, micro-optical telescopes, and microscopy.[10]

Biological eyes can be described, much like conventional imaging technologies, as an integrated set of front-end primary optics (the optical components that collect and direct light) and a back-end processor (e.g., the components that capture and process the incoming image), as well as additional optical components that serve to improve overall system function. The structures of most animal eyes can be described as a variation of one of approximately eight major eye structures, which are seen in Figure 5.2.[1] The structures that we are most interested in currently can be generally described in one of two major classifications: simple or compound eyes. Simple (or camera) eyes have a single integrated lens system, which focuses incoming light

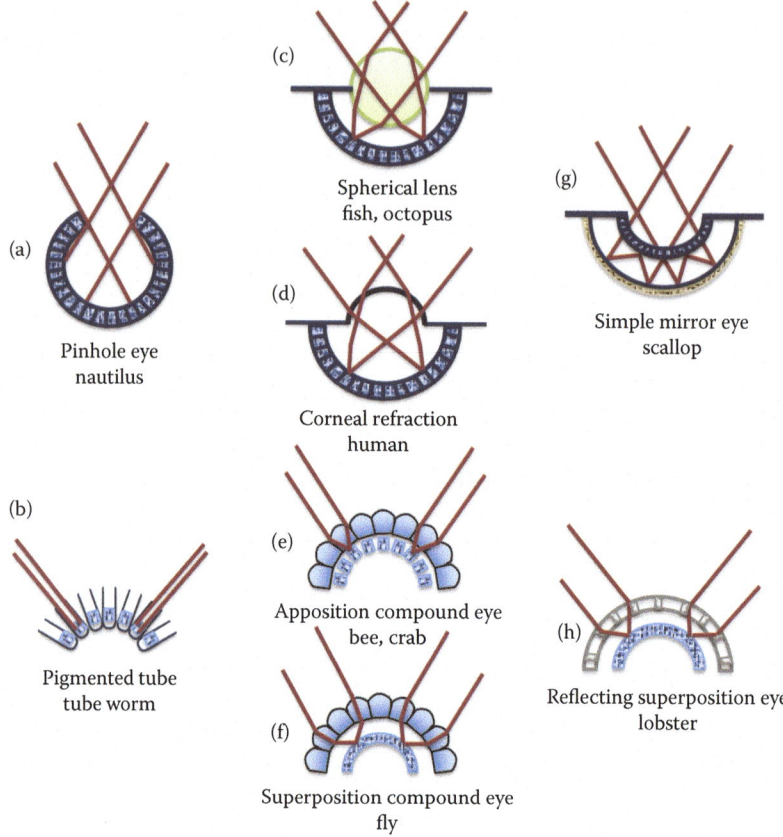

Figure 5.2 Illustrations of eight variations on the image-forming eye and examples of animals with those eyes. (a) Pinhole eye, (b) pigmented tube, (c) spherical lens, (d) corneal eye, (e) apposition compound eye, (f) superposition compound eye, (g) simple mirror eye, and (h) reflecting superposition eye. (From Land, M.F., and D. Nilsson, *Animal Eyes*, Oxford University Press, Oxford, 2002.[1])

onto common photoreceptors in the back-end structure. Compound eyes, on the other hand, have multiple lenses (in some cases thousands) per eye.

5.2 Biological Eyes: The Front-End Optics

5.2.1 Simple Eyes

Humans are very familiar with the simple eye model. It is how our eyes are structured, as well as the eyes of other land vertebrates. Other animals with simple eyes have a similar general structure, but they have species-specific design variations. An illustration of the human eye structure can be seen in Figure 5.3. Human eyes are not quite spherical and are typically about 24 mm in diameter. The eye has an outer lens called the cornea, which provides nearly two-thirds of the focusing power in the eye.[13] The pupil and iris are behind the transparent cornea. The iris acts as an aperture at the front controlling the amount of light allowed into the optical path, enabling us to see in a wide range of lighting conditions. A protein-based crystalline lens sits behind the pupil and iris. The aspherical lens is flexible and deformable. Muscles called the ciliary muscles pull on the supporting fibers (zonule fibers) attached to the lens, controlling the curvature and focal length of the lens. The cornea and crystalline lenses work in concert to create a highly focused image on the retina. The interstitial spaces between the optical components contain fluid, such as the vitreous humor between the lens and retina and the aqueous humor between the cornea and iris. Extraocular muscles rotate the eyes and ensure

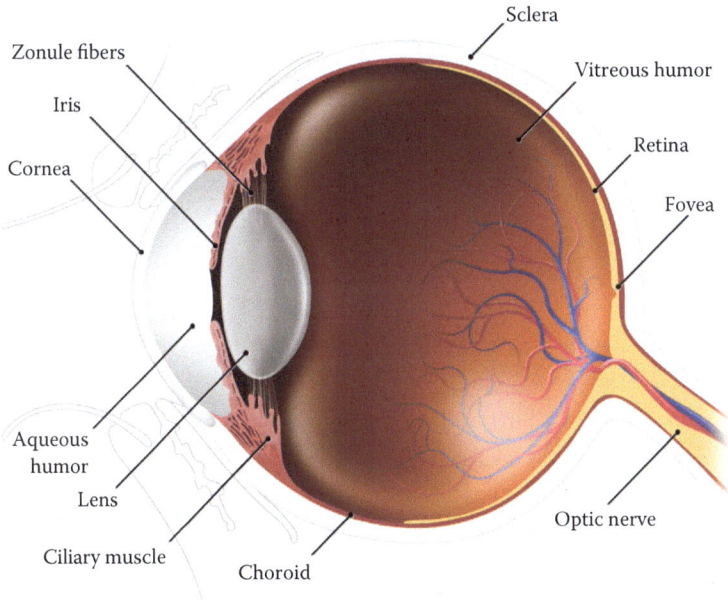

Figure 5.3 Illustration of the anatomy of the human eye.

that the image is focused on the visual axis, called the fovea, which is the central point for image focus.[14] The corneal eye is conserved among land vertebrates, meaning that the basic structural design occurs repeatedly. In certain species, the eye is adapted for resolution and high visual acuity. In others, such as nocturnal species, the eye structure may sacrifice resolution for photosensitivity. Variations exist in pupil shape and size, in the intraocular muscles that control the deformation of the cornea and inner lens, and in the overall shape of the cornea and eye.[1]

Simple eyes can range in size from the very large to the very, very small. The ostrich's eye is the largest land animal eye at 50 mm in diameter.[1] The giant and colossal squid (*Architeuthis* and *Mesonychoteuthis*) has the largest eyes in the entire animal kingdom.[15] The eye, seen in Figure 5.4, is 27 cm in diameter, with a 9-cm pupil. There are anecdotal accounts of eyes in these squids that are even larger. Spiders, perhaps, have the smallest simple corneal eyes. Eyes of the smallest spiders can be less than 500 μm in length.[16] Spiders also have many more of them than humans, typically six or eight. The eyes sit in two rows in the front of the head. The *anterior median* (front middle) eyes are called the principal eyes and point forward. The remaining eyes (*anterior lateral*, *posterior median*, and *posterior lateral*) are called secondary eyes and provide peripheral FOV.[17,18] The multiple small simple eyes (most are less than 1 mm in diameter) help save space and provide complementary functionality. Most spiders do not see particularly well, and their multiple small simple eyes give them a

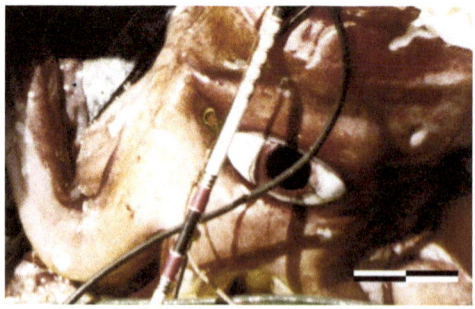

Figure 5.4 Head of a giant squid with a 90-mm pupil. The squid is likely of genus *Architeuthis*. Scale bar: 200 mm (photograph taken at the pier after catch, and scale was calibrated by a standard fuel hose across the pupil). (Reproduced from Nilsson et al., *Curr. Biol.*, 22(8), 683–8, 2012.[15])

wide FOV and the ability to distinguish light from dark. These spiders predominantly rely on vibration and air pressure to supplement their sensory capabilities to capture prey or evade predation. Some possess the capability to detect light polarization, which is used for navigation and prey capture.[19] The multiple spider eyes show differentiation in form and function. The Australian wolf spider *Lycosa leuckartii* has eight eyes. The anterior eyes seem suited for distance jumping and prey capture, whereas the posterior eyes seem suited for predator and prey detection at longer distances.[20] However, some members from the family Salticidae have been shown to exclusively use vision to hunt. The jumping spider *Portia*, seen in Figure 5.5, however, has vision that is especially well developed. Spiders of the genus *Portia* are diurnal hunters.[21] They have eight eyes, two of which are larger and more complex "principal" eyes compared to the remaining six "secondary" eyes. These spiders make accurate vision-based jumps onto prey and other targets. Their visual acuity is impressive: nearly 10× better than the dragonfly and only 5× worse than human vision. The principal eyes allow the spider to identify long-distance targets and see fine detail, whereas the secondary eyes detect short-distance movement. To fit the high-resolution simple eye into the small spider body, the principal eye has a special adaptation. Light enters the eye through the cornea and travels down a narrow tube and secondary lens before reaching a comparatively small retina. The entire system operates somewhat like a telephoto lens. Muscles attached to the eye tube enable it to sweep a larger FOV. The secondary eyes are much simpler, lacking the eye tube, and enable the spider to detect movement over multiple directions.

Marine animals also have simple eyes.[1] For example, cephalopods, such as the octopus, cuttlefish, and squid, also have a simple eye that produces high-quality images. However, marine animal eyes lack the corneas found in land animal eyes. The focusing power of the cornea seen in the eyes of land animals relies on the change in index of refraction at the air–cornea interface. Under water, there is no significant difference in index of refraction at the water–eye interface. The focusing power in this eye comes entirely from a spherical lens, which can be moved back and forth, rather than through deformation. The cephalopod eye in particular, although similar in form, seems to have evolved separately from the eyes of land vertebrates.

Figure 5.5 (a) Jumping spider *Portia* and (b) diagram of the salticid eyes. The principal eyes face forward and have high spatial acuity, and the secondary eyes function to monitor nearby space for movement. (c) Internal structure of the principal salticid eye from above. Light travels through a corneal lens down a light tube to a second lens, which magnifies the image as it reaches the retina. (From Harland, D., and R. Jackson, *Cimbebasia*, 16, 231–40, 2000.[21])

5.2.2 Gradient Index Lenses

The crystalline lenses in mammalian and marine animal simple eyes are not homogeneous. They are not made out of a single material system nor are they uniform in composition throughout the lens. The crystalline lens is made of crystalline protein and water. The protein and water concentration changes gradually through the lens. The protein has a higher index of refraction than water, and the changing protein concentration throughout the lens results in a refractive index gradient. The refractive index in these gradient index (GRIN) lenses can vary either axially or radially with the highest index in the center, and these lenses are found in all lens-based eyes. The fish eye, for example, has a radial GRIN lens with a refractive index difference (Δn) measured at 0.17 (1.54 at core to 1.37 at edge) and some show a Δn as high as 0.22.[1,22] The GRIN lens in the human eye, for comparison, has a Δn of approximately 0.03.[23] The GRIN lens in the human eye consists of nearly 22,000 layers of protein/water with an approximately parabolic refractive index gradient.[24] Figure 5.6 shows an illustration of the structure of the gradient index lens in the human eye. At the center of the lens, the refractive index is approximately 1.42, gradually changing to approximately 1.37 at the lens surface.

GRIN lenses are one of nature's solutions to the challenge of aberration. In an image-forming system, optical aberrations occur when the incoming light is not focused into a single point. The lack of focus results in poorer visual acuity and a fuzzy image. In man-made optical systems, great pains are taken to correct potential aberration. Chromatic aberration is distortion caused when the lens fails to focus light of different colors onto a single point. In biological refracting

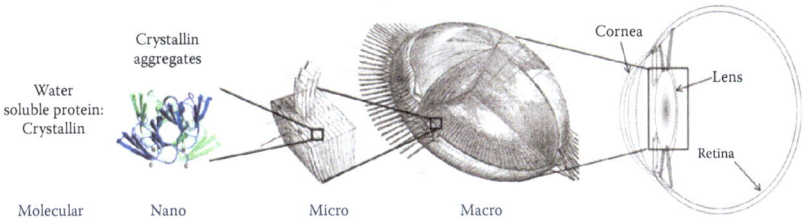

Figure 5.6 Breakdown of the hierarchical structure of the human eye lens. (From Ji et al., *Opt. Eng.*, 52(11), 2013.[24])

eyes, the optical receptors respond to an extremely narrow spectral band, allowing them to bypass the wavelengths that might be most affected.[1] Spherical aberration is distortion caused by the increased refraction experienced by light passing through the outer edge of a homogeneous lens, compared to light passing through the center. The human eye corrects for this aberration in two ways. First, the human eye is not spherical; it is somewhat flattened toward the edges of the cornea, which reduces the refraction of the rays furthest from the lens axis. Second, the gradient index lens corrects the aberration by creating a shorter optical path near the lens edges.[23] Combined, these adaptations correct for spherical aberration in the eye (as well as chromatic aberration in fish eyes) in a compact focal length.

5.2.3 Compound Eyes

In contrast to simple eyes, the compound eyes of arthropods (insects and crustaceans) seem tremendously exotic. Whereas simple eyes have a single lens, compound eyes are made up of as many as tens of thousands of individual lenses (or entire optical systems) per eye. The closely packed individual lenslets are called ommatidia; a close-up of a fly eye in Figure 5.7 shows the many ommatidia on this

Figure 5.7 An extremely sharp 25× close-up of the compound eye of a fly, showing the many lenslets (ommatidia) on this compound eye.

compound eye. (The ommatidial lenses are also sometimes called facets.) These ommatidial lens array structures result in eyes that are lower in resolution than vertebrate simple eyes, but they have a greatly increased FOV and, in some cases, higher photosensitivity. There are two main types of compound eye structures: apposition and superposition compound eyes. (Though these are the two most common, there are variations on these, as well as other less common, compound eye types.) In the apposition compound eye, the ommatidial lenses are each individually part of the front end of a self-contained and optically isolated optical system. The apposition compound eye, then, is a closely packed array of these individual optical units. In the superposition compound eye, in contrast, the ommatidia are a part of a lens array that focuses on a single common retina.[13,25] In apposition eyes, each ommatidium forms a separate inverted image. In a superposition eye, the image from each facet is overlapping with its neighbor and a single upright image is projected onto the retina. A superposition eye has the benefit of higher photosensitivity than apposition eyes; however, the overlapping images at the interfaces can result in blurring at the image edges and so the superposition eye has lower acuity.[6]

5.2.4 Apposition Eyes

The apposition eye is common among diurnal insects and crustaceans such as day-flying butterflies, wasps and bees, dragonflies, cockroaches, and some crabs. Apposition compound eyes, seen in Figure 5.8, are formed from hundreds to thousands of closely packed, optically isolated individual optical systems. The optical unit contains the lenses and waveguides that transport and focus light into individual photoreceptor bundles. Each ommatidial corneal lens focuses an image into a crystalline cone and the image is guided down to the rhabdom, which together functions as both a waveguide and a photodetector.[1,27] Each ommatidial unit is optically isolated from its neighbors, preventing optical loss. The number of ommatidia in an insect eye can vary tremendously with species. Fruit flies have only approximately 700 ommatidia per eye, whereas, on the other end of the scale, the dragonfly has perhaps the most facets with up to 30,000 ommatidia per eye.[28] A cross

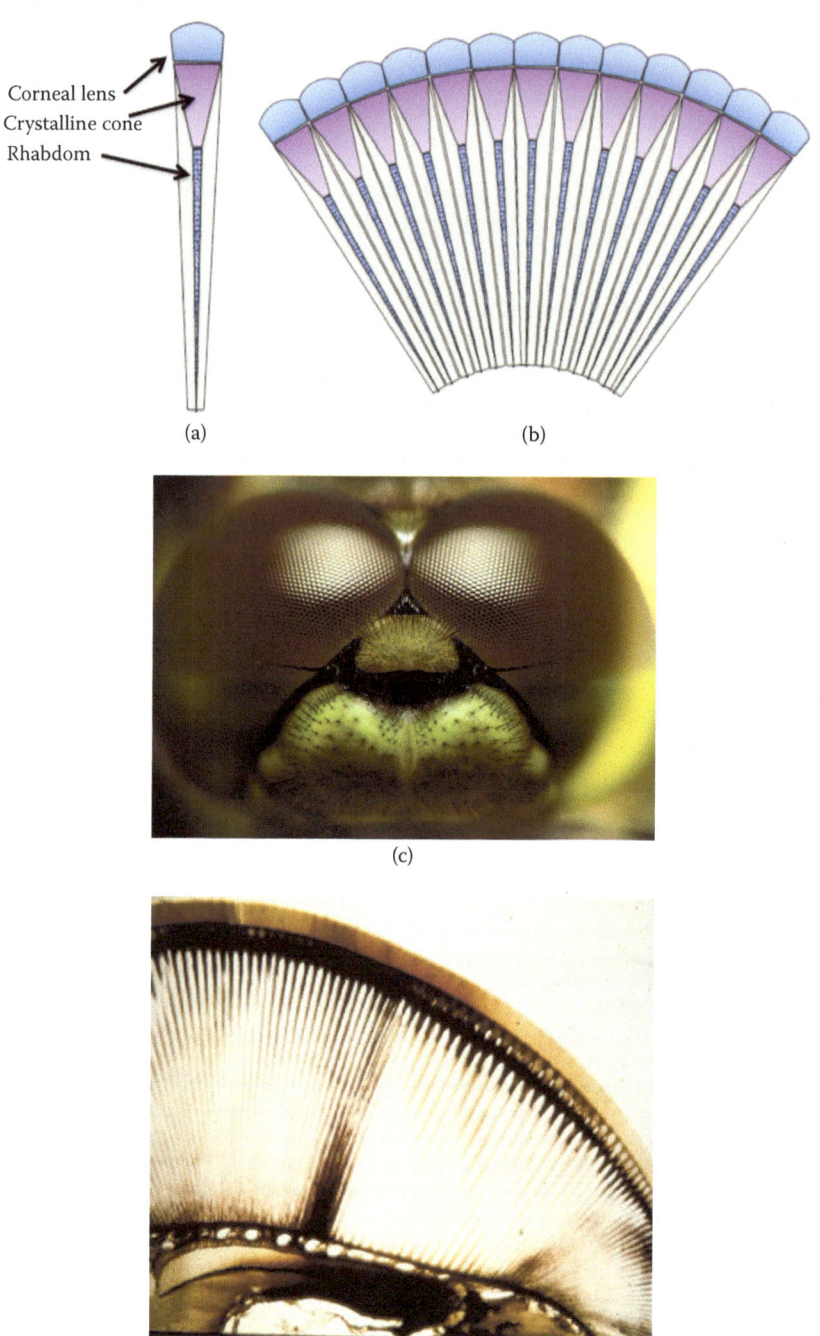

Figure 5.8 Illustration of the apposition eye. (a) Ommatidium and (b) compound eye structure. (c) Close-up of dragonfly eyes and (d) cross section of an *Hemicordulia tau* eye showing the eye's apposition structure. (From Belusic et al., *J. Exp. Biol.*, 216(11), 2081–8, 2013.[26])

section of a Tau emerald (*Hemicordulia tau*) dragonfly eye is shown in Figure 5.8, in which the individual optical segments within the eye can clearly be seen.

One significant benefit of the compound eye is that it achieves attractive imaging characteristics in a small package. Very small compound eyes are extremely interesting to technologists as they may provide clues and inspiration for shrinking synthetic optical systems. For example, the parasitoid wasp *Trichogramma evanescens*, seen in Figure 5.9, is only 3–400 μm in size.[29] This wasp has two oval apposition eyes approximately 60–70 μm long. The eyes have approximately 120 facets, each approximately 6 μm in size. In addition, *Trichogramma* also has three smaller (very simple) simple eyes called ocelli (two dorsal 11 μm in diameter, and one median 15 μm in diameter). *Trichogramma* is not the smallest insect to have compound eyes, either; another parasitoid wasp holds that honor. At <200 μm in size, the *Megaphragma mymaripenne* Timberlake is the smallest flying insect, about the size of some unicellular organisms.[30] A scanning electron microscope (SEM) image of *M. mymaripenne* can be seen in Figure 5.10, along with unicellular organisms for comparison. These extremely small insects are very interesting specifically because of their size. The eye structures in these

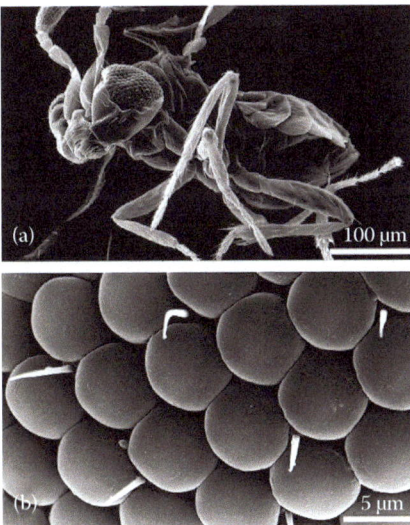

Figure 5.9 SEM images of female *Trichogramma evanescens*. (a) Overall appearance of *T. evanescens* and (b) detail on one compound eye. Individual ommatidia facets are approximately 6.4 μm in diameter. (From Fischer et al., *Visual Neurosci.*, 28(4), 295–308, 2011.[29])

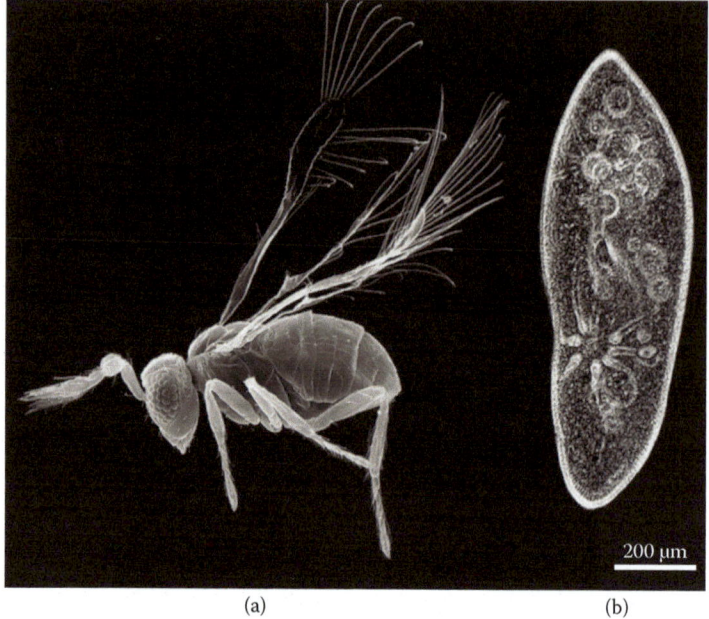

Figure 5.10 Size of the smallest insect (a) *Megaphragma mymaripenne* and a protozoan and (b) *Paramecium caudatum* for comparison. Scale bar: 200 μm. (c) SEM of head capsule of adult specimen. Scale bar: 10 μm. (From Polilov, A.A., *Arthropod Struct. Dev.*, 41(1), 29–34, 2012.[30])

organisms have to be extremely space efficient. Behavioral studies and modeling of photoreceptor behavior in *T. evanescens* imply sensitivity to shorter wavelengths, blue, violet, and ultraviolet (UV).[29] This is not unexpected because the size of the photonic system can scale with wavelength, so shorter wavelengths can enable smaller systems.

The size of the eye suggests that these wasps are likely somewhat light insensitive, and the females, who are more active, require bright light conditions. It is of significant interest to us how these ultrasmall insects with body sizes well under 1 mm cope with the performance trade-offs required of such space limitations.

5.2.5 Superposition Compound Eyes

Superposition compound eyes are found in many nocturnal arthropods (e.g., moths, fireflies, and beetles) and deep-water crustaceans, such as crayfish and krill.[1] The superposition eye is composed of closely packed lenslets, much like the apposition eye. However, in superposition eyes, ommatidia are not the front end of optically isolated individual photonic receptor systems. In superposition compound eyes, light is refracted through multiple ommatidial lenses and is focused (superimposed) on a small portion of a common retina. Between the ommatidial lenses and the photoreceptors, there is a section of unpigmented and transparent cells (called the clear zone). Figure 5.11 shows an illustration of the superposition eye cross section, an SEM of the superposition compound eye in a female owl-fly, and a histological cross section of the owl-fly eye. In the histological cross-section image, the lens, crystalline cone, clear zone, and retina can all be clearly seen. In these eyes, the superposition of the image from multiple ommatidia serves to improve the photosensitivity in light-limited environments. In the nocturnal dung beetle *Onitis aygulus*, for example, nearly 300 ommatidia contribute to the image at any one point. Some have a form of variable light adaptation to protect sensitive photoreceptors from bright light, in which dark pigments move from between the ommatidia into the clear zone to provide some optical insulation and reduce the light that reaches the retina. The superposition eye demonstrates improved photosensitivity by as much as three orders of magnitude compared to simple eyes without hampering their ability to recognize predator and prey and navigate a complex habitat.[31] This is important to eye functionality as the mean light intensity can be up to 11 orders of magnitude dimmer at night. Their increased sensitivity to light, compared to simple or apposition eyes (as well as some specific photoreceptor adaptations),

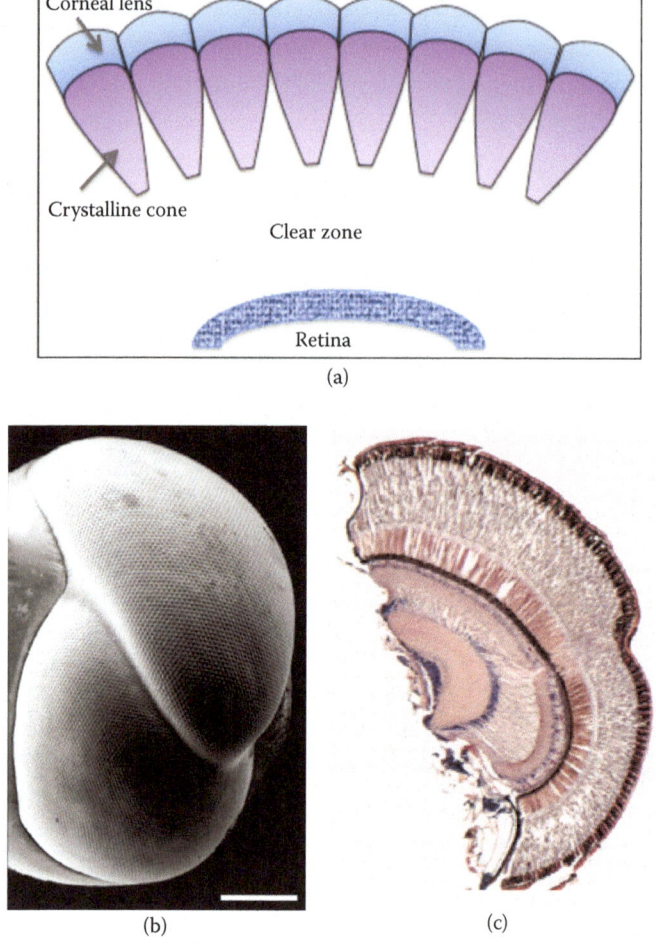

Figure 5.11 (a) Illustration of the structure of a superposition compound eye. (b) SEM of the right compound eye (lateral view, dorsal side up) of the female owl-fly *Libelloides macaronius*. The owl-fly eye is divided into the dorsofrontal (DF) and the ventrolateral (VL) parts. (c) A histological section of the eye showing its superposition structure (corneal facet lenses and the crystalline cones, the clear zone (cz), and the rhabdom layer (rl)). (From Belusic et al., *J. Exp. Biol.*, 216(11), 2081–8, 2013.[26])

allows them to trade off some spatial and temporal resolution for improved visual reliability.

5.2.6 Other Variants on the Compound Eye

The refracting superposition eye is only one variant on the superposition eye. Another variant is the reflecting superposition eyes found in decapod crustaceans like shrimp and lobster.[32] The facets of these eyes look somewhat like long rectangles of homogeneous gel, and when first discovered their behavior and function was not at all understood. As one author put it, "there was a period, between about 1955 and 1975, when shrimps and their relatives could not see," because we didn't understand how the eye structure worked.[33] By 1975, it was determined that the walls of the rectangular structures acted as biological mirrors, reflecting light to the retina. Another variation is the neural superposition eye, in which the superposition of images occurs not in the retinal plane, as in refractive eyes, but rather in the neural layer. This provides redundant sampling and increased photosensitivity and minimizes the loss of visual acuity.[34,35] Finally, there are some examples of an intermediate structure between an apposition eye and a refractive superposition. The moth *Cameraria ohridella*, or chestnut leaf miner, shows some characteristics of an apposition eye and some characteristics of a superposition eye.[36] An SEM of *Cameraria ohridella*'s head and eyes, as well as an illustration of the combination compound eye, can be seen in Figure 5.12. Like an apposition eye, *Cameraria ohridella*'s eye shows a distal rhabdom occupying the area that would contain the clear zone in a superposition eye. Like a superposition eye, this eye shows a reflecting layer behind the photoreceptors and longitudinal screening pigment migration. These adaptations seem related to its small size, enabling miniaturization with low performance impact.

5.2.7 Brittle Star: A Strange Compound Eye

A rather interesting version of the compound eye can be found in the echinoderm *Ophiocoma wendtii*, seen in Figure 5.13, which is quite frequently cited for its unusual structure.[6,37] Also called the brittle star, this marine creature is clearly sensitive to light, as it can change color based on lighting condition and move to escape predation; yet, it has no obvious eyes (or brain!). The surprise is that the body of the creature acts as one big compound eye. Within the dorsal arm plates of the organism, there is a roughly hexagonal arrangement of calcite crystals (called ossicles) in its skeletal structure, each about 40–50 μm in size. These ossicles are double-facet lenses that focus light onto nerve bundles located underneath. The top surface of the double lens is spherical, whereas the bottom

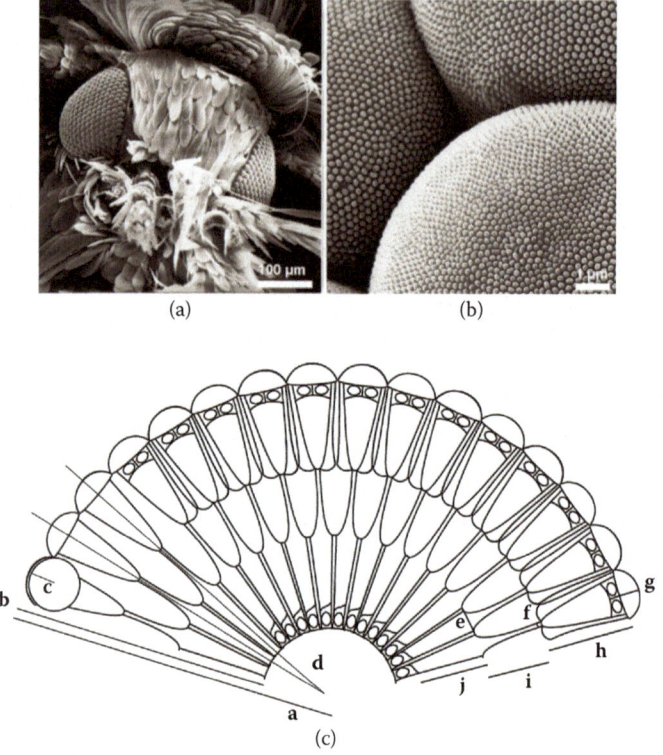

Figure 5.12 SEM of the chestnut leaf miner *Cameraria ohridella* (A) head and (B) single compound eye. (C) Schematic drawing of a longitudinal section of the eye indicating parameters that were measured for the morphometric analysis: (a) radius of curvature of the eye, (b) total length of the ommatidium, (c) outer radius of curvature of the cornea, (d) interommatidial angle, (e) maximal diameter proximal rhabdom, (f) minimal diameter distal rhabdom, (g) maximal thickness of the cornea, (h) length of the cone, (i) length of the distal rhabdom, and (j) length of the proximal rhabdom. (From Fischer et al., *Zoomorphology*, 131(1), 37–55, 2012.[36])

surface of the lens is aspherical. Surrounding the ossicles are pores with pigment-filled chromatophores that help regulate the amount of light that reaches the lens.[38] Together, these characteristics create a lens system that is intensity modulated and aberration and birefringence free.

5.3 Photoreceptors: The Imager's Back End

Like the film or charge-coupled device (CCD) in a camera, the photoreceptors in the eye serve as the biological back end of the imaging system. In simple eyes (and superposition compound eyes, for

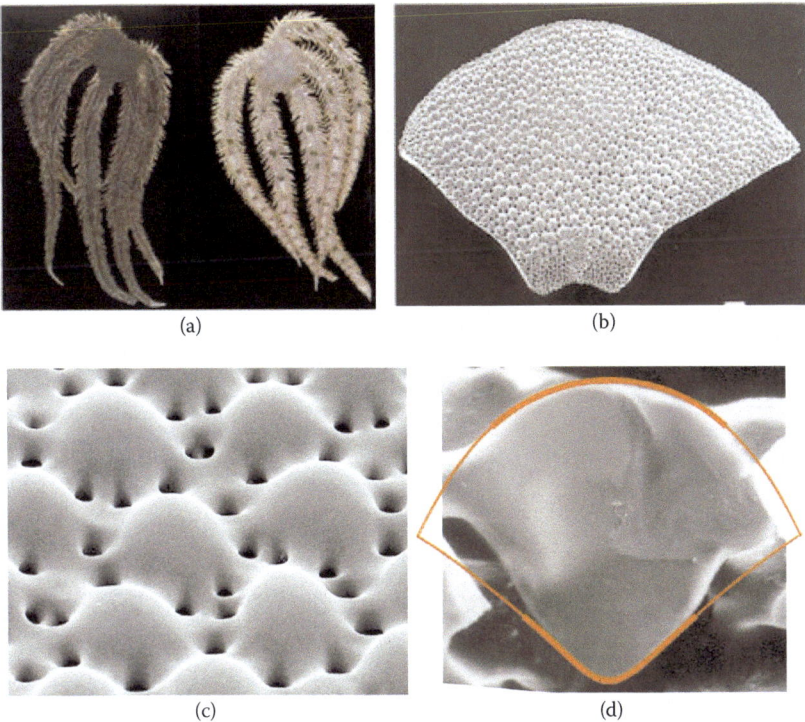

Figure 5.13 (a) Brittle star *Ophiocoma wendtii*, (b) SEM image of dorsal arm plate cleansed of organic tissue, (c) SEM of peripheral layer of dorsal arm plate with enlarged lens structures, (d) high-magnification SEM of individual lens cross section. Orange lines are the calculated profile of lens compensated for spherical aberration. (From Aizenberg et al., *Nature*, 412(6849), 819–822, 2001.[37])

example), this collection of photosensitive and optical processing molecules and cells is called the retina. Light coming from the eye's complex front-end optics is focused on the retina, which resides at the back of the eye. When light is incident on the photoreceptors in a human eye, their excitation is registered by adjacent neural ganglion cells that relay the information to the optic nerve and subsequently the brain. In mammalian eyes, the neurons reside on the anterior surface of the retina, which is the side closest to the lens, as seen in Figure 5.14. The light-sensitive photoreceptors actually lie on the back surface of the retina, meaning that the light has to travel past the neurons to reach the photoreceptors. (Within the photoreceptors are proteins called opsins, which are interesting in their own right. We will be discussing them in Chapter 6.) The ganglion

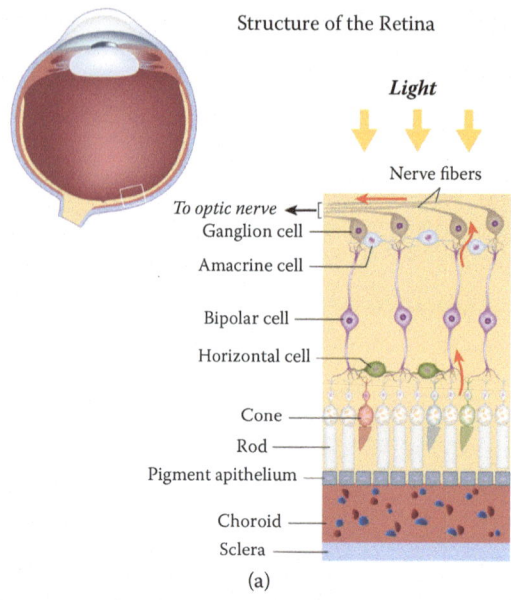
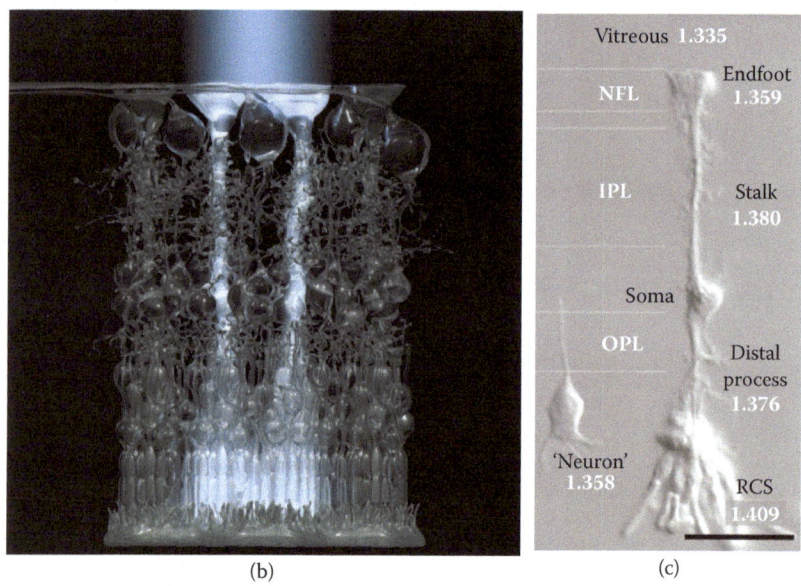

Figure 5.14 (a) Illustration of the structure of the human retina. (b) Nomarski differential interference contrast microscopy image of a dissociated guinea pig Müller cell with several adherent photoreceptor cells, including their outer segments (ROS) and a dissociated retinal neuron (bipolar cell) to the left. The refractive indices of the different cell sections are given. (c) Scale bar: 25 μm. (From Franze et al., *Proc. Natl. Acad. Sci. USA*, 104(20), 8287–8292, 2007.[39])

cells transmit their response via axons (long neural cellular fibers), which collect in one area of the retina and then stretch back into the brain (the optic nerve). The spot where these axons exit the eye leaves a portion of the retina without photoreceptors, creating the blind spot.[14] Different areas of the retina will have different packing densities of photoreceptors, or different spectral sensitivities depending on function.[40] The neural tissue in the retina does not simply register excitation, either, as there is some evidence that some initial processing is done in the eye before the information is fed back into the brain.[41] Indeed, the retina can be described as "a piece of your brain sitting on the front of your face."[42] This arrangement where the photoreceptors lie at the back of the retina is called an inverted retina and, although perhaps it seems counterintuitive, this arrangement is pretty well conserved. One exception to this is the cephalopod, which does not have an inverted retina. In the cephalopod eye, the photoreceptor cells are in front of the neurons. Because of this, cephalopod eyes do not have a blind spot.[1] At first, this inverted retinal design seems like a significant design flaw. The human retina is approximately 500 μm thick; so the optical path of the light must pass through tissue (which could potentially scatter or absorb light) before reaching the photoreceptors, reducing the photosensitivity or degrading the image quality. However, research on the neural cells that make up the vertebrate retina, called Müller cells, has uncovered some very interesting behavior.[39] Müller cells are neural glial cells. They have a long funnel shape and reach from the inner surface of the retina to the photoreceptors at the back, spanning the entire thickness of the retina, as seen in Figure 5.14. It was discovered that the Müller cells have a higher refractive index than their surrounding tissue and act like optical waveguides within the retina. These cellular waveguides enable low-loss light transmission directly to the individual photoreceptors.

The optical back-end (photoreceptor) structure in the apposition compound eye is fundamentally different from that in the simple eye retina.[1] Below the cornea and transparent cone, there are seven to eight photoreceptor cells called rhabdomeres. These cells are fused together to form the rhabdom. Pigment cells surround the bundled structure, and pigments move in and out of the bundle to control light penetration and protect photosensitive cells. Microvilli protrude from the rhabdomere and reach into the center of the bundle (microscopic protrusions of the cellular membrane). The microvilli are organized orthogonally and interdigitate, thus forming the crystalline tract down the center of the bundle. Light entering this facet travels through the lens down the cone and is focused on the distal tip of this rhabdom tract, which acts as a waveguide and photodetector. The rhabdomeres are sensitive to specific wavelengths. The ability of certain arthropods to see polarized light, such as bees, arises from the organization of the photoreceptor molecules within the microvilli.

5.4 Spectral Sensitivities

The colors and images perceived by a viewer are a measure of the intensity of light at a given wavelength relative to the observer's respective color receptor sensitivity.[1] It is the brain's reconstruction of the relative differential response of the wavelength-specific receptors. The photoreceptors in the human retina are made up of two major types, called rods and cones. Cones are wavelength specific and are responsible for most of our ability to distinguish color, whereas rods are significantly more sensitive to light and contribute to our ability to see in low-light conditions (night vision). The cones are localized near the fovea, enabling high visual acuity, whereas the rods, much more numerous, reside primarily in the surrounding area. The human eye has three types of cones that are sensitive to red (564 nm), green (533 nm), and blue (437 nm), resulting in a total sensitivity to wavelengths between roughly 400 and 750 nm, defining the visible spectrum.[14] Trichromacy in the visible wavelength range is uncommon in mammals other than humans and related nonhuman primates. There has been some recent work suggesting that a not insignificant proportion of women may actually be tetrachromats, enabling them to differentiate vastly more distinct colors in the visible spectrum than trichromats.[43] The number of different photoreceptors in an animal eye has a great evolutionary impact on their behavior and coloration. Honeybees and other insect species are trichromats; however, their photoreceptors are more sensitive in the UV through green spectrum, lacking the red sensitivity of humans.[1,40] The sensitivity to UV is thought to play a large role in the evolution of UV reflection of some plants and other insect species.

Location of food, attraction of pollinating species, and mating are behavioral adaptations to UV detection. Some web-weaving spiders create web decorations that reflect in the UV to enhance the attraction of prey with blue–violet wavelength receptor sensitivity.[44] There is also indication that UV can signify open space better than other wavelengths.[29] In addition, shorter wavelengths enable smaller optical systems before the performance becomes diffraction limited. There is some evidence that many more animals are sensitive to UV light than was previously believed, even in animals that lack a UV-specific photoreceptor.[45] Most nonprimate mammals are dichromats, whereas some animals (most birds and butterflies, for example) are tetrachromats or pentachromats (pigeons).[1,14,46] The cephalopod, always one to be different, is color blind, despite its impressive ability with camouflage.

As mentioned before, many flying and walking insects, such as dragonflies and butterflies, also have the ability to sense the polarization of light, which these animals use as a primary method of navigation by detecting the polarization pattern in the sky.[47] The skylight pattern changes relative to the position of the sun and changes with the sun's position during the day and throughout the year. The ability of insects to sense light polarization comes from the organization of the rhabdosomal microvilli of the ommatidia in the dorsal rim area in the compound eyes of these insects. Many fish and cephalopods can sense polarization as well, which also depend on the organization of microvilli in the animal retina.[48] This capability is very interesting to technologists, particularly for small navigation sensors.[47]

The retinal structure of chicken eyes has recently proved to be very interesting.[8] Chickens and other diurnal birds have very sophisticated visual systems. Their retinas have four types of single cones (red, blue, green, and violet) that are responsible for color vision, as well as double cones responsible for detecting light levels. Figure 5.15 shows the spatial distribution of the five photoreceptor types in the chicken retina. Given the visual acuity of chicken, one would expect that the arrangement would be more uniformly distributed (as it is in other animals) for optimal sampling. Yet, at first glance the arrangement looks completely random. The five cone types all lie within the same tissue layer and are all actually slightly

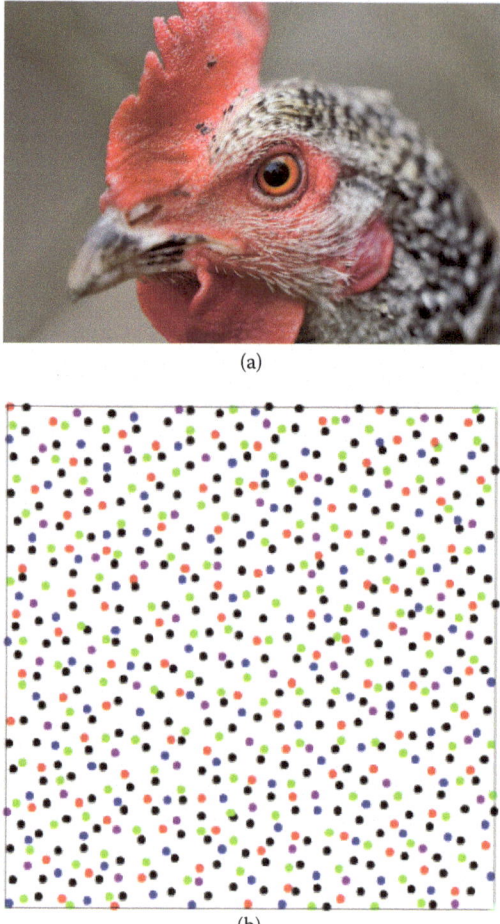

Figure 5.15 (a) Image of the head of a hen. (b) Spatial arrangements of the overall pattern of chicken cone photoreceptors: red violet, blue, green, and double species (experimentally obtained). (From Jiao et al., *Phys. Rev. E*, 89(2), 2014.[8])

different in size. Modeling shows that the cones arrange themselves so that they are not too close to any other cone of the same type. Because of this "exclusion" requirement, organization into a more periodic or uniform distribution is frustrated. The arrangement of the cones that at first looked random is actually ordered over long distances. This type of arrangement is known as "disordered hyperuniformity" and is the first known occurrence of this novel state of matter to be found in biology. This type of organization acts like a crystal in that it lacks large-scale variations and is highly uniform, but it also acts like a liquid in that it has the same physical properties in all directions. In fact, not only is the overall pattern hyperuniform but also each individual cone type simultaneously exists in its own hyperuniform spatial pattern independent of the overall organization, which the authors call "multihyperuniformity." Understanding states of matter like this may enable the development of novel and advanced materials, such as self-organizing colloids, hyperuniform optical circuits, light detectors, and other optical metamaterial-like systems.

5.5 Secondary Structures

Most adult winged arthropods have additional light-sensitive organs, which provide additional separate optical functionality from their more primary image-forming eyes. Ocelli, such as those seen in Figure 5.16, are extremely simple corneal eyes, which supplement the function of a compound eye in insects (or the primary eyes in spiders). Ocelli have a static cornea and a layer of photoreceptors, sometimes separated by vitreous humor. Dragonflies and other flying insects typically have three ocelli, two focused to either side of the head and one facing forward. Dragonflies, in particular, seem to have highly developed ocelli, which enable them to detect correlated changes in light levels enabling impressive flight control capabilities, particularly tracking and stability.[47] These organs can also be sensitive to infrared or UV wavelengths, providing additional visual information. Pit organs with infrared sensitivity have been found in some snakes, bats, beetles, and butterflies.[6,49–51] Intriguingly, there is some behavioral evidence for infrared detection in the juveniles of the giant jellies *Nemopilema nomurai*. Infrared detection in aquatic animals is unexpected, as water absorbs most infrared wavelengths.[52] More work is needed, however, to determine the mechanism of infrared sensitivity.

In both simple and compound eyes, additional mirrorlike structures may be found beyond the retina or rhabdom. These reflective films are called the tapetum lucidum, meaning "shining carpet," and reflect light back to the photoreceptors to increase the capture of available photons. They are particularly common in vertebrates and arthropods active in light-constrained environments. Tapeta are

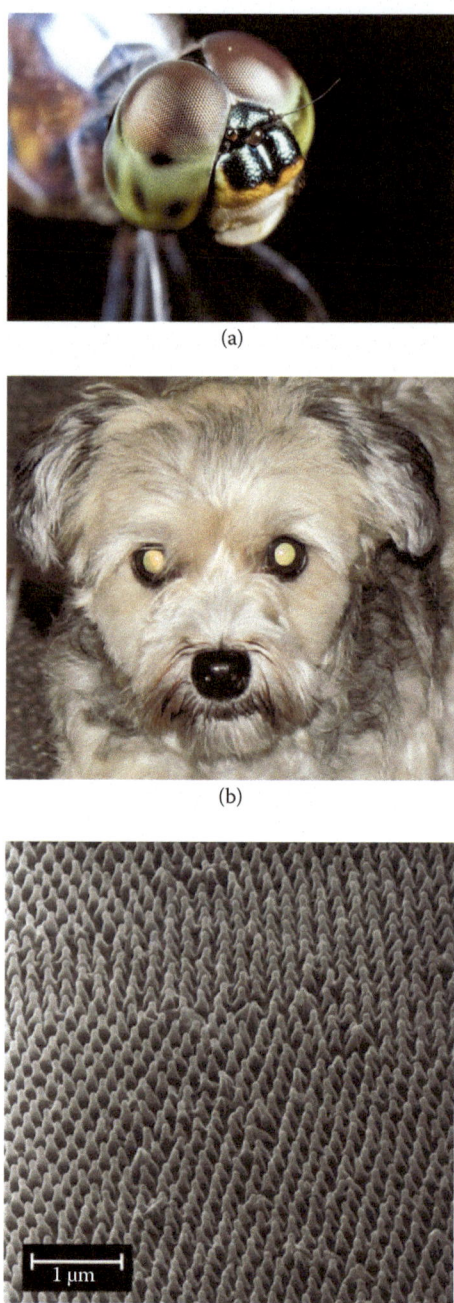

Figure 5.16 (a) Photograph of dragonfly showing the three ocelli used for navigation. (b) Photograph of the eyeglow in the eyes of a dog. (c) Helium ion microscope image of nanoscale pillar arrays on the surface of transparent sections of the wings of *Cephanodes hylas* (sample tilted by 45°). (From Stavroulakis et al., *Opt. Express*, 21(1), 1–11, 2013.[53])

multilayer reflectors, which may be made of, for example, guanine crystals in sharks, rods of zinc salt crystals in cats, or chitin and air in moths. The reflected light, also called eyeshine, often appears to the viewer like the eyes are glowing, as in Figure 5.16, and is visible when viewed from the direction of original illumination.[33] Arctic reindeers (*Rangifer tarandus*), such as those seen in Figure 5.17, live in far northern climates, such as northern Europe and North America.[54] The amount of environmental light in these climates changes dramatically between winter and summer. Reindeers have an adaptation that may increase light sensitivity during the long hours of darkness in winter while preserving their vision during the sunshine of the summer. The reindeers have a tapetum lucidum behind their retina, which shines golden in the summer months. However, during the winter months the eyeshine color changes to blue. The reindeer tapeta are composed of spaced collagen fibers. In winter, the spacing between the fibers is compressed and the wavelength of reflection of the multilayer shifts to blue. The mechanism for compression is thought to be an increase in intraocular pressure that occurs as the pupil of the eye is dilated for

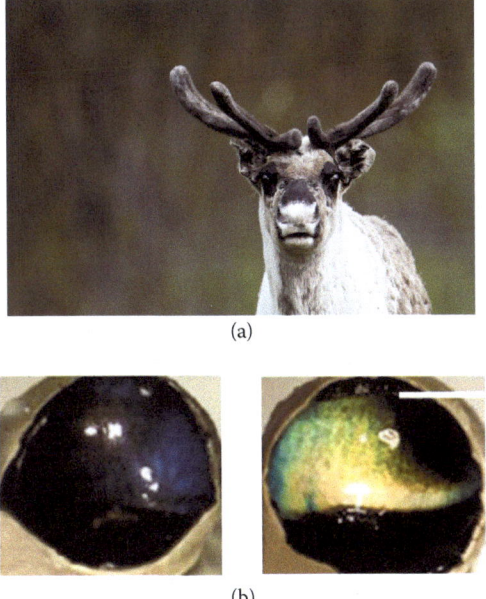

Figure 5.17 (a) Photograph of an Arctic reindeer. (b) Photographs of a reindeer eye tapetum lucidum in winter (left) and summer (right) (cornea, lens, and vitreous have been removed). Scale bar: 1 cm. (From Stokkan et al., *Proc. R. Soc. B-Biol. Sci.*, 280(1773), 2013.[54])

extended periods of time in winter. It is possible that this adaptation enables the eye to capture more of the UV wavelengths, which are more abundant in winter light.

Many insects have additional nanostructures on the outer surface of the corneal lenses that look like arrays of small, rounded cones.[55-60] These structures are often called "nipple arrays" or "moth-eye arrays." As you may guess from the name, these structures are found on the ommatidial surfaces of arthropods, such as butterflies, moths, and mosquitoes. Figure 5.16 shows the surface of the ommatidial array of the owl-fly, and the moth-eye array is clearly visible. They are also found on the wings of moths such as *Cephonodes hylas* and *Cacostatia ossa* as well as cicada wings. The small chitin structures are closely packed and have dimensions smaller than the wavelengths of visible light. For example, the (*Cephonodes hylas*) hawkmoth's structures are approximately 250 nm in height and approximately 200 nm tip to tip. They are narrower at the top and wider at the bottom. The tapered structure changes the air to chitin ratio from the outer surface of the structure to the corneal surface, producing a gradual change in the index of refraction. This reduces spurious reflections at the surface and can result in as much as two orders of magnitude reduction in reflection. For the structures on eyes, it improves vision in low-light conditions. It improves transparency and aids with camouflage, making the animal less visible to predators. Structures like these are useful as antireflective (AR) coatings for applications where the efficiency of light absorbance and emission is important, such as light-emitting diodes and solar cells.

BOX 5.1 FOSSIL EVIDENCE

We have fossil evidence of complex optical eye structures dating back to the Cambrian explosion, over 500 million years ago.[1] The familiar trilobite fossils show evidence of compound eyes. Recently, significant specimens of the preserved anomalocaridid *Anomalocaris* compound eyes were found. *Anomalocaris* was a shrimp-like arthropod relative dating back to approximately 515 million years ago.[61] Figure 5.18 shows an image of the eye

BOX 5.1 (*Continued*)   FOSSIL EVIDENCE

(a)

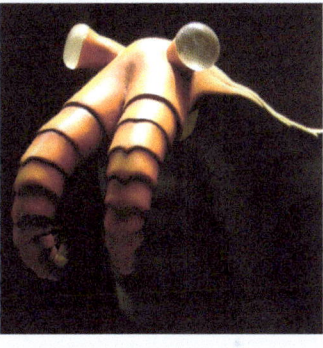

(b)

Figure 5.18 (a) *Anomalocaris* eyes. Scale bar: 2 mm. Inset: detail of ommatidial lenses. Scale bar: 300 μm. (b) Artist's renderings. (From Paterson et al., *Nature*, 480(7376), 237–240, 2011; Anomalocaris model by Gaetan Lee. Licensed under Creative Commons Attribution 2.5 via Wikimedia Commons. http://commons.wikimedia.org/wiki/File:Anomalocaris_model.jpg; Anomalocaris predator model by Yinan Chen. Licensed under the Creative Commons Public Domain Dedication. http://commons.wikimedia.org/wiki/File:Gfp-anomalocaris-predator.jpg.[61–63])

fossil remnant, as well as an artist's rendering of the animal. Do not let the description "shrimp-like" fool you, though. This animal could be up to 2 m long! The eye specimens found in these fossils showed well-preserved, albeit flattened, compound eyes that were 2 to 3 cm across and contained more than 16,000 ommatidia each, rivaling the dragonfly for complexity. There is speculation that visual guided predation resulted

(*Continued*)

BOX 5.1 (*Continued*) FOSSIL EVIDENCE

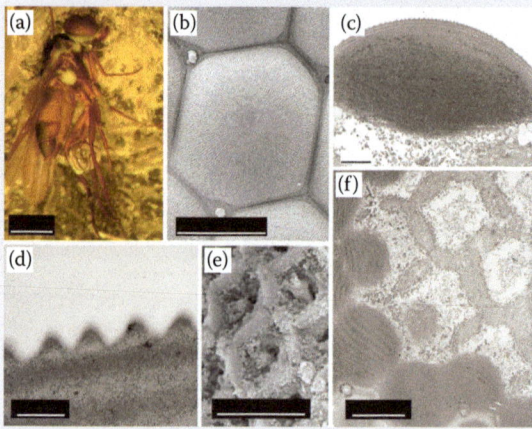

Figure 5.19 Dolichopodid fly eyes, mid Eocene, Russia. (a, b, c) Specimen SUM-CA0001: (a) photomicrograph, right lateral view of specimen in amber; (b) SEM image of the surface of ommatidia showing the fly eye grating structure (fg). (c, d, f) Specimen SUM-CA0002: (c) transmission electron microscope (TEM) image of a cross section of a single lens, showing the exocuticle (ex) and procuticle (pr); (d) TEM image of the exocuticle in figure (c), showing the eye grating structure (fg; Miller [1979]). (e) SEM image of pigment screen. (f) TEM image of a cross section of the eye showing cuticular lenses (cl), primary pigment cells (ppc), and crystalline cone (cc). (From Tanaka et al., *Proc. R. Soc. B-Biol. Sci.*, 276(1659), 1015–1019, 2009.[65])

in macroscopic animals. It is believed that these creatures were apex predators of their age, and it is speculated, in part because of the complexity of their eyes, that they did so with a high degree of visual acuity.[64] Good specimens of eye structures in fossils are hard to find as the soft tissues typically leave little evidence behind. Amazingly, however, soft tissue preservation can be found in some amber-trapped specimens. For example, a dolichopodid fly found preserved in Baltic amber dating back to the Eocene, about 45 million years ago, has recently been examined.[65] The insect and its preserved eye structures can be seen in Figure 5.19. In this fossil, it was possible to examine not only the corneal structure and preserved AR structure but also the delicate ultrastructure below. The images show a rhabdom organization indicative of a neural superposition eye, with seven rhabdomeres, which is similar to modern extant flies.

5.6 Applications

The camera is a very familiar piece of technology whose design is technically bioinspired. It is the modern handheld simple eye. Light enters through an aperture, travels through a lens or lenses, and then registers on photosensitive receptors such as the photodetectors on a CCD. Modern cameras even mimic some of the more dynamic properties of the biological eye, like autofocus, dynamic aperture control, and so on, and include memory and some processing; however, this comes with significant system complexity. Biological optics has the advantage of packing a significant amount of optical functionality in a very small space, and the potential savings in size, weight, and power are driving application interest in biological vision systems. High-performance, ultraminiature cameras have applications in commercial, medical, and military spaces. In some applications, we may be willing to accept some reduction in acuity in return for small, inexpensive, and rugged systems. There is demand for ever-improving performance, which competes with the extreme space and power limitations. Cameras in modern cell phones, for example, must often fit their camera systems in spaces only a few square millimeters in area. Multiaperture cameras and GRIN lenses are of interest for medical applications such as endoscopes. There is also a demand for lightweight, high-performance cameras for reconnaissance, surveillance, and navigation.[22] UAVs are often small and light and extremely limited in their payload capacity. Inspiration from biological imaging systems can improve the performance of synthetic optical systems and eliminate the need for high-cost, heavy, complex, and multielement optics.

5.6.1 GRIN Lenses

The development of gradient index lenses is particularly interesting because they have the potential to enable improved focusing power and FOV and correcting aberration in a compact, single-lens system. In addition, the ability to make deformable lenses, such as those found in nature, would enable rapid, dynamic multifocus lens systems with all of those benefits. Many current GRIN lenses are made of rigid materials (e.g., glass) in which the gradient index is induced via processes such as ion exchange or bombardment and are widely available and the design flexibility is limited.[66] To achieve similar performance to the single biological GRIN lenses, these state-of-the-art synthetic optical designs require multiple (often five or more) homogeneous refractive index lens elements to form images. If we can improve lens performance via bioinspiration, we can potentially reduce the number, size, and weight of optical system components without sacrificing performance. We are not yet capable of matching the performance of natural lenses. In particular, current commercial GRIN systems are not deformable and lack the ability to reach the high Δn seen in nature. Researchers from the Case Western and the U.S. Naval Research Laboratory (NRL) have been developing nanolayer polymer composite GRIN lenses that can overcome the Δn challenge.[23,24] These lenses were created from stacked fabricated films of alternately layered polymethyl methacrylate (PMMA) and styrene-*co*-acrylonitrile copolymer (SAN17). To create a lens with a specific index gradient, 50-µm-thick films were fabricated via coextrusion. Each film had up to approximately 4000 layers (each individual layer was ~12 nm thick), which were then forced together using elevated temperatures and compression (coextrusion) to form a composite film. Creating thin layers well below the wavelengths of visible light and a layer thickness < $\lambda/4$, the polymeric composite actually transmits light instead of acting as a multilayer reflector. Using coextrusion, the volumetric percentage of individual polymers could be controlled to within 2%. Films with gradually varying volume percentages of the two polymers were created and stacked sequentially to create a thick composite sheet with the desired refractive index profile. Thermoforming was used to consolidate the entire 3–7 mm thick stack (sheet). These gradient index sheets could then be formed into a variety of lens shapes, with (for example) axial, spherical, or radial refractive index distributions, and then shaped, polished, and diamond turned to the desired lens geometry. An illustration of the process can be found in Figure 5.20. To demonstrate the power of this technique, human eye lens mimics and spherical lenses were created. The intraocular lens (IOL) in the human eye has a parabola-shaped refractive index gradient with $n = 1.42$ at the center and $n = 1.37$ at the surface. To demonstrate the human eye lens mimic, two aspherical planoconvex lenses were created using the aforementioned techniques and reversibly assembled using optical gel. This GRIN lens contained

Figure 5.20 (a) Illustration of the procedure to build a gradient index (GRIN) lens: stacking, consolidation, shaping, and diamond turning. (b) Spherical GRIN ball lens forming two hemispherical GRIN lenses adhered together with optical gel. (c) A Case Western Reserve University logo image that was focused using a bioinspired "Age = 5" human eye GRIN lens and captured with a charge-coupled device camera. (From Ji et al., *Opt. Eng.*, 52(11), 2013.[24])

thousands of nanolayers and had a designed refractive index differential (Δn) of approximately 0.04, very similar to the human IOL. The eye lens mimic showed good optical performance, and the shape of the response matched that of the model lens (however, the response in the mimic was wavelength shifted due to different refractive indices of the constituent polymers compared to protein and water found in the natural lens). The next challenges are to increase the number of potential polymers that can be used with this technique to increase the refractive index differential even further and to explore deformable polymers to create deformable lenses. Thus far, several examples of deformable lenses have been demonstrated using approaches such as liquid crystals, microfluidics, and liquid lenses, but none matches the GRIN profile developed at NRL and Case Western. Creating deformable lenses may enable the creation of variable focal length optics that is responsive to an external trigger, such as pressure or electrical or chemical stimulation. Understanding the deformable optical performance in both natural and synthetic systems will be crucial.

5.6.2 Artificial Eye Prosthetics

Perhaps, the most obvious entry into the discussion of bioinspired applications is the eye prosthetic. It is estimated that nearly 200 million people have impaired vision (or no vision) in both eyes, with additional hundreds of millions more having debilitation in a single eye.[67] Solutions are being developed to replace damaged components, but a completely self-contained high-performance replacement of the entire eye is not yet achievable. Obviously, the capability to replace damaged components of the eye, or develop a fully functional full-eye prosthetic, could have profound impact. We cannot even come near to creating a functional prosthetic eye that matches the performance of the human eye. One interesting recent advancement has been the development of a wearable system that enables the wearer to switch between normal and magnification mode. The system consists of contact lenses with telescopic lenses, and a set of eyeglasses. When triggered on, the lenses can magnify up to 3×. The team that developed this technology is now hoping to help people with age-related macular degeneration.[68] Even in the case of component-level prosthetics, such as IOLs used in cataract surgery, the functionality still does not match what is achieved regularly in biological eyes. A cataract is clouding of the IOL within the eye, and modern cataract surgery removes that lens and implants a synthetic replacement. Deformable GRIN lenses could have a profound impact on the performance of IOLs. Current IOL implants are made of flexible polymers, making them easier to implant than rigid lenses. However, these lenses are primarily fixed monofocal lenses. Multifocal and toric lenses are also available, but none have the dynamic focusing capability of natural lenses. Accommodating IOLs attempt to mimic the deformable capabilities of a natural lens, although thus far performance is still lacking. A theoretical examination of gradient index lenses for implants has been done, but no practical versions are on the market yet.[69]

Artificial retinas are an extremely complicated system that requires overcoming challenges in fabrication, biocompatibility, resolution, power, and integration, to name a few. Conventional electronic processing techniques are planar. Sensors require power; many approaches require wires or inductive power sources and external hardware. Unlike many of the other technologies we have discussed in this book, approaches to this problem have to solve additional challenges of interfacing with neural tissue and the more general problem of biocompatibility. In the future creating ocular prosthetics that have capabilities beyond the natural eye, such as additional spectral bands, or higher photosensitivity, will be possible.

5.6.3 Inspired Compound Eye Lens Arrays

Apposition compound eyes serve as a source of inspiration due to their potential for wide FOV (>180°) and fast motion detection. Designs inspired by these biological photonic systems offer potential advantages for, for example, microcameras in extremely space-limited applications. There is significant work going on toward fabricating both individual components and systems.[70,71,95] Fabricating the lenses alone is challenging. Biological systems have a facility for creating nonplanar, flexible, and deformable systems, and fabricating curved synthetic structures is a not insignificant challenge. Many of the technologies and processing techniques used in the fabrication of micro- or nanoscale structures, such as various lithographic techniques, ink-jet printing, and etching, are planar in nature. These techniques can also be expensive, require complex processing and long processing times. Initial efforts to create curved lenses utilize these and other planar techniques in clever ways to create curved structures.

One technique that shows promise to create compound eye lens structures is the fabrication of lens-on-lens arrays.[72] Lens-on-lens arrays are small (2 to 3 µm) concave polymer structures that sit on top of larger (100 µm) concave structures, as seen in Figure 5.21. In a clever and fairly simple approach, the small structures were created first. To start, 2- to 3-µm cylindrical microchannels were formed in silicon. A polycarbonate film was sealed to the microchannel array and heated, and a pressure differential between the inside and outside of the microchannels was applied. This process was then repeated with a larger set of microchannels (this time

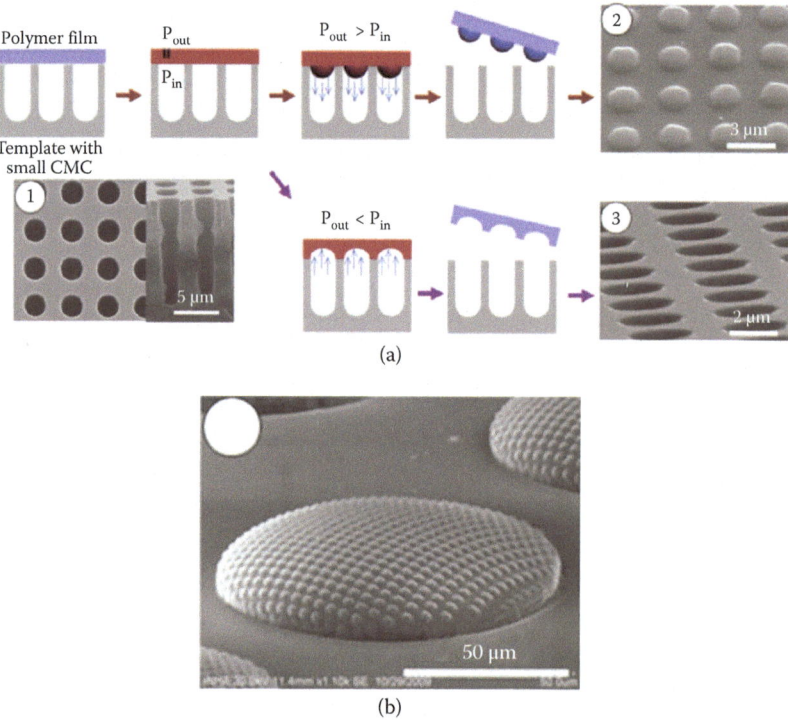

Figure 5.21 (a) Schematic diagram of lens array fabrication process, (b) oblique SEM image of convex lens-on-lens (c) images formed by the small lens in the lens-on-lens array ($d = 100$ µm). (From Park et al., *Soft Matter*, 8(6), 1751–1755, 2012.[72])

~100 µm). The resulting lenses are not as closely packed as natural lens structures, and the small lenses are significantly larger than biological examples. However, basic imaging with the array was demonstrated. The organization, dimensions, and height of the small and large lenses can be controlled through fabrication of the template, and the pressure applied during the fabrication process. The template fabrication limitations (such as the size and space between channels) and the material choice of the lens are potential challenges. The resulting lens-on-lens array can be fabricated with any combination of concave and convex lenses as well, adding interesting design flexibility. While this overall lens-on-lens array is planar, the arrays are fabricated out of the polymer and could be made deformable.

Another approach uses irreversible thermomechanical deformation to form a curved microlens array.[73,74] A planar array of concave silica microlenses was fabricated using femtosecond laser pulses, as seen in Figure 5.22. Silica was then used as a mold to form a planar array of convex lenses out of PMMA. The array was then heated and bent around a glass hemisphere. Using this method, the team was able to fabricate a 7600-microlens array with a 2° angle between nearest neighbor ommatidia. The resulting array was able to demonstrate a 140° FOV, which is a significant improvement over planar lens array structures with FOVs around 90°.

Another interesting example is the oil and water interface-based camera system developed at the University of Wisconsin-Madison.[75] In this effort, they are attempting to address the issue of fixed focus, which is true of most current fabricated lenses. A planar polymeric lens structure is fabricated and wrapped around a hemispherical dome, as seen in Figure 5.23. Polymer bridges between the lens elements serve to reduce the deformation stress. The lenses are created by filling the lens cavity with water, capped with oil. The lenses are pinned at the hydrophobic–hydrophilic boundary. A thermally responsive hydrogel acts as an actuator, allowing tunable deformation of the water–oil meniscus. Initial arrays have shown rudimentary imaging. However, the liquid lenses and thermally driven performance dependence may limit potential applications.

Several groups have demonstrated templating techniques, which allow replication of the curved eyes of arthropods. Conformal

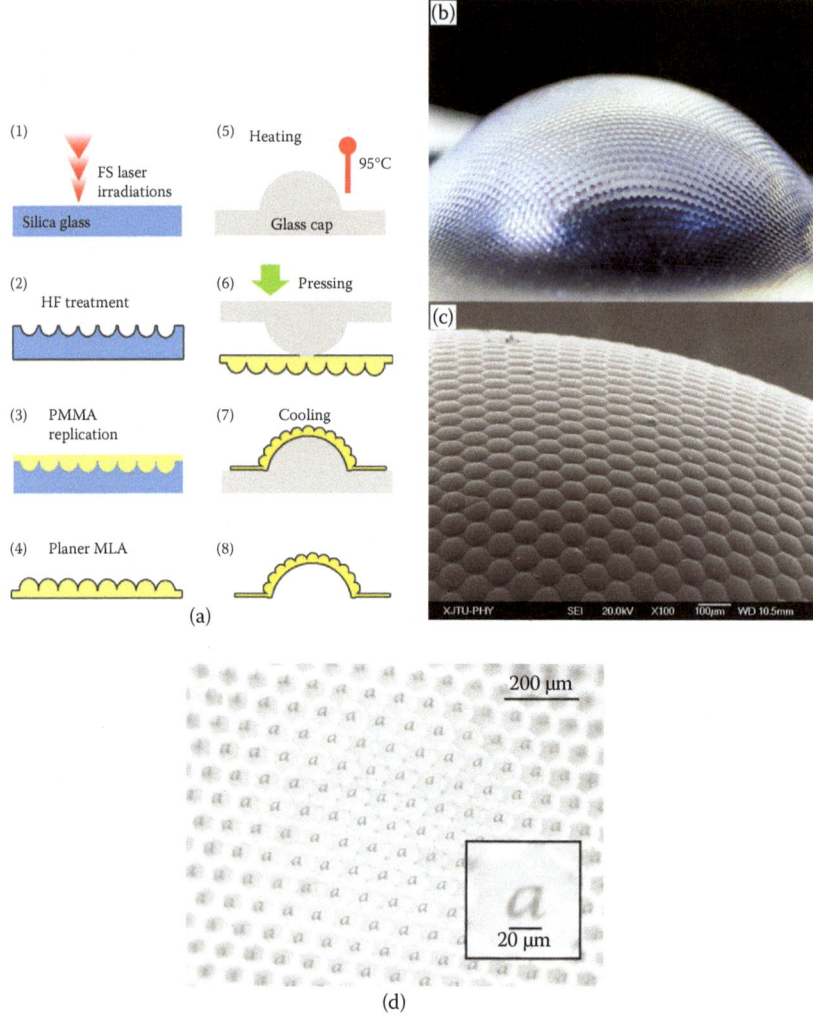

Figure 5.22 (a) Illustration of microlens array thermomechanical fabrication process. (b) Macrophotograph of the artificial compound eye. (c) Field emission Scanning electron microscope image of microlens array. (d) Imaging performance of the microlens array. The result was obtained via a 5 objective lens. (From Liu et al., *Appl. Phys. Lett.*, 100(13), 2012; Qu et al., *Opt. Express*, 20(5), 5775–5782, 2012.[73,74])

evaporation of chalcogenide glasses were used to replicate the micro- and nanoscale features of the fruit fly eye.[76] These materials have large indices of refraction and a high infrared transmittance. The eyes of the *Attacus atlas* moth were templated using soft lithography to create molds, as seen in Figure 5.24.[77] With both of these techniques, faithful three-dimensional structures can be fabricated in a range of

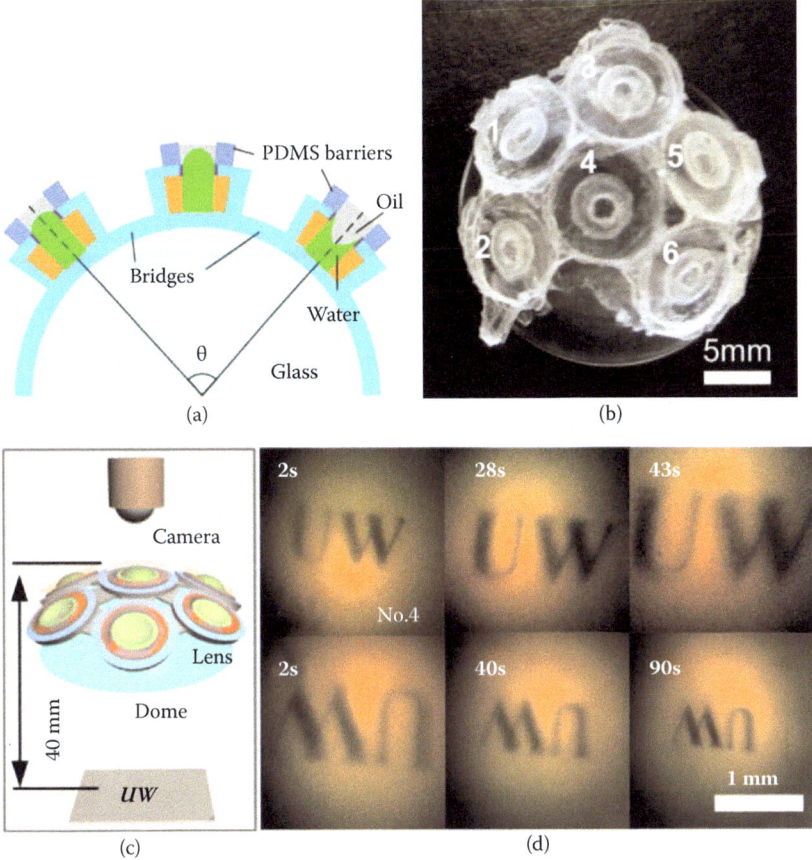

Figure 5.23 (a) Illustration of the domed tunable-focus microlens array, showing both concave and convex water–oil interface lenses. (b) Image of the six-element array on a dome with a diameter of 18 mm. Scale bar: 5 mm. (c) Illustration of imaging through the microlens array on a glass hemisphere. A logo of 1-mm-tall UW on a transparency film was placed 22 mm below the glass dome. (d) Images of "UW" taken through the number 4 microlens element. The images change as a change in temperature over time causes the water–oil meniscus to shift from convex to concave. (From Zhu et al., *Appl. Phys. Lett.*, 96(8), 2010.[75])

materials, lending some flexibility of material choice, but the design will be predetermined. The structure will be that of the natural eye and cannot be changed or optimized.

Brittle star microlens structures are interesting in part because of the distributed organization of the lenses and nanopores that can produce images with modulated light intensity without aberration or birefringence.[38] They are also interesting because in this particular case the lenses are made out of a biomineral, calcite (calcium carbonate),

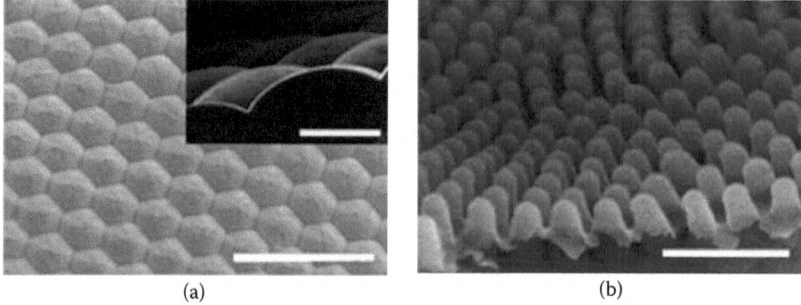

Figure 5.24 (a) SEM of Attacus atlas moth eye replica in ultraviolet curable polyurethane. Inset shows curved surfaces of the replica. Scale bar: 100 µm, inset scale bar: 20 µm. (b) Close-up of the moth-eye structures in the replica. Scale bar: 500 nm. (From Ko et al., Soft Matter, 7(14), 6404–6407, 2011.[77])

rather than an organic protein or a biopolymer. How biology fabricates some of these structures and the materials used in them is the focus of increasing study.[78] For example, researchers at Harvard were able to develop synthetic microlenses inspired by brittle star microlens structures as well as the pigment-transporting holes in organic materials, such as photoresist, using multibeam interference lithography.[38] Similar microlens array structures are being fabricated directly from calcite crystals.[79] Amorphous $CaCO_3$ particles are precipitated out of saturated $Ca(OH)_2$. The hemispherical nanoparticles form at the air–solution interface. Being in fluid allows nanocrystals the freedom to self-assemble as their size increases, and they form a close-packed two-dimensional array. An organic surfactant regulates the growth uniformity and size. The close-packed particles grow together to form a film, which is then heated to crystallize the amorphous material into calcite. These arrays can then be glued to curved surfaces using chitosan to form a curved lens array. Figure 5.25 shows the polymer array, as well as the calcite array fabrication approach and microlens array image.

Recently, a group from the Korea Advanced Institute of Science and Technology (Kaist) fabricated planar, lithographically defined synthetic compound eye structures coupled with a fluorescent photodetector array to probe the behavior and transport of light through these structures.[80] The planar array consisted of 115 ommatidia with an interommatidial angle of 1.6°, which enabled the array to span 180°. Simple apposition-type eye structures were composed of cylindrical polymer microlenses, waveguides, and photodetector elements.

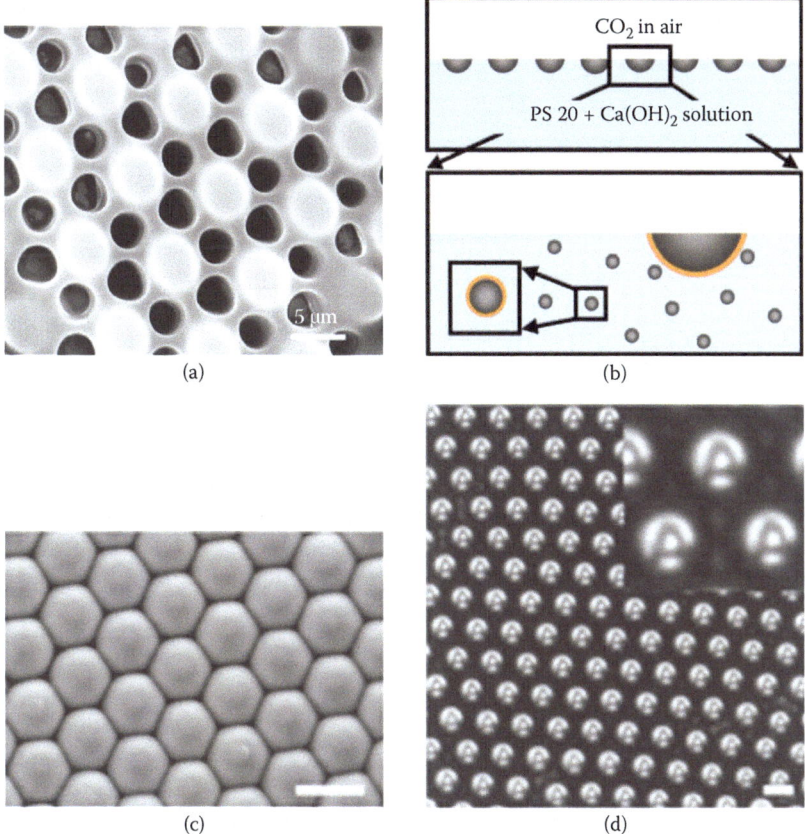

Figure 5.25 (a) SEM of a synthetic biomimetic microlens array with integrated channeled pores fabricated via multibeam lithography. (b) Illustration of the $CaCO_3$ microlens fabrication process, and (c) SEM top view of a well-ordered $CaCO_3$ microlens array. (d) Optical microscopy image of $CaCO_3$ microlens array imaging inversely projected 'A' array. Scale bars: 5 µm. (From Aizenberg, J., and G. Hendler, *J. Mater. Chem.*, 14(14), 2066–2072, 2004; K. Lee, *Nat. Commun.*, 3, 725, 2012.[38,79])

The microlenses were each 2.6 µm in thickness and 25 µm in diameter, with a 50-µm focal length. The waveguides were 2 µm wide, 25 µm in diameter, and 500 µm long. The fluorescence photodetector array (FPDA) was constructed of fluorescent dye–doped polymer resin, and the elements were aligned with the ends of the waveguides. Figure 5.26 shows examples of the planar eye structures in this scheme. To image the light propagation within these structures, the microlenses and waveguides could also be fabricated out of the dye-doped polymer. Structures based on the reflecting superposition eye were also fabricated and tested. Finally, by selectively defining FPDA

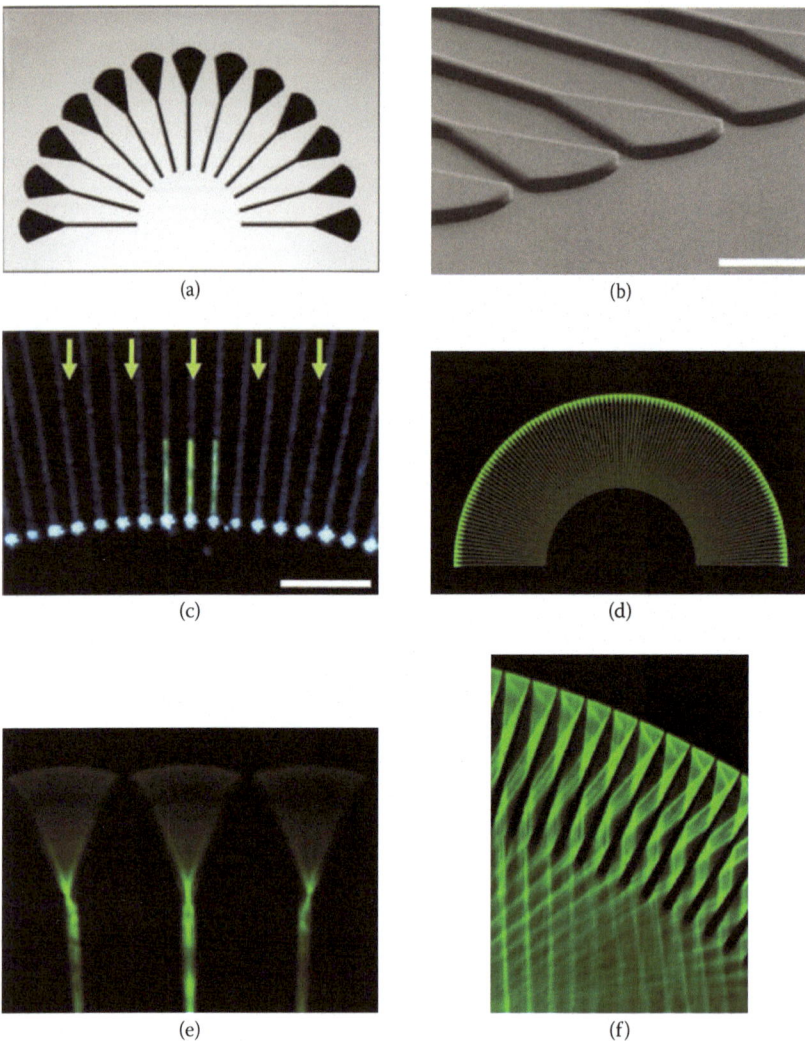

Figure 5.26 (a) Schematic of the simple apposition-type planar compound eye structure. (b) SEM image of the simple apposition microlenses. (c) A dark-field image of fluorescently activated fluorescence photodetector array integrated with the waveguides of the simple apposition type under the side-coupled light illumination. (d, e, f) Direct visualization of light propagation inside the artificial ommatidia, defined with Rhodamine 6G–doped photosensitive polymer resin by photolithography. (d) An epifluorescence image of the simple apposition type. Fluorescence image shows that the dye is distributed uniformly throughout the device. (e) Light propagation through a single ommatidium of the simple apposition type. The amount of light coupling from the microlens to the waveguide depends on the orientation of the ommatidium with respect to the propagation direction of the illuminated light. (f) Conical structures of the reflecting superposition type redirect the coupled light by internal reflection. The propagation loss inside the conical structure increases with the incident angle of the side-coupled light due to multiple internal reflections. Scale bars: 20 μm (b), and 50 μm (c), 200 μm (d), 25 μm (e), and 50 μm (f). (From Keum et al., *Small*, 8(14), 2169–2173, 2012.[80])

the angular sensitivity of the ommatidial array could also be probed. There are many different compound eye architectures and material choices that could be examined using this method. Planar synthetic compound eyes offer a comparatively simple method of probing different ommatidial architectures that may impact both planar and nonplanar optical systems.

5.6.4 Compound and Simple Eye Imaging Systems

There is also research beyond fabricating stand-alone lens systems to create whole optical systems (integrating synthetic lenses and photodetectors). Replication and fabrication are extremely challenging as each synthetic ommatidium requires the entire optical path from cornea to receptor and processor to be well aligned, and each synthetic ommatidium must be optically isolated from its neighbor. Synthetic, high-performance imaging systems such as these may enable the development of systems with reduced number and size of optical components compared to conventional optical systems. Applications for these high-performance systems may widen a path for compact cameras, and novel imagers for micro-optical telescope and microscope applications, medical applications, and machine vision.[10] Terrestrial and aerial vehicles, medical instruments, prosthetic devices, home automation, surveillance, motion capture systems, smart clothing, and robotics explore the fundamental principles of animal sensory-motor control.[7]

At the macroscale, the University of Wyoming in collaboration with the United States Air Force Academy is developing vision sensors based on the neural superposition compound eye of the *Musca domestica*, or housefly.[81,82] These eyes are interesting because the fly demonstrates hyperacuity, that is, the ability to detect motion better than one would expect from the size of its optics alone. *Musca domestica* eyes and eyes of this type are able to detect small, fast moving objects and small object movements. A seven-pixel array of lenses connected to a multimodal optical fiber demonstrated increased signal to noise ratio and motion acuity compared to other fly eye-based sensors. The array was formed with off-the-shelf components, and each lens is fairly large, a few millimeters across.

At the microscale, Duparre et al. have demonstrated an ultrathin microlens array camera.[83] In this work, a microlens array was fabricated and attached to a silica substrate with a pinhole array and a photodetector array. The group was able to demonstrate an effective superposition eye-type detector system. However, the potentially more interesting development was that of a stacked lens array called the "cluster eye," which uses a stack of three lenses in each ommatidium to achieve extremely high resolution for a compound type eye. However, it does so at the cost of significant increase in fabrication complexity and overall device thickness. In addition, fabricated eye cameras of this type may be optimized for anticipated viewing angles, and therefore corrected for aberration.

One challenge of integrating curved ommatidial lens arrays with planar sensor arrays is that distance between the lens and the sensor changes. The lens will be farther away from the sensors in the middle of the array than it will be at the array edges. However, to obtain a clear image the sensor must be at the focal plane of the lens. Most of the approaches to creating curved microlens arrays create lenses with a single, uniform focal length across the array. That is, each lens in the array has the same focal length. If the sensor is placed in the focal plane of the center lenses, then the images at the edge of the array will be out of focus. Researchers at Xiamen University in the People's Republic of China have developed an approach to creating microlens arrays with variable focal lengths via electrostatic deformation.[10] In this approach, a hexagonal close-packed array of electrodes and spacers is fabricated to act as a first template and addressable in such a way that different voltages can be applied to different groups of electrodes (e.g., the electrodes at the center of the array vs. those around the edges). A UV curable resin is poured over the concave template array, and the applied voltage pulls the material toward the electrode. The amount of deflection for any given microlens is determined by the amount of voltage applied. A polydimethylsiloxane (PDMS) lens is created using the resin as a template. The PDMS lens array is then formed into a hemispherical shape via negative pressure, and more UV resin is applied. This approach creates a hemispherical microlens array with varying focal lengths. This type of array should demonstrate a larger FOV and better imaging characteristics than flat arrays or curved arrays with uniform focal lengths.

Researchers at the University of Illinois at Urbana-Champaign have developed a simple eye–mimicking camera with both a curved lens and a curved photodetector array approximately the size of the human eye that enables tunable zoom-like features.[84,85] The imager was first demonstrated with a rigid planoconvex lens and a hemispherically curved optoelectronic array.[84] Creating curved optoelectronic arrays is an extremely significant challenge to the creation of some bioinspired photonic systems, deformable arrays even more so. This curved array was fabricated using planar fabrication techniques, but it was then transfer printed onto the hemispherical substrate. A 16 × 16 array of photodetectors, diodes, and wavy interconnects was fabricated onto a Si-on-insulator wafer. This array was then transferred onto a PDMS film and subsequently transfer printed onto a shaped glass. When coupled with a hemispherical cap with a simple lens, the assembly creates a simple-eye camera. Figure 5.27 shows the

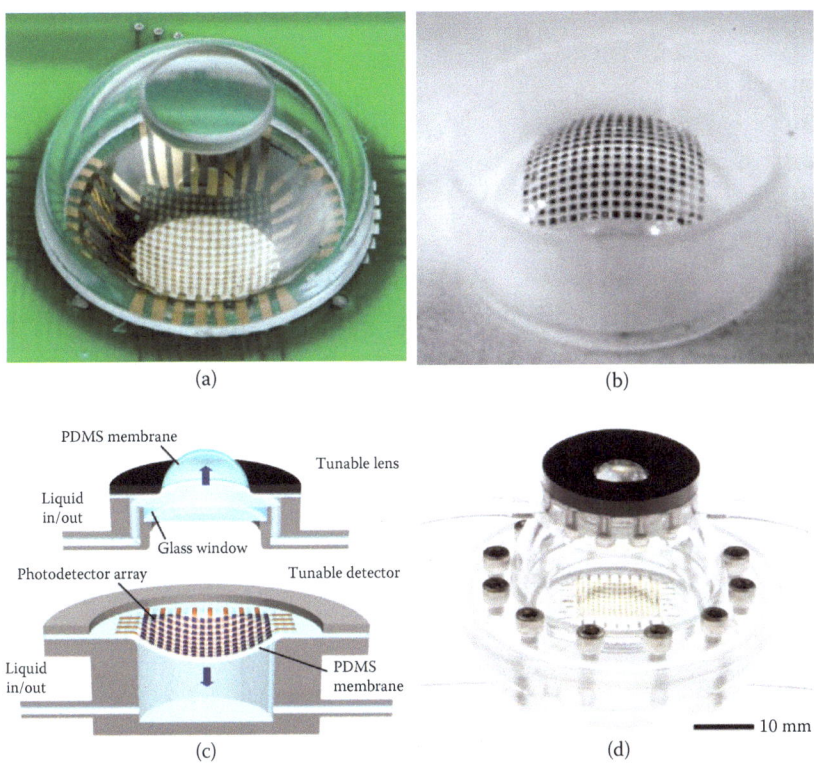

Figure 5.27 (a) Photograph of the camera with a transparent hemispherical cap. The cap has a simple, single-component imaging lens. (b) Photograph of a hemispherical polydimethylsiloxane transfer element with a compressible focal plane array on its surface. (c) Illustration of the two components of the hydraulically actuated simple eye camera. Top section is the tunable lens module and bottom section is the tunable detector module. (d) Photograph of a complete hydraulically tunable camera. (From Ko et al., *Nature*, 454(7205), 748–753, 2008; Jung et al., *Proc. Natl. Acad. Sci. USA*, 108(5), 1788–1793, 2011.[84,85])

hemispherical optoelectronics array and the completed eye system. Follow-on work on the bioinspired simple eye then demonstrated an integrated, dynamically deformable optoelectronics and lens system.[85] The eye was created in two parts. To create the dynamically deformable optoelectronics array, the 16 × 16 array was fabricated on a planar Si-on-insulator and then transferred to the PDMS film. The PDMS film was elastomeric and is reversibly deformable, creating a flexible optoelectronics array capable of high strain without loss of performance. The detector sheet was then placed over a fluid-filled cavity, and the fluid pressure was used to actuate the shape of the array. The lens was created by placing a PDMS film over a fluid-filled space and a glass window; the shape and curvature of the PDMS film were controlled by the controllable fluid pressure. The hydraulically deformable polymer film acts as a lens and, when coupled with the deformable optoelectronics array, created a high-performance hemispherical electronic eye with a pathway to adjustable zoom. Figure 5.27 shows an illustration of the eye structure and a picture of the assembled system. While a 16 × 16 array is insufficient to match the resolution of the human eye, for example, this and other efforts like this enable the potential for integration in biological systems (or on them) for a host of applications such as health monitoring and prosthetics.

In a natural extension of the simple eye work, the team has since created a hemispherical compound eye camera inspired by arthropod apposition compound eyes, which can be seen in Figure 5.28.[4] An elastomeric lens array was created from molded PDMS into a 16 × 16 array of tall, rounded cylindrical structures. The result is an array of convex microlenses (~180 individual ommatidia in the active region) that sit atop cylindrical supporting posts. This number of ommatidia is comparable to the eyes of fire ants (*Solenopsis fugax*) and bark beetle (*Hylastes nigrinus*). A flexible photodetector array was fabricated and aligned to the lens array. The photodetector array was then bonded directly to the back of the lens film, integrating the two components. A perforated sheet of black silicon placed atop the lens array and a bottom black silicon support serve to block any stray light. The height of the posts (~400 μm) and the thickness of the base film (~550 μm) separate the lenses from the photodetectors by

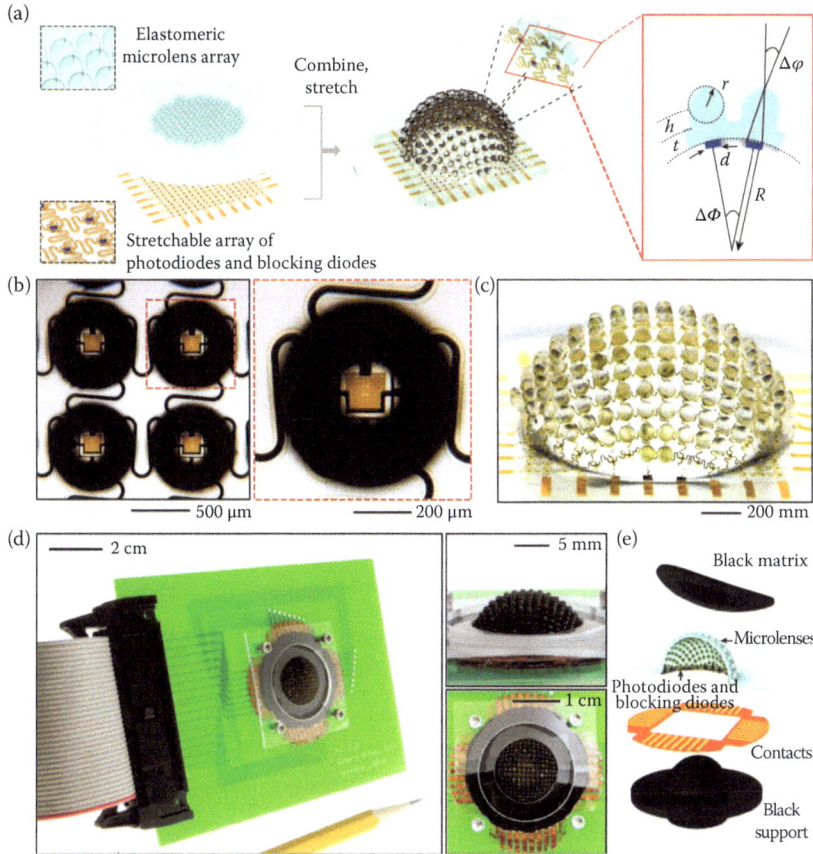

Figure 5.28 (a) Illustration of the elastomeric microlens array and supporting posts on the base membrane (above) and the optoelectronics array (below). These components are fabricated in planar geometries (left) with magnified views of four adjacent unit cells (artificial ommatidia) in the upper and lower insets. Center image shows the elastically deformed hemispherical camera ("combine, stretch") with an exploded view of four adjacent unit cells center inset. Right-most image is a cross-sectional illustration highlighting key parameters: the acceptance angle ($\Delta\phi$) for each ommatidium, interommatidial angle ($\Delta\Phi$), radius of curvature of the entire device (R) and of an individual microlens (r), height of a cylindrical supporting post (h), thickness of the base membrane (t), and diameter of the active area of a photodiode (d). (b) Image of the hemispherical compound lens system. (c) Optical micrograph of an integrated artificial ommatidium. (d) Illustration of the components of this system: perforated sheet of black silicone (black matrix), hemispherical integrated microlens and optoelectronics arrays, thin film contacts layer for external interconnects, and hemispherical supporting substrate of black silicone. (e) Photographic close-up of the tilted and top views of the completed camera mounted on a printed circuit board. (From Song et al., *Nature*, 497(7447), 95–9, 2013.[4])

approximately 1 mm and place the detectors in the focal plane of the lenses. The flexible integrated optical system can be deformed from their planar form into a nearly full hemispherical shape, giving the array a 160° FOV without loss of alignment or damage to the arrays or system performance. Like the previous work on the simple eye cameras, these films can be deformed dynamically via hydraulic actuation, enabling full control of the curvature.

In a final example, a complete bioinspired optical imaging system is the integrated three-layer artificial compound eye from researchers at École Polytechnique Fédérale de Lausanne in Switzerland.[7] In this approach, Figure 5.29, three separate planar array layers—microlenses, photodetectors, and electromechanical interconnect layer—were fabricated and integrated into a completely curved optical system. The first layer, the microlens array layer, was 500 μm thick with 630 individual lenses (42 × 15 array) that were molded onto a glass carrier. A 300-μm-thick Si-based photodetector layer was the second layer, and a 100-μm-thick flexible interconnect layer was the third. This last layer supported the system physically and enabled the transfer of signal output from individual ommatidia to the processing units. Below the microlens array, an opaque metal layer that had low reflection served to suppress cross talk. A second metal layer was placed close to the focal plane ahead of the photodetector. The layers were then aligned and integrated. The ommatidial layers were cut (diced) down to the interconnect layer using high-precision dicing. As the interconnect layer was flexible, this step allowed the entire array to be flexible and bendable. The final imager was light, with the prototype weighing only 0.36 g, and demonstrated good hemispherical FOV (180° × 60°), high temperature resistance, and local adaptation to illumination. Fabricating the layers separately, and in this manner, allows fabrication using planar techniques, as well as much more complete optimization and customization of the final product (with properties such as the number of elements, size, focal length, acceptance and interommatidial angles, etc.). Future challenges include developing a more cost-effective method of mass production of artificial ommatidia, as well as a more complex dicing method to be able to cut other additional bending patterns.

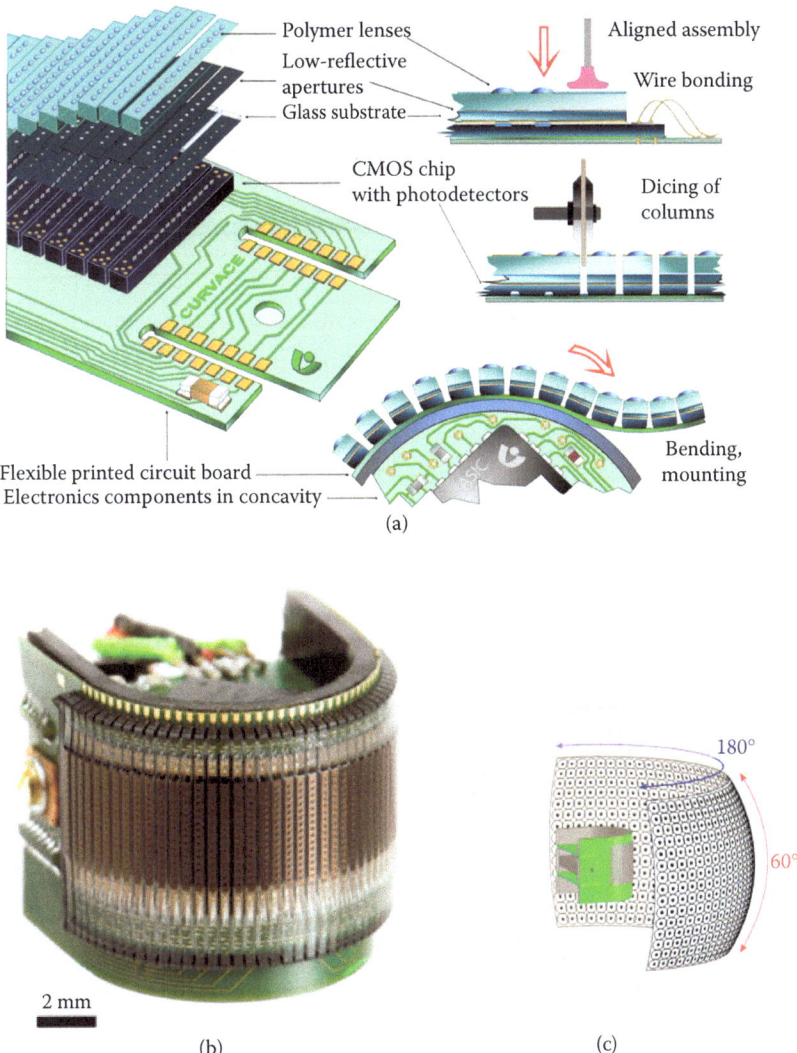

Figure 5.29 (a) "CurvACE" compound eye design and assembly. (Left) illustration of the multiple layers that compose the artificial ommatidia: optical (microlenses and apertures), photodetector (complementary metal oxide semiconductor chip), and interconnection (printed circuit board). (Top right) accurate alignment and assembly process of the artificial ommatidia layers in planar configuration. (Middle right) dicing of the assembled array in columns down to the flexible interconnection layer, which remains intact. (Bottom right) curving of the ommatidial array along the bendable direction and attachment to a rigid semicylindrical substrate with a radius of curvature of 6.4 mm to build the CurvACE prototype. (b) Image of the CurvACE compound eye prototype. The entire device occupies a volume of 2.2 cm^3, weighs 1.75 g, and consumes 0.9 W at maximum power. (c) Illustration of the panoramic field of view of the fabricated prototype. The dots and circles represent the angular orientation and acceptance angle $\Delta\rho$ of every ommatidium. (From Floreano et al., *Proc. Natl. Acad. Sci. USA*, 110(23), 9267–9272, 2013.[7])

5.6.5 Polarization Sensors

Inspired by the ability of some arthropods, particularly flying insects, to detect polarization patterns, researchers are creating integrated sensors capable of measuring polarization.[9,86–89] Nanofabrication technology is making it possible to integrate polarization light filtering and image processing into the individual pixel to create an active, monolithic polarization imaging sensor. These sensors would be useful for a wide variety of applications. The small size of arthropod eyes, their polarization sensors, makes them particularly attractive for applications where size and weight are extremely restricted, such as in UAVs. Polarization sensors can act as an automatic polarization compass for navigation, guidance, and flight control. This is particularly true in environments where more conventional technologies, such as geographical positioning system, may be limited, for example, in cluttered and obstructed environments such as urban canyon environments. They can be useful for imaging in conditions where traditional imagers are unable to capture detail, such as under water or in foggy or low-light conditions. Polarization imaging sensors are potentially useful for remote sensing, guidance, and surveillance applications, as well as in fingerprint detection.

5.6.6 Antireflective Structures

Biologically inspired antireflective nanostructures offer several performance and design benefits that make them attractive from an application perspective. Other antireflection approaches involve multilayer or multielement structures. Moth eye structures more significantly reduce unwanted reflections over a broad range of wavelengths and increase optical transmission, and they do so over a larger FOV than other techniques.[90,91] The structures do so without adding significantly to system size or complexity.

AR structures are being used for a wide range of optical and electro-optical applications, where performance is increasingly demanding. AR coatings on glass and plastic are being used for windows and other lens applications.[55] They are also being integrated into other bioinspired attempts at imaging optics.[77] These structures are also very useful in alternative energy applications, such as solar panels.

The highest-efficiency solar panels are fabricated with single-crystal silicon, which when untreated has very high surface reflectance and high reflection, approximately 35% at approximately 200–1000 nm. The ability to fabricate high-quality antireflecting surfaces and reduce the reflectance to less than approximately 1% would enable a significant improvement in efficiency.[92] AR coatings can also reduce the occurrence of ghost images in displays and improve detector performance.[77]

AR coatings can also be used to improve the efficiency of quantum cascade and other surface emitting lasers, light-emitting diodes (LEDs), and organic LEDs by reducing surface reflection-related optical losses. One lovely example is the application of these AR structures and microlens arrays to enhance the optical output power of aluminum gallium indium phosphide (AlGaInP) LEDs.[93] In these structures, the output efficiency is hampered by internal reflection at the surface–air interface. This means that the light generated within the device is reflected back at the surface and cannot escape, so the amount of light coming out of the laser (the optical output power) is reduced significantly (30%–100% depending on the angle). Researchers from the Gwangju Institute of Science and Technology demonstrated that an integrated ommatidium-like microlens and subwavelength antireflection structure monolithically fabricated on the top layer of an LED could significantly reduce this total internal reflection.

The fabrication process for these structures is illustrated in Figure 5.30. First, a hexagonally patterned photoresist mask was defined on the gallium phosphide (GaP) surface of the LED using conventional lithography and developed. The device was then briefly heated to enable thermal reflow of the photoresist, which softens and rounds the photoresist edges and allows controlled tapered profiles from the subsequent etch. Dry etch is used to define the approximately 2-μm-diameter microlenses, with approximately 1 μm spacing. A similar technique had previously been used on ultraviolet LEDs, for example, with larger microlens structures (~12 μm) and no additional AR structure.[94] After contacts were deposited, silver nanoparticles were grown on the microstructure surface.[93] These nanoparticles were used as a mask in a subsequent etch stage to define tapered nanostructures on the microstructure surfaces. Any residual Ag was then removed. The final nanostructures were randomly distributed across the microlens, with a nanostructure period of less than 200 nm

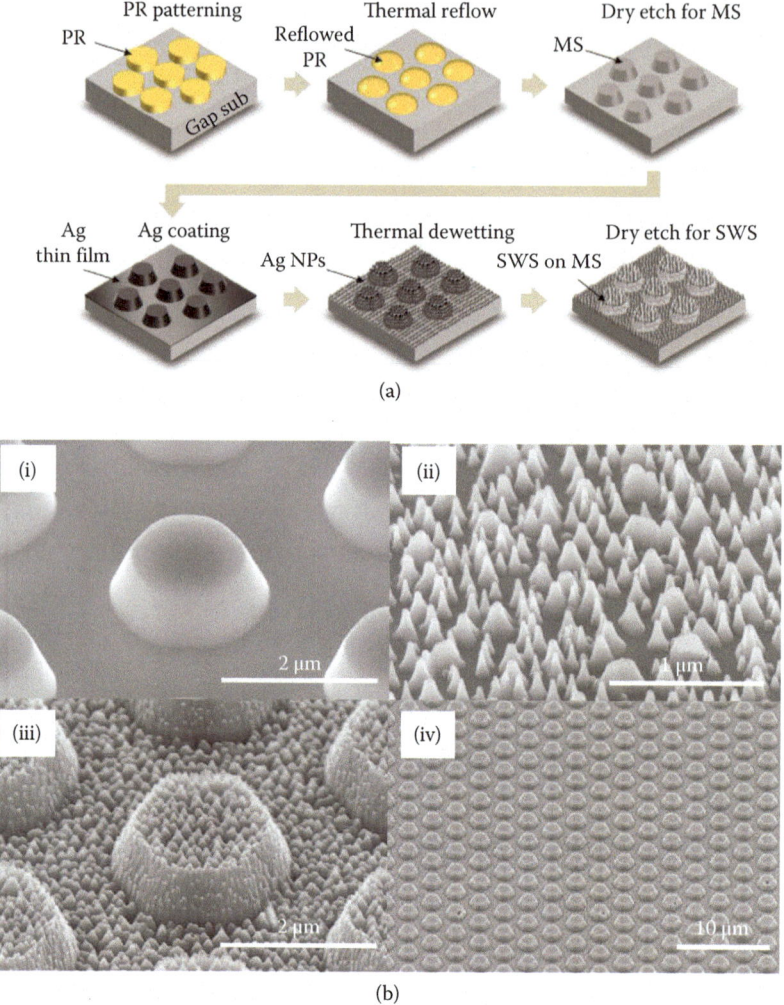

Figure 5.30 (a) Schematic illustration of the fabrication procedure for the microstructure/subwavelength structure architecture. (b) Tilted-angle view of SEM images for the fabricated samples with (i) microstructure, (ii) a subwavelength structure, and (iii) a combined micro- and subwavelength structure on a gallium phosphide substrate. (iv) A lower magnification image of (iii). (From Song et al., *Opt. Express*, 19(6), A157–A165, 2011.[93])

and a height greater than 100 nm. Demonstrated on a 635-nm (red) AlGaInP LED, these structures improved optical power output by over 70% without any loss of electrical performance.

Most fabrication approaches for these structures are top-down and involve some sort of lithographic, colloidal, or mask and etch technique. Many examples of these approaches can be found in the literature and in several thorough reviews.[60,91,95] The size scale and structure of these nanostructures provide additional potential functionalities: hydrophobicity, self-cleaning, and antifogging. Capturing multifunctionality at the level of nature remains an as yet unattainable goal. Mechanical stability and durability are a challenge as many of these structures are mechanically not robust.

References

1. M. F. Land and D. Nilsson, *Animal Eyes*. (Oxford University Press, Oxford, 2002).
2. P. Lariguet and C. Dunand, *J. Mol. Evol.* **61** (4), 559–U559 (2005).
3. W. R. Briggs and M. A. Olney, *Plant Physiol.* **125** (1), 85–88 (2001).
4. Y. M. Song, Y. Z. Xie, V. Malyarchuk, J. L. Xiao, I. Jung, K. J. Choi et al., *Nature* **497** (7447), 95–99 (2013).
5. S. F. Barrett and C. H. G. Wright, *Bioinspiration, Biomimetics, and Bioreplication* **7975** (2011).
6. L. P. Lee and R. Szema, *Science* **310** (5751), 1148–1150 (2005).
7. D. Floreano, R. Pericet-Camara, S. Viollet, F. Ruffier, A. Bruckner, R. Leitel et al., *Proc. Natl. Acad. Sci. USA* **110** (23), 9267–9272 (2013).
8. Y. Jiao, T. Lau, H. Hatzikirou, M. Meyer-Hermann, J. C. Corbo, and S. Torquato, *Phys. Rev. E* **89** (2) (2014).
9. G. C. Giakos, T. Quang, T. Farrahi, A. Deshpande, C. Narayan, S. Shrestha et al., *Sensing Technologies for Global Health, Military Medicine, and Environmental Monitoring Iii* **8723** (2013).
10. H. D. Sun, S. F. Deng, X. B. Cui, and M. Lu, *J. Micromech. Microeng.* **24** (6) (2014).
11. X. Zeng, C. T. Smith, J. C. Gould, C. P. Heise, and H. Jiang, *J. Microelectromech S.* **20** (3), 583–593 (2011).
12. S. A. Boden, A. Asadollahbaik, H. N. Rutt, and D. M. Bagnall, *Scanning* **34** (2), 107–120 (2012).
13. M. F. Land, *Contemp. Phys.* **29** (5), 435–455 (1988).
14. H. Kolb, R. Nelson, E. Fernandez, and B. Jones, *Webvision—The organization of the Retina and Visual System*. Webvision, 2015, http://webvision.med.utah.edu/.
15. D. E. Nilsson, E. J. Warrant, S. Johnsen, R. Hanlon, and N. Shashar, *Curr. Biol.* **22** (8), 683–688 (2012).
16. P. Cardoso and N. Scharff, *Zootaxa* (2246), 45–57 (2009).
17. R. DeVoe, *J. Gen. Physiol.* **59**, 247–269 (1972).
18. M. F. Land, *J. Exp. Biol.* **119**, 381–384 (1985).
19. M. Dacke, T. A. Doan, and D. C. O'Carroll, *J. Exp. Biol.* **204** (14), 2481–2490 (2001).
20. C. J. Clemente, K. A. McMaster, E. Fox, L. Meldrum, T. Stewart, and B. Y. Main, *J. Arachnol.* **38** (3), 398–406 (2010).
21. D. Harland and R. Jackson, *Cimbebasia* **16**, 231–240 (2000).

22. G. Zuccarello, D. Scribner, R. Sands, and L. J. Buckley, *Adv. Mater.* **14** (18), 1261 (2002).
23. G. Beadie, J. S. Shirk, A. Rosenberg, P. A. Lane, E. Fleet, A. R. Kamdar et al., *Opt. Express* **16** (15), 11540–11547 (2008).
24. S. Z. Ji, K. Z. Yin, M. Mackey, A. Brister, M. Ponting, and E. Baer, *Opt. Eng.* **52** (11) (2013).
25. S. Exner, "Die Physiologie der facettirten Augen von Krebsen und Insecten: eine studie" Leipzig und Wein, Franz Deuticke (1891).
26. G. Belusic, P. Pirih, and D. G. Stavenga, *J. Exp. Biol.* **216** (11), 2081–2088 (2013).
27. M. Srinivasarao, *Chem. Rev.* **99** (7), 1935–1961 (1999).
28. R. Ranganathan, D. M. Malicki, and C. S. Zuker, *Annu. Rev. Neurosci.* **18**, 283–317 (1995).
29. S. Fischer, C. H. G. Mueller, and V. B. Meyer-Rochow, *Visual Neurosci.* **28** (4), 295–308 (2011).
30. A. A. Polilov, *Arthropod Struct. Dev.* **41** (1), 29–34 (2012).
31. E. J. Warrant and P. D. McIntyre, *Prog. Neurobiol.* **40** (4), 413–461 (1993).
32. M. F. Land, *Curr. Biol.* **19** (2), R78–R80 (2009).
33. M. F. Land, *J. Opt. A-Pure Appl. Opt.* **2** (6), R44–R50 (2000).
34. J. Zeil and M. M. AlMutairi, *J. Exp. Biol.* **199** (7), 1569–1577 (1996).
35. C. Bruckner, B. Pradarutti, O. Stenzel, R. Steinkopf, S. Riehemann, G. Notni et al., *Opt. Express* **15** (3), 779–789 (2007).
36. S. Fischer, C. H. G. Muller, and V. B. Meyer-Rochow, *Zoomorphology* **131** (1), 37–55 (2012).
37. J. Aizenberg, A. Tkachenko, S. Weiner, L. Addadi, and G. Hendler, *Nature* **412** (6849), 819–822 (2001).
38. J. Aizenberg and G. Hendler, *J. Mater. Chem.* **14** (14), 2066–2072 (2004).
39. K. Franze, J. Grosche, S. N. Skatchkov, S. Schinkinger, C. Foja, D. Schlid et al., *Proc. Natl. Acad. Sci. USA* **104** (20), 8287–8292 (2007).
40. A. D. Briscoe, *Integr. Comp. Biol.* **49**, E21–E21 (2009).
41. H. Wassle, *Nat. Rev. Neurosci.* **5** (10), 747–757 (2004).
42. S. Palmer, Personal communication, (2012).
43. S. S. Deeb, *Clin. Genet.* **67** (5), 369–377 (2005).
44. S. J. Blamires, D. F. Hochuli, and M. B. Thompson, *Biol. J. Linnean Soc.* **94** (2), 221–229 (2008).
45. R. H. Douglas and G. Jeffery, *Proc. R. Soc. B-Biol. Sci.* **281** (1780) (2014).
46. I. C. Cuthill, J. C. Partridge, A. T. D. Bennett, S. C. Church, N. S. Hart, and S. Hunt, *Adv. Stud. Behav.* **29**, 159–214 (2000).
47. J. Chahl, S. Thakoor, N. Le Bouffant, G. Stange, M. V. Srinivasan, B. Hine et al., *J. Robot. Syst.* **20** (1), 35–42 (2003).
48. L. Cartron, N. Josef, A. Lerner, S. D. McCusker, A. S. Darmaillacq, L. Dickel et al., *J. Exp. Mar. Biol. Ecol.* **447**, 80–85 (2013).
49. V. Moiseenkova, B. Bell, M. Motamedi, E. Wozniak, and B. Christensen, *Am. J. Physiol.-Regul. Integr. Comp. Physiol.* **284** (2), R598–R606 (2003).
50. H. Schmitz, H. Soltner, and H. Bousack, *IEEE Sens. J.* **12** (2), 281–288 (2012).

51. T. C. Pappas, M. Motamedi, and B. N. Christensen, *Am. J. Physiol.-Cell Physiol.* **287** (5), C1219–C1228 (2004).
52. K. Ohtsu and S. I. Uye, *Biol. Bull.* **221** (3), 243–247 (2011).
53. P. I. Stavroulakis, S. A. Boden, T. Johnson, and D. M. Bagnall, *Opt. Express* **21** (1), 1–11 (2013).
54. K. A. Stokkan, L. Folkow, J. Dukes, M. Neveu, C. Hogg, S. Siefken et al., *Proc. R. Soc. B-Biol. Sci.* **280** (1773) (2013).
55. A. R. Parker and H. E. Townley, *Nat. Nanotechnol.* **2** (6), 347–353 (2007).
56. C. G. Bernhard, W. H. Miller, and A. R. Moller, *Acta Physiol. Scand.* **58**, 381–382 (1963).
57. S. A. Boden and D. M. Bagnall, 2009 (unpublished).
58. A. Yoshida, M. Motoyama, A. Kosaku, and K. Miyamoto, *Zool. Sci.* **13** (4), 525–526 (1996).
59. O. Deparis, N. Khuzayim, A. Parker, and J. P. Vigneron, *Phys. Rev. E* **79** (4) (2009).
60. K. L. Yu, T. X. Fan, S. Lou, and D. Zhang, *Prog. Mater. Sci.* **58** (6), 825–873 (2013).
61. J. R. Paterson, D. C. Garcia-Bellido, M. S. Y. Lee, G. A. Brock, J. B. Jago, and G. D. Edgecombe, *Nature* **480** (7376), 237–240 (2011).
62. Anomalocaris model by Gaetan Lee. Licensed under Creative Commons Attribution 2.5 via Wikimedia Commons. http://commons.wikimedia.org/wiki/File:Anomalocaris_model.jpg.
63. Anomalocaris predator model by Yinan Chen. Licensed under the Creative Commons Public Domain Dedication. http://commons.wikimedia.org/wiki/File:Gfp-anomalocaris-predator.jpg.
64. G. Kuhl, D. E. G. Briggs, and J. Rust, *Science* **323** (5915), 771–773 (2009).
65. G. Tanaka, A. R. Parker, D. J. Siveter, H. Maeda, and M. Furutani, *Proc. R. Soc. B-Biol. Sci.* **276** (1659), 1015–1019 (2009).
66. http://www.grintech.de/gradient-index-optics.html; http://www.edmundoptics.com/optics/optical-lenses/aspheric-lenses/gradient-index-grin-rod-lenses/3145; http://www.newport.com/Gradient-Index-Micro-Lenses/141056/1033/info.aspx; http://www.agiltron.com/GRINLENSandTubes.htm.
67. http://www.who.int/blindness/en/.
68. Kelly Servick, Telescopic contact lenses could magnify human eyesight, http://news.sciencemag.org/technology/2015/02/telescopic-contact-lenses-could-magnify-human-eyesight.
69. D. Siedlecki, H. Kasprzak, and B. K. Pierscionek, *J. Mod. Opt.* **55** (4–5), 639–647 (2008).
70. J. Y. Kim, K. H. Jeong, and L. P. Lee, *Opt. Lett.* **30** (1), 5–7 (2005).
71. K. H. Jeong, J. Kim, and L. P. Lee, *Science* **312** (5773), 557–561 (2006).
72. B. G. Park, K. Choi, C. J. Jo, and H. S. Lee, *Soft Matter* **8** (6), 1751–1755 (2012).
73. H. W. Liu, F. Chen, Q. Yang, P. B. Qu, S. G. He, X. H. Wang et al., *Appl. Phys. Lett.* **100** (13) (2012).

74. P. B. Qu, F. Chen, H. W. Liu, Q. Yang, J. Lu, J. H. Si et al., *Opt. Express* **20** (5), 5775–5782 (2012).
75. D. Zhu, C. Li, X. Zeng, and H. Jiang, *Appl. Phys. Lett.* **96** (8) (2010).
76. R. J. Martin-Palma, C. G. Pantano, and A. Lakhtakia, *Nanotechnology* **19** (35) (2008).
77. D. H. Ko, J. R. Tumbleston, K. J. Henderson, L. E. Euliss, J. M. DeSimone, R. Lopez et al., *Soft Matter* **7** (14), 6404–6407 (2011).
78. S. Yang and J. Aizenberg, *Mater. Today* **8** (12), 40–46 (2005).
79. K. Lee, *Nat. Commun.* (2012).
80. D. Keum, H. Jung, and K.-H. Jeong, *Small* **8** (14), 2169–2173 (2012).
81. R. S. Prabhakara, C. H. G. Wright, and S. F. Barrett, *IEEE Sens. J.* **12** (2), 298–307 (2012).
82. G. P. Luke, C. H. G. Wright, and S. F. Barrett, *IEEE Sens. J.* **12** (2), 308–314 (2012).
83. J. W. Duparre and F. C. Wippermann, *Bioinspir. Biomim.* **1** (1), R1–R16 (2006).
84. H. C. Ko, M. P. Stoykovich, J. Song, V. Malyarchuk, W. M. Choi, C.-J. Yu et al., *Nature* **454** (7205), 748–753 (2008).
85. I. Jung, J. Xiao, V. Malyarchuk, C. Lu, M. Li, Z. Liu et al., *Proc. Natl. Acad. Sci. USA* **108** (5), 1788–1793 (2011).
86. R. C. Hardie, *Curr. Biol.* **22** (1), R12–R14 (2012).
87. S. Thibault and P. Desaulnier, in *Photonic Applications for Aerospace, Transportation, and Harsh Environment Iii*, edited by A. A. Kazemi, N. Javahiraly, A. S. Panahi, S. Thibault, and B. C. Kress (Spie-Int Soc Optical Engineering, Bellingham, WA, 2012), Vol. 8368.
88. J. Van der Spiegel, X. T. Wu, M. L. Zhang, N. Engheta, and Ieee, *Polarization Image Sensors: Learning from Biology to Make the Invisible Visible*. (IEEE, New York, 2012).
89. J. Chahl and A. Mizutani, *IEEE Sens. J.* **12** (2), 289–297 (2012).
90. Y. Li, J. Zhang and B. Yang, *Nano Today* **5** (2), 117–127 (2010).
91. S. Chattopadhyay, Y. F. Huang, Y. J. Jen, A. Ganguly, K. H. Chen, and L. C. Chen, *Mater. Sci. Eng. R-Rep.* **69** (1–3), 1–35 (2010).
92. X. C. Li, J. S. Li, T. Chen, B. K. Tay, J. X. Wang, and H. Y. Yu, *Nanoscale Res. Lett.* **5** (11), 1721–1726 (2010).
93. Y. M. Song, G. C. Park, S. J. Jang, J. H. Ha, J. S. Yu, and Y. T. Lee, *Opt. Express* **19** (6), A157–A165 (2011).
94. M. Khizar, Z. Y. Fan, K. H. Kim, J. Y. Lin, and H. X. Jiang, *Appl. Phys. Lett.* **86** (17) (2005).
95. L. Yao and J. H. He, *Prog. Mater. Sci.* **61**, 94–143 (2014).
96. B. M. Ross, L. Y. Wu and L. P. Lee, *Nano. lett.* **11** (7), 2590–2595 (2011).

6
BIOMATERIALS FOR PHOTONICS

Many of the biomaterials nature uses to create its photonic systems are themselves very attractive for future applications. Biological materials range from biopolymers, such as chitin, to vivid fluorescent proteins and more. Scientific interests in these materials, how they are formed and how we might leverage their useful properties, form the basis of a growing area of development. One reason that biomaterials are of interest is the potential for greatly reduced cost compared with other inorganic materials. Many biomaterials are easily obtainable on large scales from the waste products of other industries or historical domestication. Organic optical systems, such as organic light emitting diode displays, are a huge area of technology development because of the lower cost compared with complex semiconductor fabrication requirements. Fabrication of organic devices uses less energy than their semiconductor counterparts. Biomaterials also enable fabrication of flexible or curved devices and systems. The mechanical properties of biomaterials can be tuned according to their composition, which enables their deposition, casting, and coating into and onto a wide variety of substrate conformations. Use of these materials may be less environmentally hazardous, as well. As our society moves further and further toward viewing even high-performance systems as disposable technologies (one could argue that we have already reached that point), the lifetimes over which these systems must perform become shorter, and biodegradability, lack of environmental toxicity, and recyclability become increasingly important.

Another reason we look to these materials is their biological compatibility. Applications in biological sensing and imaging, implanted devices, prosthetics, medical devices, and more can all benefit from fabrication or incorporation of biological materials. Photonic approaches in general are attractive for medical applications, such as implantable probes, because of reduced size and lower parasitic losses compared with electronic devices. In addition, devices tuned to the NIR (800–1000 nm) may enable readout without the potential side effects of other biological probes such as implantable electrodes, or ionizing radiation, as near infrared (NIR) light travels comparatively well through tissue with a moderate penetration depth. Similar devices made instead from biological materials can be not only lightweight, flexible, at lower cost, but also potentially with lower potential toxic or immunogenic effect. Finally, biomaterials are making huge impacts on our ability to understand biology itself. Because of their compatibility, and our increasing ability to incorporate it into the systems we study, development of tools and techniques by using biomaterials such as green fluorescent protein (GFP) have resulted in vast improvements in fields such as microbiology.

These materials and their use are not without limitations. Organic materials are not always able to match the performance characteristic of inorganic devices. Biomaterials have a smaller set of available refractive indices, control, and uniformity of the resulting material properties; the need for high-performance computing capability means that all-organic systems are still a long way away. It is unlikely that, for some applications, we will ever completely eliminate or negate the need for inorganics. Another limitation on our current ability to create all-biomaterial devices is the lack of true localized control over the material composition. Tailoring the material properties for desired optical and mechanical properties often requires chemical modification of the native or purified biomaterial. Nature demonstrates the ability to change material properties over extremely small length scales and close proximity. Viable methods of heterogeneous integration, where materials and devices fabricated with incompatible processing are integrated into the same device or circuit, are currently being explored, but these often require additional fabrication techniques for both the bio- and inorganic components. This is one reason why exploration of natural photonic structures is so vital. Nature has found ways to incorporate design features and fabrication techniques that may enable us to overcome some of these limitations.

Natural photonic structures are created out of only a handful of primary biomaterials, such as chitin in arthropods and crustaceans; keratin in birds and higher order vertebrates; and silicates in diatoms. Yet despite this potential limitation, a tremendous variety of properties can be achieved. These biomaterials, particularly the structural biomaterials such as chitin, do not often appear in pure form in nature.[1] They are often nanocomposites—complex combinations of proteins, lipids, polymers, minerals, pigments, and more. The specific combination of materials can have a significant effect on the behavior of the material. The same primary material can be hard, soft, opaque, transparent, stiffer, less brittle, and more, all due to the composition of its constituents. In many cases, we still do not understand the functionality of the constituents in the nanocomposite. Identifying the critical pieces is still a challenge, and there has been insufficient investigation into the complex structure/function relationships seen in the wild, particularly for structural color-related materials.

6.1 Chitin

Chitin is one of the most abundant biomaterials on the planet.[2] It is a polymer—a polysaccharide—and it functions as a structural material for shells and bodies of crustaceans and arthropods. It is a white, hard, inelastic material that is highly insoluble. It is amenable to chemical modification which, however, makes it enticing for manufacturing.[3] One of the most common modifications is to turn chitin into chitosan by deacetylation, making it somewhat more soluble and easier to work with, while retaining the attractive properties of chitin. The structure of chitin and chitosan can be seen in Figure 6.1. Other modifications may include attaching sugars, dendrimers, nanoparticles, and so on; the resulting material can be made not only into fibers, yarns, threads, colloids, and more, but also amenable to multiple fabrication techniques, such as spray coating, molding, and laser irradiation. Chitin is also relatively easily acquired, as it can be obtained from crab and shrimp shells from food industry waste.[2]

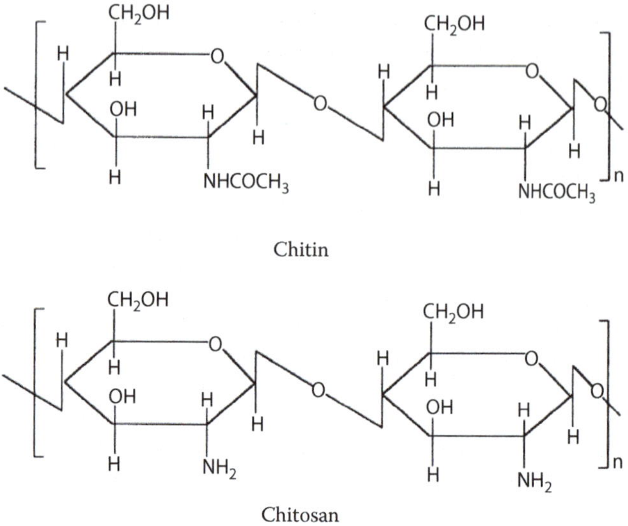

Figure 6.1 The structures of chitin and chitosan. (From Kumar, M., *React. Funct. Polym.*, 46(1), 1–27, 2000. With permission.[2])

Chitin and its chemically modified cousins have very attractive material properties for biomedical applications.[4] It is biodegradable, nonimmunogenic, and has demonstrated antibacterial and wound healing properties. Applications are wide ranging in areas of tissue engineering, drug delivery, and agriculture. Our interest is in the photonic and optical applications, and there are several interesting ones. Chitin can be optically transparent, and composites are being developed for use as contact lenses, sensors, and optical screens for, for example, flexible flat-panel displays, e-paper devices, and solar cells.[5–7]

Researchers at Kyoto University and colleagues have investigated nanofibrous chitin and resin composites for exactly these latter examples.[8] Inspired by the investigation of cellulose and other biomaterials for optical composites, this team has demonstrated that chitin–acrylic resin nanocomposites can be fabricated, and that they show attractive optical and thermal properties. Chitin–acrylic composites were first demonstrated using the shell of a *Chionoecetes opilio* (snow crab), removing all of the nonchitin components, and impregnating the remaining structure with UV-cured acrylic resin. The resulting transparent crab shell showed that the resin penetrated the clean nanofibrous chitin matrix in the shell. One key result from this process demonstrated that even though chitin comprised only

~20%–25% of the natural shell, the transparent crab kept its structural integrity, and retained many of the structural and morphological details of the whole crab (such as its eyes). Images of the crab at various stages of processing can be seen in Figure 6.2. Rather than using the whole crab for subsequent work (which may make fun pictures, but is completely impractical), chitin in microparticle form was then subjected to the same treatment. After removing the non-chitin components, and roughing the microparticles to form surface nanofibers, the team then formed chitin powder "paper" sheets and impregnated those with the acrylic resin. After curing, the result was a flexible, optically transparent sheet (as shown in Figure 6.3) with

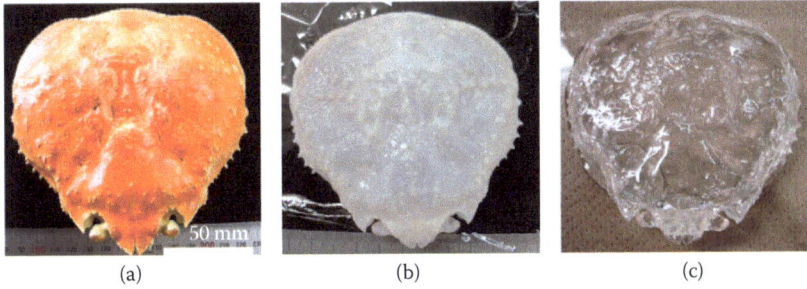

Figure 6.2 Transforming a crab shell—Prep and process of transparent crab shell. (a) Original crab shell. (b) Crab shell after removal of matrix substances. (c) Transparent crab shell after immersion in acrylic resin. (From Shams, M.I. et al., *Soft Matter.* 8(5), 1369–1373, 2012. With permission.[8])

Figure 6.3 (a) (Left) Appearance of chitin powder paper. (Right) Transparent and flexible chitin powder sheet with acrylic resin ABPE 10, 22 wt% fiber content.

(Continued)

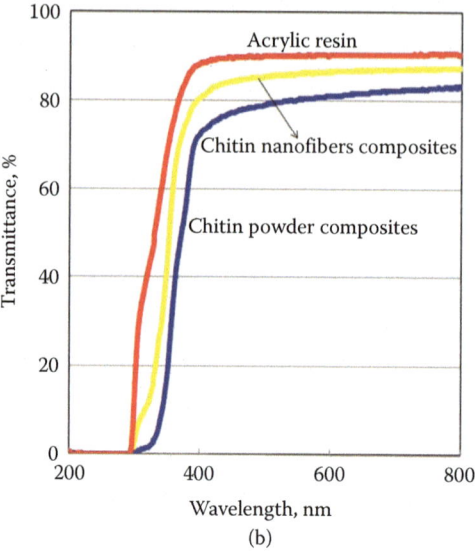

(b)

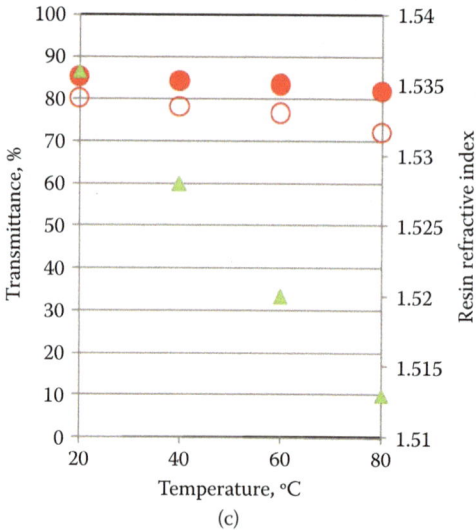

(c)

Figure 6.3 (*Continued*) (b) Regular transmittance of the chitin powder reinforced acrylic resin sheet (thickness 190 mm, fiber content: 22%) and chitin nanofibers reinforced acrylic resin sheet. (Thickness 100 μm and FC: 36%). (c) Temperature dependence of the refractive index of the neat acrylic sheet (triangle) at 600 nm and the regular transmittance of chitin powder acrylic resin sheet (open circles) and chitin nanofibers acrylic resin sheet (filled circles). (From Shams, M.I. et al., *Soft Matter.* 8(5), 1369–1373, 2012. With permission.[8])

BOX 6.1 COMPLEXITY IN BIOLOGICAL MATERIALS

The materials from which biological photonic structures are made are extremely complex.[1] We often describe them as made of a single material (e.g., chitin in insects or keratin in birds) but actually these structures are extremely heterogeneous. The photonic structure can contain not only a large percentage of the primary material, but often pigments and other materials such as lipids, polysaccharides, and amino acids as well. Chitin, for example, may comprise as little as 20%–30% of the total composite material in the biological photonic structure, with the remainder consisting of proteins, minerals, and other organic compounds. Like chitin, keratin is a common structural biomaterial. It is a protein and is found in most land vertebrates in hair, hooves, claws, feathers, and so on. Unlike chitin structures, keratinaceous structures may only contain 10% additional material, such as pigments. Figure 6.4 shows a scanning electron microscopy (SEM) image of the photonic structure found

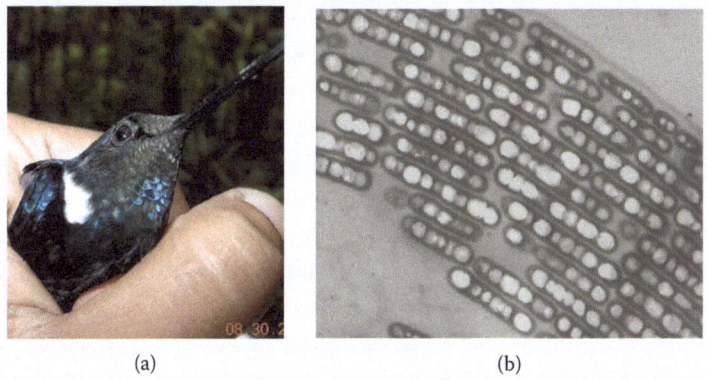

(a) (b)

Figure 6.4 (a) The hummingbird *C. prunellei*. The box indicates the colored region from which the tunneling electron microscopy (TEM) images and spectral data were taken. (b) TEM image of an iridescent green barbule from the hummingbird *Coeligena iris*, showing highly ordered layers of air filled melanin platelets in a keratin matrix.

(Continued)

BOX 6.1 (*Continued*) COMPLEXITY IN BIOLOGICAL MATERIALS

in the hummingbird *Coeligena iris* as an example, highlighting the composition of air, keratin, and pigment (melanin). Other materials that may also exist in the structure were not identified. Also like chitin, these proteins tend to be biocompatible, processable, and also have the interesting ability to maintain the biological function of live biomaterials encased within. The exact composition of any given photonic structure may vary with species, may vary with location on the body, and even locally within the photonic structure itself. The challenge with inhomogeneous materials is that their composition has a direct effect on the optical and physical properties of the structure, so understanding local composition is vital to our ability to untangle the effects, both dominant and subtle, on the structure's functional properties.

The refractive index (RI) of a material is the key property that determines the material's optical behavior. The RI is actually a complex value, with a real and imaginary component. The real component (n) of the RI is defined as the phase velocity of light in the material relative to the phase velocity of light in a vacuum. The imaginary component (κ) is called the extinction coefficient and it describes how light is absorbed in the material. The complex RI can be written as $n = n + i\kappa$ and, both of these components may depend on wavelength. In many published articles on biological photonic structures, the imaginary component of the RI is often assumed to be negligible, and only the real component is used in calculations. (There are exceptions.)[9] In addition, the real RI values quoted in articles may not have been measured directly from the sample, but instead may be assumed to be a generic, constant value. Values for chitin in the literature, for example, range from 1.4 to 1.73 (most use $n = 1.56$). The extinction coefficient, while small, may not

BOX 6.1 (*Continued*) COMPLEXITY IN BIOLOGICAL MATERIALS

truly be negligible. The challenge with this is that local variations in structure composition may result in varying RI within the structure. A change in the real or imaginary components may alter the hue, brightness, reflected wavelength, and so on, of the structure. Although using an inexact value for the RI may be acceptable to model the primary features of the optical properties of a structure, understanding the details will require more accurate measurement of the true local structural material composition and complex RI.

very encouraging optical properties. The chitin sheets demonstrated good transparency, ~80% transmittance between ~400 and 800 nm, only about 10% less than the acrylic alone. The benefit of the chitin nanocomposite becomes apparent when you examine the thermal properties of the sheet. The optical properties of the sheet remain relatively immune to temperature variation, particularly compared with the acrylic alone.

6.2 Silk

Whether from spider, silkworm, or wasp, the silk fibroin protein shares many of the same positive attributes of chitin and keratin. It is biocompatible, with well-known, impressive mechanical properties.[10] Spider silk in particular is known for its strength and toughness, malleability and flexibility. Some spider silks are also known to be UV reflective and their optical properties for interest in applications have recently begun to be explored.[11–15] Silk films have demonstrated >90% transparency in the visible wavelength range, which is slightly higher than other biopolymers.[16] It is, however, unlikely that spider silk will be attractive for large-scale processing simply owing to the problem of production or harvesting of the silk itself.[17] Spiders do not get along well with each other and domestication is a significant challenge. Silk from the *Bombyx mori* silkworm (Figure 6.5), in contrast, offers many of the same positive material properties and the silkworm has been domesticated for millennia and is easily acquired in large volumes.

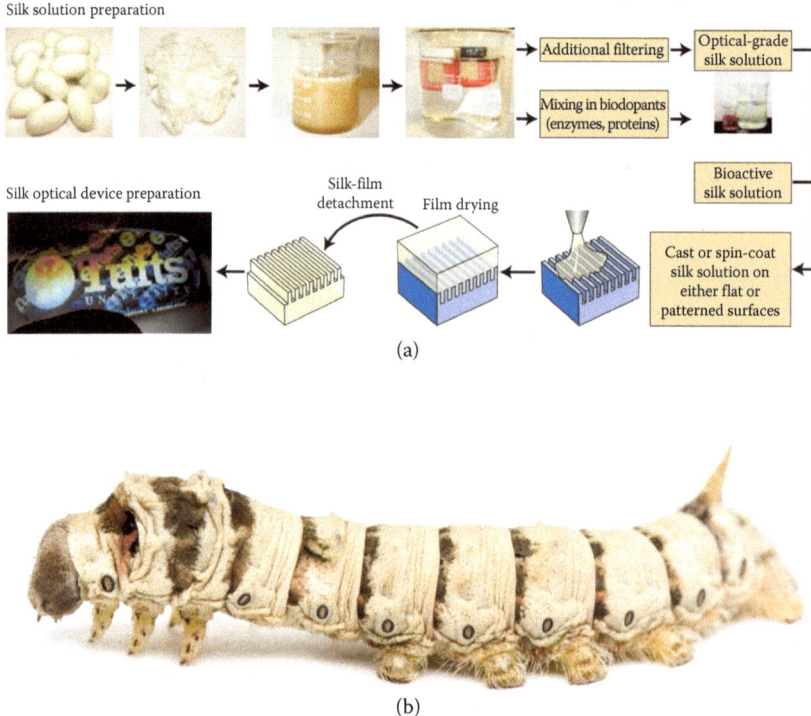

(a)

(b)

Figure 6.5 (a) The transformation process of silk optics. Realization of silk optics starts from the processing of silk cocoons, which are cut, cleaned, boiled, and dried, and subsequently dialyzed to remove the contaminating sericin. Filtering yields a clear water-based silk-fibroin solution, which can either be used to realize undoped or "activated" silk optics by simply mixing biodopants into the solution. Either the undoped or doped silk solution can be deposited on appropriate substrates and detached after its transition from the liquid to the solid phase, yielding an optical element, such as the silk white-light hologram shown. (b) The *Bombyx mori* silkworm. (From Omenetto, F.G. and Kaplan, D.L., *Nat. Photonics*, 2(11), 641–643, 2008. With permission.[18])

Professors at Tufts University are leading the way in silk-based biomedical and photonic applications.[19] Silk from the *B. mori* silkworm consists of the proteins fibroin and sericin; fibroin functions as the structural component, whereas sericin acts like glue to hold the fibers together.[20] The silk is not woven into these photonic systems. Instead, it is extracted from the natural cocoons (the sericin and other impurities removed) and processed into an aqueous silk fibroin solution.[19] Figure 6.5 shows the purification and photonic structure fabrication process. Materials, devices, and structures are then fabricated from this liquid silk in a wide variety of methods and formats, depending on the needs of the application. It can be deposited, sprayed, or printed; optical structures can be stenciled, imprinted, or lithographically defined (for example) with high-resolution feature sizes. This silk is engineerable into a multitude of interesting material structures, such as hydrogels, colloids, films, fibers, and coatings, and can be designed with long-term stability, or controllable degradation. All of this has been demonstrated either with the silk in its pure form, or in combination with organic and inorganic additives, including live active biologicals. The result is an incredibly flexible material foundation for applications development. That flexibility has driven the demonstration of many interesting optical structures in recent years. Many examples of silk-based metamaterials, microlens arrays, waveguides, gratings, sensors, and plasmonics are found in the literature, defining the state of the art of what organic photonic materials can achieve.[20–29] Figure 6.6 shows some of the variety of photonic structures fabricated in silk.[18] Silk is also greatly of interest due to its biocompatibility and biodegradability, opening the possibility for implantable and bioresorbable photonic structures. Silk materials can also be doped with organic and biological dopants, enabling the potential for integration with living tissues, theranostics, biosensors, and more.[30] Finally, silk can be integrated with flexible microelectronics for a host of bioresorbable and biocompatible optoelectronic devices.[31–34] Recently, the fabrication of silk inverse opal photonic structures has been reported.[35] Silk inverse opals (SIOs) were fabricated via the process shown in Figure 6.7. Poly(methyl methacrylate) (PMMA) colloids were stacked in a face-centered cubic (FCC) structure on a silicon substrate. Multiple structures were fabricated using spheres of 250, 320, and 350 nm diameter. Aqueous fibroin solution was then allowed to infiltrate the PMMA template and solidify, resulting in an amorphous

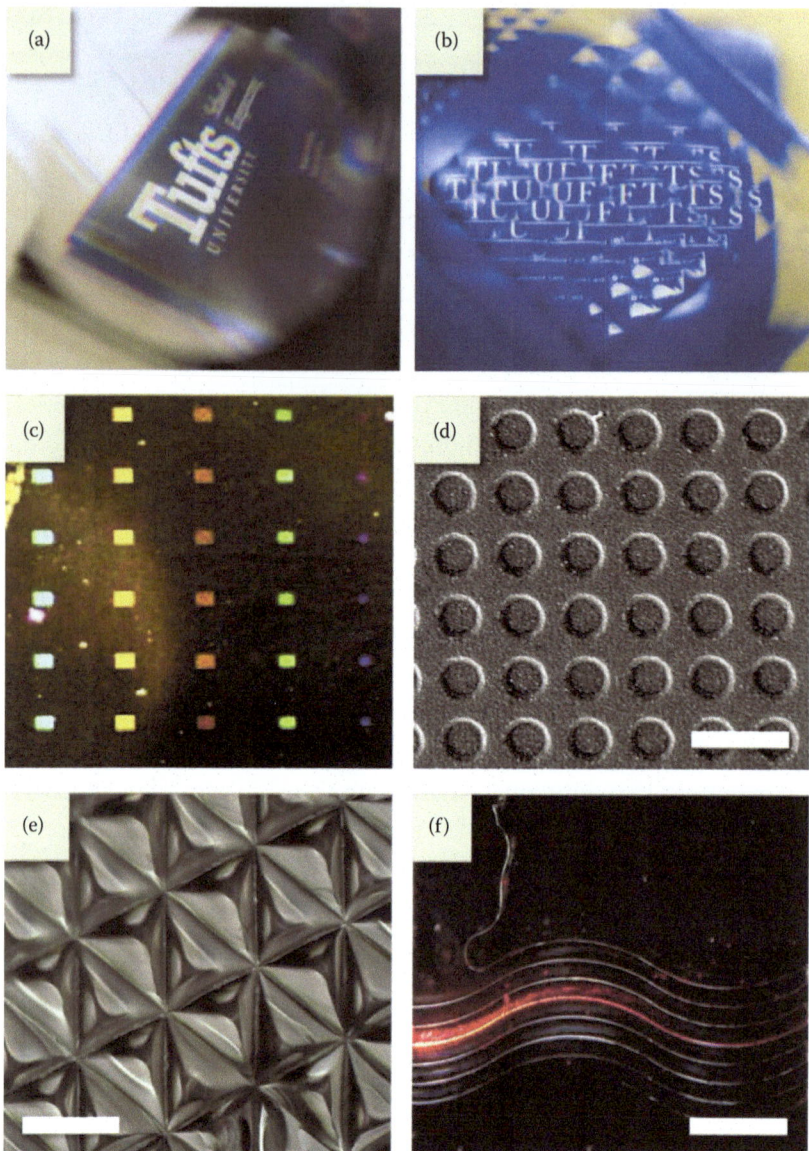

Figure 6.6 Examples of silk-based optical and photonic elements and devices. (a) An image through a 1-cm-diameter Fresnel lens. (b) An image through a 6 × 6 microlens array. (c) Periodic two-dimensional array of nanoholes in silk under dark field illumination. (d) SEM image of the periodic nanohole array with lattice constant equal to 400 nm. (e) SEM image of a silk microprism array. (f) An image of inkjet-printed silk waveguide. (From Omenetto, F.G. and KapLan, D.L., *Nat. Photonics*, 2(11), 641–643, 2008; Tao, H. et al., *Adv. Mater.*, 24(21), 2824–2837, 2012. With permission.[18,19])

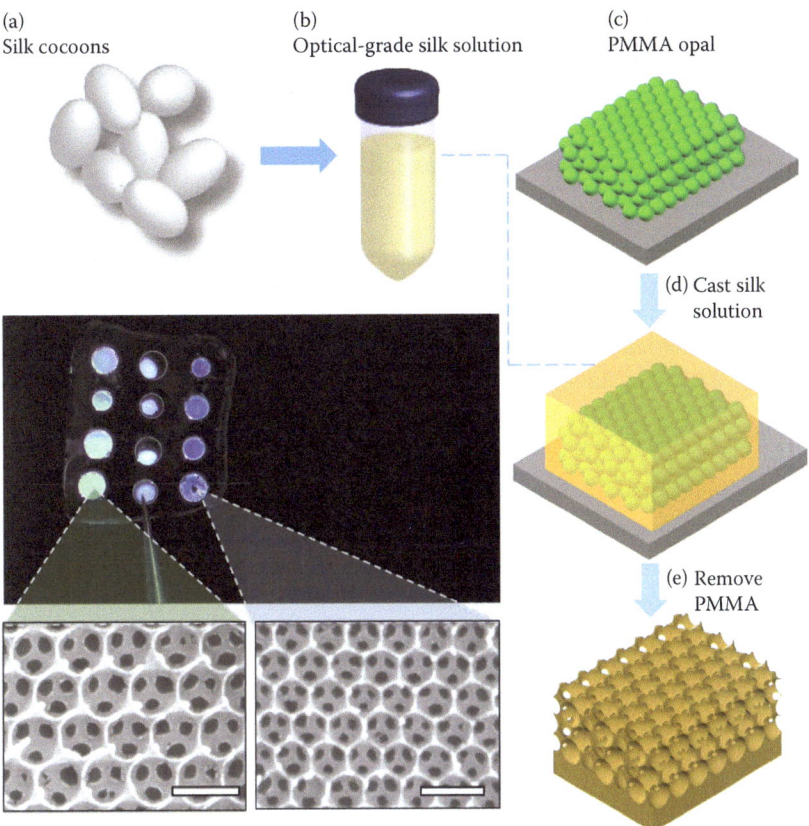

Figure 6.7 (a, b) Optical-grade silk fibroin solution is obtained from the cocoons of the *B. mori* silkworm. (c) A PMMA opal, with PMMA sphere diameters of 250, 320, and 350 nm is generated on a silicon substrate. (d) The opal on silicon is infiltrated with silk fibroin solution. (e) After drying, the silk film is removed from the silicon wafer. PMMA spheres in the free-standing silk film are dissolved in acetone. A nitrogen blower is used to remove acetone in the voids of the SIO film. The image is a photograph of a 2.5 × 1.5 cm free-standing silk film with patterned inverse opals. Corresponding SEM images show the green (left) and blue (right) silk opals. Scale bars: 500 nm. (From Kim, S. et al., *Nat. Photonics*, 6(12), 817–822, 2012. With permission.[35])

freestanding silk film. To create the inverse opal structure, the substrate was then removed from the silk film and the PMMA spheres dissolved by immersion of the structure in acetone. The reflectance spectrum of the SIO film shows reflectance peaks appropriate for the lattice spacing. These structures offer the potential for all-protein-based contrast agents and sensors for use within biological tissues. The air gaps in the inverse opal structure can be filled with a contrasting fluid that changes the RI profile of the structure. The silk can also be doped, for example, with gold nanoparticles, which extends the functionality of the structures (in this example, laser plasmon resonance-induced photonic band gap-enhanced heating for biomedical applications). Silk offers a very attractive platform for photonic applications, particularly in the biomedical space.

6.3 Biosilica

Biosilicate organisms, such as diatoms and glass sponges, are able to take up environmentally available, soluble silicon and store it in their cells, where they can then catalyze its transformation into silica.[36,37] The process creates nano- and microstructured hierarchical assemblies that form the organism's skeleton and structural supports. Silica and silicates are widely used in many applications in medicine, micro- and nanophotonics, and electronics. Silica nanostructure fabrication processes are also compatible with the fabrication processes of conventional semiconductor device technologies, which make them comparatively easy to integrate into conventional devices. However, the fabrication processes are also extreme; they require high temperature, high pressure, caustic chemicals, and more. Biological silica production, in contrast, is able to create highly complex nanostructures under mild processing conditions, such as ambient temperature and pressure. Fabrication occurs inside the organism, in conditions compatible with life. We describe these conditions as ambient, environmental, or physiological conditions, and the ability to create micro- and nanomaterials at ambient, physiological conditions is something of a holy grail in materials science. Inspired by the biological silica nanostructures found in diatoms and glass sponges, several approaches have been taken to leverage nature's ability to create these complex biosilica structures. In the case of the diatoms, the biosilica structures often serve as templates for patterned inorganic nanomaterial growth. The glass sponges have inspired a new way of fabricating fiber optic waveguides, particularly for use in short-range communication. Inspired by these organisms and their structures, exploration in this area may enable us to fabricate highly complex photonic structures in ways that are significantly more environmentally friendly, and more cost effective, than today's technology.

Diatoms are unicellular microalgae. They are ubiquitous across the planet, and found every place that has water.[38–40] Diatoms consist of an outer biological coating of polysaccharides and proteins that surround a nanostructured skeleton, which is made of amorphous hydrated nanoporous silica. The diatom skeleton consists of two halves called frustules that overlap and protect the diatom protoplasm within. The frustules are intricately patterned with pores and grooves that are typically periodically organized. Diatoms can range in size from 2 to 200 µm or so and the features on the frustule can range from 10 nm up to several hundred nanometers in size. There seems to be some evidence that the nanoscale patterns of pores on the frustule are the result of fusing or aggregation of biologically formed silica nanoparticles. Diatoms are fascinating and beautiful to look at, as the examples shown in Figure 6.8, and come in a variety of structures, shapes, and symmetries. Centric diatoms, for example, have radial symmetry, whereas pennate diatoms have bilateral symmetry. These characteristics are used by biologists to identify the more than 200,000 estimated extant species.[41] Diatoms are useful for a host of potential applications due to their porous frustule's large surface area,

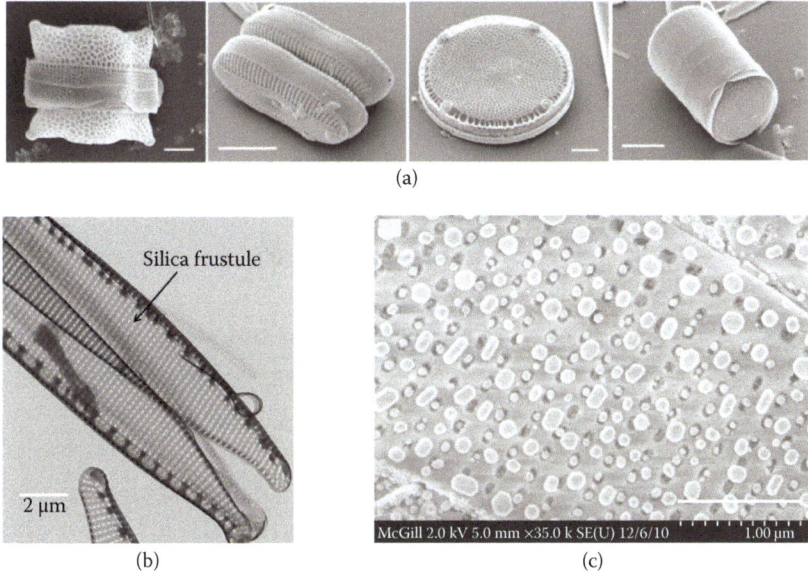

Figure 6.8 (a) SEM examples of diatom structures. (b) TEM of *Nitzschia closterium* frustules showing the rectangular pores (foramina) periodically organized on the silica surface. (c) SEM of *N. closterium* frustules coated with 10 nm gold film and annealed at 700°C. Scale bar: 1 µm. (From Andrews, M.P. et al., In A. Adibi, S.Y. Lin, and A. Scherer [eds.], *Photonic and Phononic Properties of Engineered Nanostructures*, 2011. With permission.[40])

high pore volume, periodically ordered pore channels, and high thermal and mechanical stabilitiy. Frustules of diatoms are photoluminescent. They are also widely functionalizable, that is, you can bind many other functional molecules and materials to their surface. It has been discovered that centric frustules can focus light onto a focal plane within the interior of the structure, where the cytoplasm is encapsulated.[39] Photonic applications range from sensor systems (gas or biological), bioimaging, theranostics, photonic crystals, microlenses, waveguides, and more.[38–40,42–46]

The frustule structure, the geometry of their pores, and the ability to grow or harvest large quantities of the diatoms inexpensively, combined with the functionalizability of the surface, makes them attractive as templates from which to grow nanostructures of other desirable materials. *Nitzschia closterium*, for example, is a pennate diatom frustule that has been used to template arrays of silver and gold nanoparticles.[40] Researchers from McGill University took two approaches: topochemical functionalization for deposition of silver and gold nanoparticles from solution and also thermal wetting/dewetting of evaporated gold films. Figure 6.8 shows a tunneling electron microscopy (TEM) image of the *N. closterium* diatom with regular arrays of pores. The pores on *N. closterium*'s frustule are spaced ~100 nm apart with the rows spaced ~150–200 nm apart. Overall the pore pattern is something like a triangular lattice. The topochemical approach resulted in selective deposition of silver and gold nanoparticles in and around the array of pores. The resultant structure acted as a nanoplasmonic array and demonstrated surface-enhanced Raman effects. The thermal dewetting approach produced arrays of faceted gold nanoparticles both within the pores and on the surface of the frustule, as shown in Figure 6.8. The surface particles were larger, and seemed to show some degree of uniaxial growth. Within the pores, the faceted particles were smaller and more uniform. Finite difference time domain calculations were performed based on an idealized nanoplasmonic frustule (with a gold nanoparticle in the pore, and a uniform film of gold on the surface of the frustule) and show strong coupling between the localized surface plasmon modes of the nanoparticles in the pores and overlaid gold film.

Another example of a biosilicate organism that is inspiring application is the deep sea "glass" sponge *Euplectella*, which has a

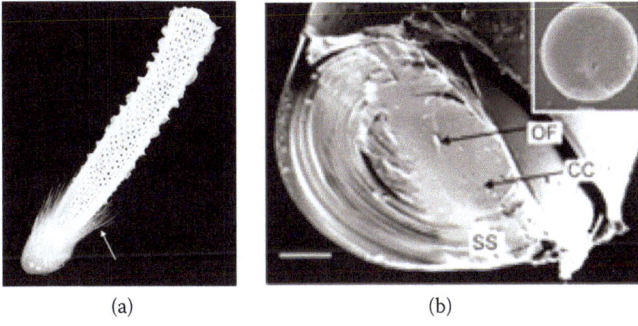

(a) (b)

Figure 6.9 Structure and fiber-optical properties of spicules in the sponge *Euplectella*. (a) The glass sponge, showing the basket-like cage structure and basalia spicules (arrow). Scale bar: 5 cm. (b) Mechanically cleaved spicules show three structural regions. OF, organic filament; SS, outer striated shell; CC, central cylinder. Inset: Smooth cross section of a stress-free spicule. (From Sundar, V.C. et al., *Nature*, 424(6951), 899–900, 2003. With permission.[47])

skeletal support structure made from high-purity silica (shown in Figure 6.9). Harvard researchers have examined these "spicules" and discovered that they contain structures that demonstrate fiber optical properties similar to the optical fibers used in the telecom industry.[48] The spicules, which are between 5 and 15 cm long, and about 40–70 µm in diameter, show intriguingly complex structure. The fiber is multilayered, with three different regions with differing refractive indices. The inset in Figure 6.9 shows a cross section of one of these fibers. In the center is a ~2-µm-diameter silica core and organic filament, the middle region is a cylinder with a high organic content, and the outer region is a multilayer with gradually decreasing organic content. These are all glued together by organic films. The index of refraction of the entire structure changes radially with the material content, being highest in the center, then a region of low index, followed by the multilayer of alternating high and low index of refraction, gradually increasing toward the outer layer. These spicule fibers are able to carry one or more optical modes with the light mostly confined to the core. The biosilica fibers are interesting not only because of their mild environmental fabrication requirements, but also because they are crack and fracture resistant compared with commercial fibers, due in part to the organic material incorporated into the structures. Insights into the biomineralization and formation of the complex sponge and diatom structures may inspire new composite materials, new designs, and new fabrication techniques that avoid or overcome the challenges of current approaches, or enable entirely new structures.

Researchers from Università del Salento in Italy have demonstrated synthetic production and organization of biosilica, through the use of recombinant silicatein-α.[37] Sponges, such as the glass sponges, use the enzyme silicatein to catalyze the biosilicification of environmentally available silicon into biological silica structure. The natural sponge spicule fibers have a graded composition ranging from mostly protein in the core to mostly silica at the outer edges. The waveguiding effect seen in these biological fibers is promising but shows a fairly high degree of optical loss (10^{-4}–10^{-1} cm^{-1}), about four orders of magnitude higher than commercial fibers. Synthetic formation of silica waveguides inspired by the biosilica formation process would enable optimization of these waveguide structures to improve the optical losses, at the same time scale up production at more friendly process requirements without sacrificing biological organisms to do so. Silicatein-α is a subunit of the sponge proteins that catalyze the biosilicification reactions. Recombinant silicatein-α from demosponge *Suberites domuncula* was used in conjunction with soft microlithography to create synthetic biosilica photonic waveguides. The process can be seen in Figure 6.10. The silicatein-α was patterned using microfluidics on a hydroxylated Si/SiO$_2$ substrate and incubated in a silica precursor solution, resulting in a structure with very similar protein

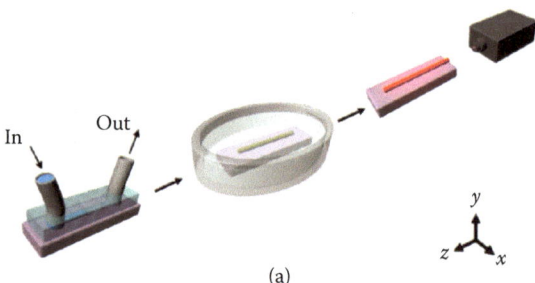

(a)

Figure 6.10 (a) Silicatein patterning using pressure-driven microfluidic flow, within a device assembled by a poly(dimethylsiloxane) (PDMS) elastomeric element and a Si substrate. Small black arrows indicate the microflow direction. After peeling off the elastomer, incubation in a tetraethoxysilane (TEOS) solution results in biosilicification directed by the patterned protein. Finally, the waveguiding properties of coupled laser light by the biomineralized microstructures are investigated. The system of coordinates (x, y, z) used to identify the longitudinal axis of the realized fibers is also shown in the lower right corner.

(Continued)

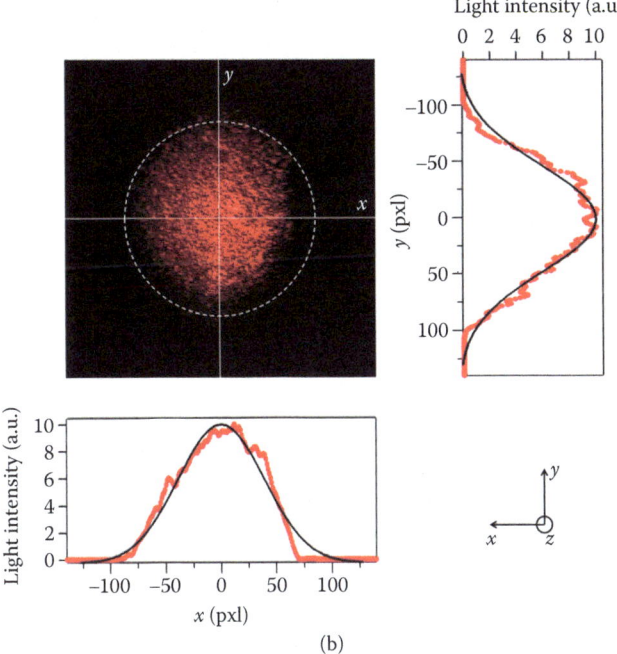

Figure 6.10 (*Continued*) (b) Multimode light spot (l5,633 nm) transmitted by a fiber. A laser is coupled into the fiber, and transmitted photons are collected through an objective lens (N.A. 0.10) and imaged by a CCD (bottom left and top right). Light intensity profiles (red dots), measured along the vertical and cross-section lines in (top right) and (bottom left), respectively, with corresponding best fits by Gaussian curves (superimposed lines). (From Polini, A. et al. *Sci. Rep.*, 2, 2012. With permission.[37])

fiber and silica profile compared with biological spicules. The optical properties of the biosilica fiber are similar to that of amorphous quartz ($n \sim 1.46$ at $\lambda = 633$ nm). The optical losses were improved over the biological fibers, 5–10 cm^{-1}, but still higher than conventional optical fibers. However, the optical properties are already good enough to be attractive for short-range communication applications, such as lab-on-chip systems and other microscale on-chip waveguides. There are several potential strategies to improve the fiber performance, and reduce the optical losses, such as improving the microfluidic process, improving the roughness of the surface, and adding cladding layers.

6.4 Reflectins

The Bobtail Squid are just about the cutest darn squid you have ever seen, as you can see from Figure 6.11.[49] Living in the very shallow waters off the shores of Hawaii, *Euprymna scolopes*, is a tiny fellow measuring only

(a)

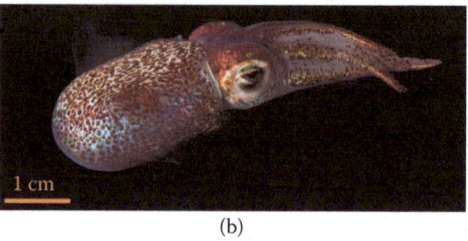

(b)

Figure 6.11 (a) Bobtail squid *Euprymna berryi*. (b) Hawaiian Bobtail Squid *Euprymna scolopes*. (From PLoS Biology, image courtesy of Chris Frazee and Margaret McFall-Ngai. With permission.[49])

a few centimeters long in adulthood.[50] This squid has a very active camouflage system that helps it escape predation during its short life. It has variably reflective tissues, statically reflective tissues, and bioluminescent light organs called photophores. These reflective tissues regulate the direction of reflected sunlight or self-generated bioluminescence. The light organs consist of a bed of symbiotic bioluminescent bacteria surrounded by silvery reflecting iridophores, the multilayer reflectors that have been described in Chapters 2 and 4. In the last decade, the composition of the reflector of *E. scolopes* has been examined, and it was discovered that within it was a previously unknown family of proteins.[51] These proteins have been named "reflectins." Since their discovery in the static reflectors of the bobtail squid, reflectins have been found in other squid such as the *Doryteuthis opalescens*, *Loligo pealeii*, and *Lolliguncula brevis*.[52–55] In *D. opalescens*, reflectins were found in both the dynamically adaptive iridophores, as well as the adaptive leucophores. Although the three-dimensional protein structure of the protein has not yet been reported, the mechanism behind its activation is beginning to become clear. In the iridosomes of the bobtail squid, the protein is assembled into platelets that act as thin-film reflectors, as shown in Figure 6.12.[52,53] In the active squid iridophores, acetylcholine (a neurotransmitter) release drives phosphorylation of the protein causing a reversible aggregation and conformation change. The altered spacing, thickness, and density strongly affect the RI of the material.

Our relatively new understanding of reflectins is incomplete, but there are already several approaches inspired by the interesting properties of the protein.[54,55,57] Scientists at Air Force Research Laboratory (AFRL) have studied the behavioral and self-assembling properties of recombinantly expressed reflectin *in vitro*, demonstrating that a variety of protein-based structures can be created.[54] Reflectin 1a was expressed and purified from *Escherichia coli*. The recombinantly expressed protein was then studied under a variety of processing conditions to attempt to understand the relationship

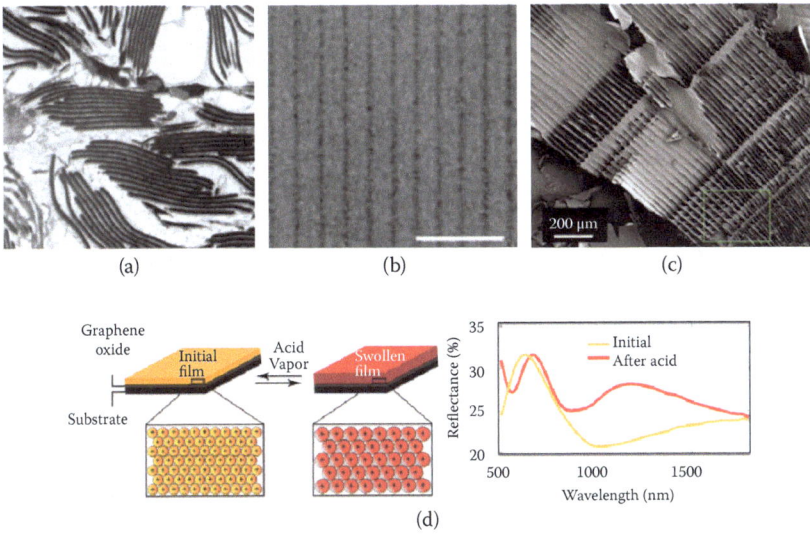

Figure 6.12 (a) Cross-sectional TEM of reflectin platelet stacks from the light organ of *Euprymna scolopes*. Scale bar: 5 μm. (b) Surface diffraction gratings produced from reflectin/ionic liquid thin films using dipping velocity of 15 mm/s. Scale bar: 20 μm. (c) SEM of a film of self-assembled lyophilized recombinant refCBA. (d) Illustration of acetic acid vapor-induced swelling for a film composed of net positively charged reflectin A1 nanoparticles (left). Effect of acetic acid vapor on the reflectance spectrum of a reflectin A1 film. Swelling of the film leads to the appearance of a large reflectance peak centered at 1200 nm. The curves have been smoothed for clarity (right). (From Kramer, R.M. et al., *Nat. Mater.*, 6(7), 533–538, 2007; Qin, G. et al., *J. Polym. Sci. Pt. B-Polym. Phys.*, 51(4), 254–264, 2013; and Phan et al. With permission.[54–56])

between protein structure and the assembly and material properties. In one example, iridescent surface diffraction gratings were assembled in a reflectin thin film that was created by dissolving reflectin in an ionic liquid. When the ionic liquid was removed from the film by submerging it in a water bath, grating structures spontaneously formed parallel to the dipping direction. SEM of a reflectin grating structure can be seen in Figure 6.12. The spacing of the grating depends on the velocity of the dip process. In addition to the gratings, thin films and fiber structures were also fabricated, demonstrating that reflectins have the potential to become a viable option for fairly simple bottom-up fabrication of photonic structures and optoelectronic devices. Understanding the assembly of reflectin proteins may also inform assembly of and act as fabrication templates for inorganic materials. Researchers at AFRL and colleagues from Tufts University have also created single and multilayer thin films from recombinantly expressed reflectin-based domain, a recombinant peptide, to understand the structural and photonic properties of this comparatively less-complex reflectin domain.

Researchers at the University of California, Irvine, have developed an approach for dynamical and reversible change in reflecting wavelength of a reflectin film for squid-inspired infrared camouflage.[56] Loliginid squids, such as *Loligo (Doryteuthis) pealeii* can change the optical reflectance of their skin over all visible wavelengths and out into the NIR (λ~800 nm). Thin films of reflectin A1 (expressed in *E. coli*) were deposited by blade coating (or doctor blading) the protein on top of a substrate of graphene oxide on silica. Two films were fabricated, one blue (125 nm thick and λ = 431 nm), and one orange (207 nm, and λ = 625 nm), and were fabricated. These films were exposed to vapors, such as humidity and other chemical stimuli, to swell the films and shift the reflectance to longer wavelengths. Concentrated acetic acid vapor exposure (Figure 6.12) caused a change in the reflectance wavelength out to 1200 nm. (This wavelength corresponds to a 394-nm film thickness.) The films were stable and the optical response could be shifted back and forth repeatedly between visible and IR reflectance, as many as 140 times. These films are comparatively easy to fabricate and offer the potential for dynamically tunable color on conformal or unconventional substrates. The approach offers an interesting proof of concept for infrared camouflage. The University of California,

Santa Barbara, in partnership with Raytheon, are interested in developing polymers that are inspired by the conformational changes seen in reflectin proteins on activation for polymer-based imaging components, hoping to avoid the large mechanical shutters and other bulky components and reduce the size, weight, and potentially power requirements versus conventional imaging systems.[58] To do this, the team is developing thin films of conjugated polymers that can change their morphology (and therefore RI) when activated electrochemically. Finally, researchers from the University of California, Irvine have reported that, in addition to its photonic properties reflectin is actually quite a good proton conductor as well, comparable with synthetic state of the art materials.[59] Proton conductors are important for many renewable energy and bioelectronics technologies, such as fuel cells, batteries and sensors, and protein-based conductors may launch a new class of organic bio-optoelectronic systems.

6.5 Luciferins and GFP—Bioluminescence and Fluorescence

Bioluminescence is the hauntingly beautiful phenomenon of light produced by living organisms. More than 700 different genera (plant and animal) have species that are bioluminescent, a few of which can be seen in Figure 6.13.[60,61] On land, organisms such as the firefly, railroad worms, and click beetles, fungus are bioluminescent.[64] However, far and above the number of bioluminescent land species, the ocean is where most of the bioluminescent species reside. Our identification, understanding, and development of the molecules and mechanisms of light production in these organisms have paved the way for a host of revolutionary applications, particularly in biology and medicine.

Organisms use bioluminescence to enhance their ability to find food, find mates, and avoid predation.[61] In oceanic species, it is thought that the prevalence of bioluminescence is due to adaptations necessary to overcome the challenging environment. In the open ocean, there are fewer environmental cues to locate prey, lack of light, and a distinct lack of places to hide. The bioluminescence may function to illuminate or lure prey. It may be used to temporarily blind or distract a predator. Some annelid worms (shown in Figure 6.13) discovered recently in the northeast and western Pacific Ocean can release bioluminescent "bombs" from their bodies, which can glow brightly for

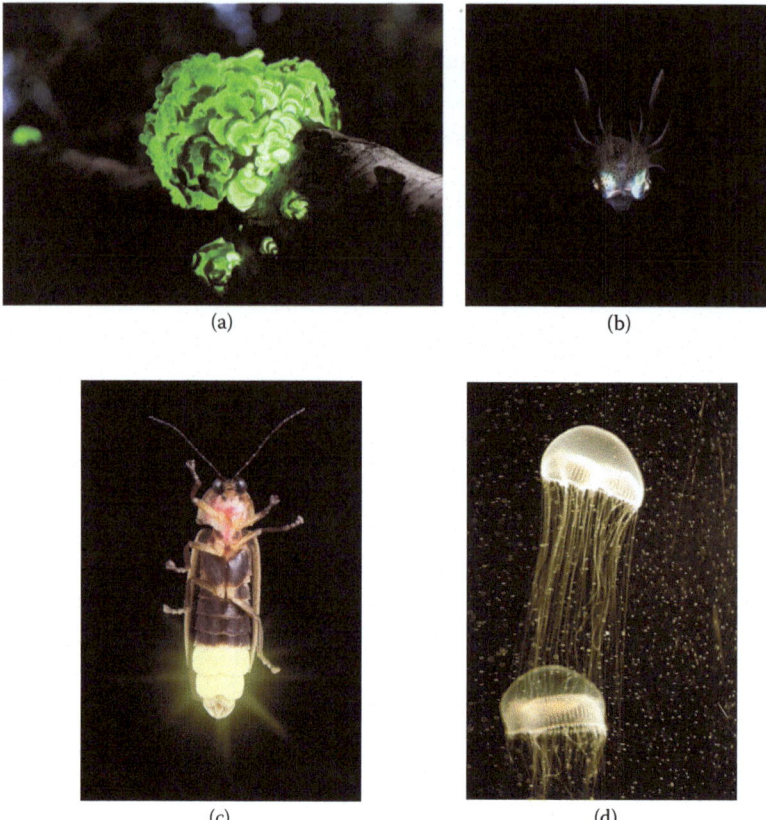

Figure 6.13 Examples of bioluminescent organisms. (a) The fungus *Panellus* stipticus. (b) Bioluminescent squid. (c) Firefly (d) *Aequorea victoria* (crystal jellies). (From Wikimedia commons.[63])

several seconds after release.[62] Behavior such as this may distract the predator and allowing the worm to escape. Similar behavior has been seen in the brittle star and some squid. Bioluminescence may be able to provide counter illumination. For example, the bioluminescence in sharks may be localized on the belly region, and is thought to provide counter shading and edge blurring for predators and prey that swim below. The emitted light from the belly matches the color and intensity of the downdwelling sunlight, which distorts or eliminates the animal's silhouette when seen from below.[65] Others, like the firefly, send species specific indicators; blinking patterns that attract mates of the right species.[66,67] The glow may also signal to predators that the organism will taste bad.

Bioluminescence is a chemiluminescent process; light is the result of a chemical reaction within the organism. The wavelengths of bioluminescence are solidly in the visible range, with no heat created in the chemical process. Bioluminescence, marine bioluminescence especially, comes in a tremendous variety of colors, patterns, and kinetics.[60,61,63] Colors range across the visible, even into the longer wavelengths that do not travel well in water. On land yellowish-green seems most common, and in the ocean the most common color is blue. Reds, oranges, yellows, and violets are much rarer. The prevalence of yellow/green on land may be due to its ability to stand out from leaves. In the ocean, blue travels well in seawater. Regarding kinetics, some organisms emit a persistent glow, whereas others have a high degree of control over the time and duration of emission. New imaging technologies are making a big impact on our ability to study these organisms in their natural habitats with minimal disturbance.

6.5.1 Luciferase and Luciferin

In the land insects and in most marine species, bioluminescence is the result of a chemical reaction between an enzyme, known as a luciferase, and its partner molecule, a luciferin. Luciferase in the presence of species-specific cofactors will catalyze its substrate (luciferin) and produce a luciferin molecule in an electronic excited state. When that excited electronic state relaxes back down to the ground state, that energy is released as light. The particular structure of the luciferin/luciferase molecules, as well as the necessary cofactors, is species specific, though there are examples of common luciferins. A list of some luciferin structures and the species that use them are shown in Figure 6.14. In fireflies, for example, the enzymatic reaction requires the presence of cofactors ATP, Mg^{2+}, and oxygen. The organism will either synthesize the necessary molecules, or extract them from the food they eat. The variation in the emitted wavelengths comes from the differences in the amino acid sequences in the varying luciferase molecules. Bioluminescent bacteria, such as the *Vibrio fischeri* that live in symbiosis with the bobtail squid (and are responsible for the squid's bioluminescent appearance), shares the gene sequence that encodes for luciferin/luciferase production, called the lux operon.[61,68]

Luciferins		
Chemical structure	Common descriptor	User groups
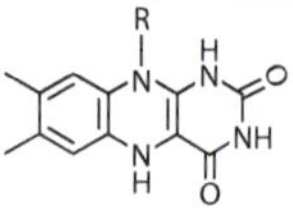	Bacterial	Bacteria some fish some squid pyrosomes?
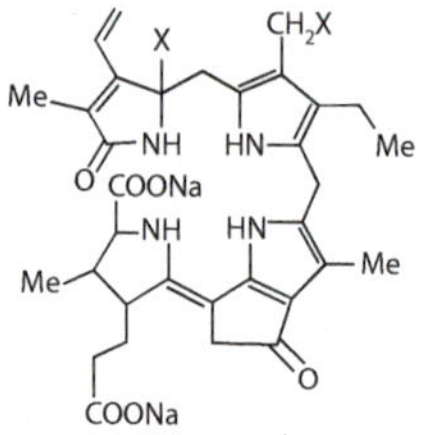	Dinoflagellate	Dinoflagellates euphausiid shrimp
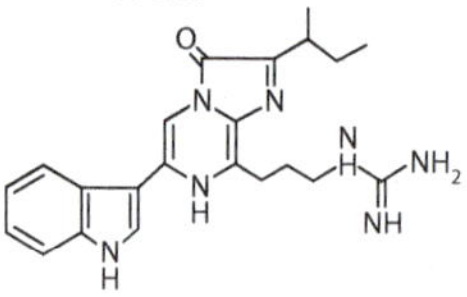	Cypridina	Some ostracods midshipman fish some other fish
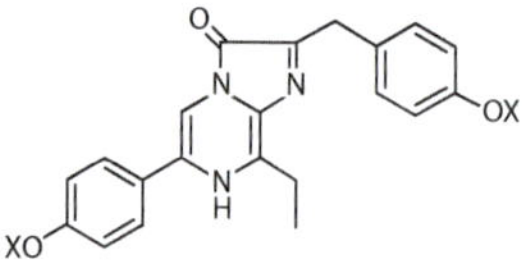	Coelenterazine	Radiolarians ctenophores cnidarians squid some ostracods copepods decopod shrimp mysid shrimp some ophiuroids chaetognaths larvaceans some fish
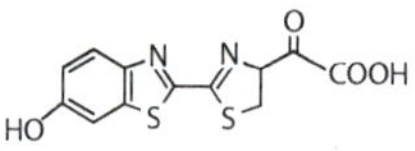	Firefly	Fireflies railroad worms click beatles
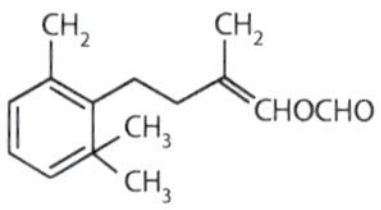	Latia	Limpet
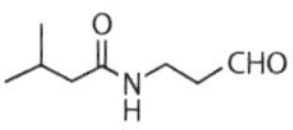	Diplocardia	Earthworm

Figure 6.14 Chemical structures of known luciferins and the taxonomic groups known to use them. (From Widder, E.A. and Falls, B., *IEEE J. Sel. Top. Quantum Electron.*, 20(2), 10, 2014. With permission.[61])

6.5.2 Green Fluorescent Proteins

Although most organisms create bioluminescence via the enzymatic reaction described above, some organisms create light via biolumines-cent photoproteins.[60,61] One example of this is the jellyfish *Aequorea victoria*, shown in Figures 6.13 and 6.15, which is also called the crystal jellie, and can be found in the Pacific, off of the coast of North America.[69] This jellie has a few hundred bioluminescent photoorgans on the edge of its umbrella, which gives off green light when stimulated. In *A. victoria*, two proteins are involved in the bioluminescence process: aequorin, which is a calcium-binding photoprotein, and an accessory protein called GFP, or green fluorescent protein. Images of *A. victoria*, aequorin, and GFP structures are shown in Figure 6.15. When calcium binds to the aequorin, the bound chromophore, the coelenterazine chromophore is oxidized into an excited state. When it relaxes back to the ground state, the protein gives off blue light at ~480 nm. (Coelenterazine is a luciferin, a substrate for luciferase reactions in other organisms.) The chromophore in GPF absorbs the energy directly from the aequorin-excited state via a process called bioluminescent resonance energy transfer and that energy is emitted from the GFP as green light (~509 nm). GFP is an 11-stranded β-barrel with an α-helix running through the center where the chromophore lies. GFP is extremely useful as it is very stable and entirely encoded in a single gene; it requires no other cofactors except molecular oxygen. Since GFPs discovery and characterization, the ability to isolate and

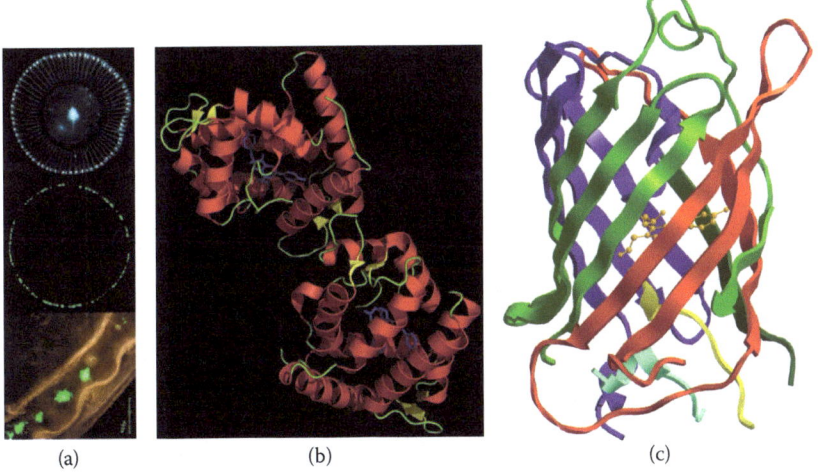

Figure 6.15 (a) *Aequorea victoria* under visible light (top), the whole jellyfish in the dark (middle), and photo organs (bottom). (b) The structure of aequorin. (c) The structure of GFP. (From Zimmer, M., *Chem. Soc. Rev.*, 38(10), 2823–2832, 2009 and Wikimedia commons, http://upload.wikimedia.org /wikipedia/commons/archive/6/6b/20090721180920%21Aequorin_1EJ3.png and http://lifesci.ucsb .edu/Bbiolum/. With permission.[69–71])

clone these other photoproteins and accessory proteins has led to a wide range of available fluorescent proteins for use in the laboratory, either from mutations of the original wild-type GFP, or discovered in other marine creatures, such as red from corals (Figure 6.16). Many of these mutations optimize not only for emission wavelength, but also for other beneficial properties, such as brightness, photostability, and excitation spectral characteristics.

6.5.3 Applications for Bioluminescence

It is hard to overstate the impact that the discovery and development of bioluminescent luciferin–luciferase and fluorescent proteins have had on science, particularly medical research and molecular biology.[69] The impact has been so profound that GFP's discovery and development won the scientists responsible the 2008 Nobel Prize in chemistry. These molecules are now extensively used as luminescent and fluorescent markers to probe cellular biology, gene expression, and disease progression. They can be used for monitoring protein–protein interactions, genetic inheritance, cell viability, cellular growth

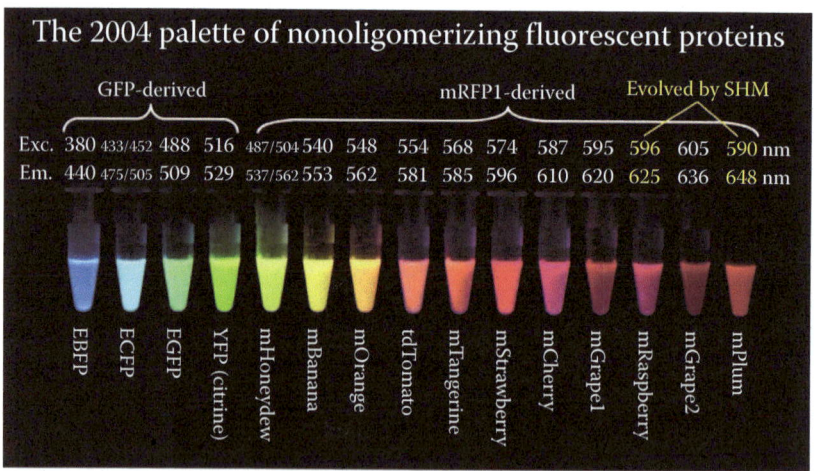

Figure 6.16 Monomeric and tandem dimeric fluorescent proteins derived from *Aequorea* GFP or *Discosoma* RFP, expressed in bacteria, and purified. This photo is a time exposure of fluorescence excited at different wavelengths and viewed through different cutoff filters. SHM, somatic hypermutation. (From Tsien, R.Y., *Angewandte Chemie-International Ed.*, 48(31), 5612–5626, 2009. With permission.[72])

and apoptosis, tumor growth, protein synthesis, neuronal behavior, diagnostics, and therapeutics development.[60,64,69,73] It has also been suggested that these molecules could be used to track and evaluate environmental pollution.[61]

GFP expression is entirely encoded in a single gene. Similarly, the bacterial lux operon encodes everything required of the luciferin/luciferase reaction. These systems are now being utilized as reporter genes that can get inserted into transfected DNA, near a target gene sequence. We can then identify whether the target gene sequence is being expressed as the two genes will be expressed together. If expression is successful, the reporter protein, for example, will fluoresce. Other luciferase genes can also be inserted, and subsequent gene expression can be identified by adding luciferin and its required cofactors.

Bioluminescent imaging can be used to track disease progression in cells and *in vivo* in small animal models.[64] Mammalian cells lack intrinsic bioluminescence and produce a very low optical background. (Although the human body does demonstrate some ultraweak photon emission.[74]) This low background means that for both *in vitro* and *in vivo*, bioluminescence is a useful and sensitive tool for imaging. In small animal models, the fluorescence produced by the fluorescent and bioluminescent markers can be externally detected. This enables rapid, real-time *in vitro* and *in vivo* identification, monitoring, tracking and analysis of disease, and infection development at molecular level, without sacrificing the animal. It can enable us to monitor and understand potential therapeutic efficacy. Fluorescent markers that emit in the longer wavelengths are particularly useful, as tissue is comparatively transparent to NIR light. These bioluminescent molecules are also being utilized in new super-resolution microscopy techniques.[69,75] In addition, transgenic animals such as mice, rats, dogs, cats, and marmosets, have been created for scientific study of disease processes such as Huntington's and AIDS.[77-78] Transgenic animals have also been made for other purposes. Glowing rabbit was created for use as social commentary in art and glowing fish are for sale in pet stores (see the Glofish in Figure 6.17).[79-81]

These systems are also used to provide insight into intracellular biology.[61] The luciferin/luciferase reactions rely on the presence of cofactors, and the light created in the enzymatic reaction will be

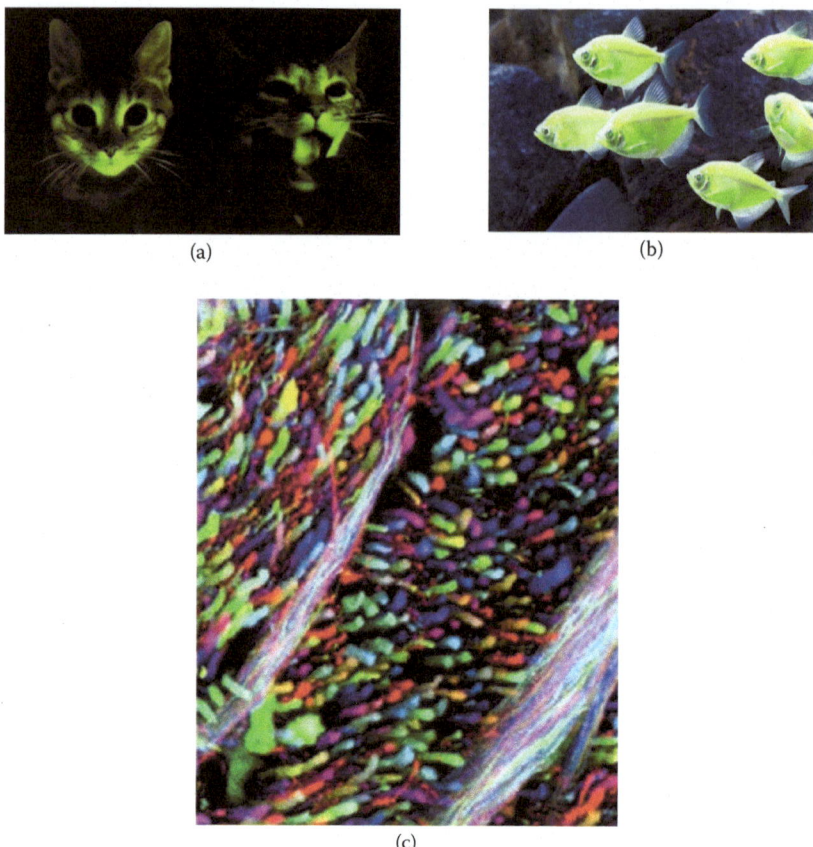

Figure 6.17 (a) Fluorescence image of RFP expression in two transgenic cloned cats 60 days after birth. (b) Transgenic glowing fish available in pet stores.(c) Multicolor neuronal labeling of an axon tract in the brainstem in Brainbow transgenic mice. In the Brainbow mice from which these images were taken, up to ~160 colors were observed as a result of the cointegration of several tandem copies of the transgene into the mouse genome and the independent recombination of each by Cre recombinase. The images were obtained by the superposition of separate red, green, and blue channels. (From Sasaki, E. et al., *Nature*, 459(7246), 523–U550, 2009; Lichtman, J.W. et al., *Nat. Rev. Neurosci.*, 9(6), 417–422, 2008; and Wikimedia commons, http://www.glofish.com/[glofish .com]. With permission.[76,78,82])

proportional in some manner to the amount of cofactor molecules available. If the target molecule we want to measure (e.g., enzyme or metabolite) is linked to a luciferin/luciferase cofactor, then we can use the luminescence as a measure of the quantity of the target molecule. One example given for this is the production of ATP by a kinase-catalyzed reaction. Firefly luciferin/luciferase reactions require ATP, and produce light in proportion to the amount of

ATP in the cell. So the light produced by the luciferin/luciferase reaction would, in this particular case, be proportional to the kinase available. There are a variety of luciferins available that require different cofactors, meaning that there are many molecules that can be used as probes. These assays have become a mainstay in the field due to their ease of use, speed, reproducibility, high sensitivity, and low cost.

XFPs, as the variants of GFP have been called, have been used to study the neural connections in the brain, by expressing different colored fluorescent proteins in individual neurons resulting in something now called the "Brainbow."[73,82] Multiple reporter genes with different wavelengths, cofactor requirements, kinetics, and so on, can be used simultaneously to track concurrent expression of more than one gene. In this work, researchers from Harvard and colleagues used a combinatorial approach to create expression of more than 100 colors in a single system. In this approach, genetic constructs were created and multiple copies inserted into the mouse genome to generate a random mixture of three fluorescent proteins at varying intensities in each neuron. Figure 6.17 shows an axon tract in the brainstem of a "brainbow" mouse with individual neurons randomly colored.

Another intriguing recent application of GFP has been its use as a gain medium for lasing in living cells and microorganisms. Harvard physicists transfected kidney cells to produce GFP.[83] To create the laser, a single cell was placed between two highly reflective Bragg reflectors (as shown in Figure 6.18), spaced approximately a cell's diameter apart (~20 μm). This microcavity was pumped through a microscope objective with pulsed (nanosecond) 465 nm light. The resulting emission showed clear narrow-band peaks, indicative of lasing behavior. Amazingly, these cells remained alive even after repeated stimulation at significantly higher power above the lasing threshold (50 nJ/pulse vs. ~850 pJ/pulse). Not long after, in vivo lasing in colonies of E. coli bacteria was demonstrated.[84] Stable E. coli bacteria were developed which could express GFP. A single colony was placed between highly reflective dielectric mirrors, forming a Fabry-Perot type cavity. Spacer beads prevented the bacteria from rupturing and controlled the cavity length. In this configuration, the cells were closely packed, forming a compact living gain medium. The GFP-laden bacteria were then pumped with nanosecond-pulses of 465 nm

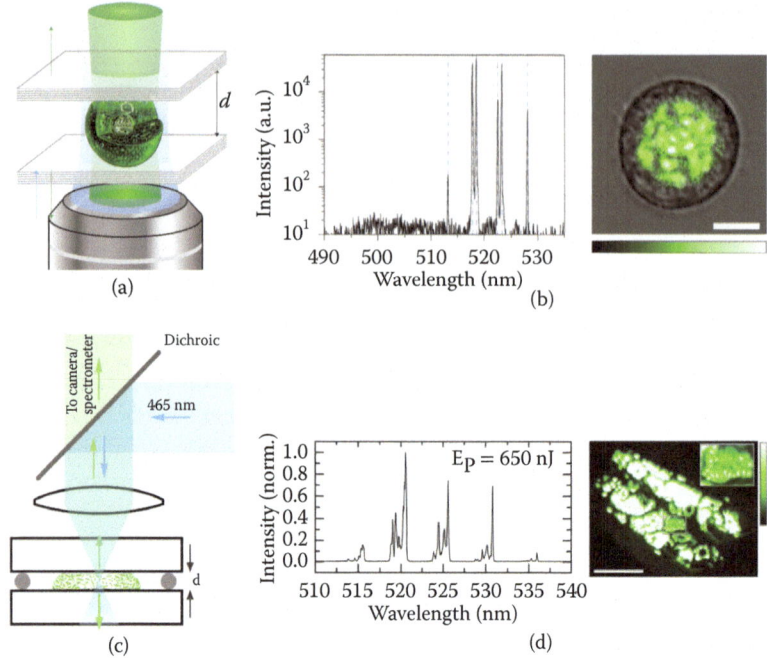

Figure 6.18 (a) Illustration of the single-cell laser. A live eGFP-expressing 293ETN cell is placed inside a high-Q resonator consisting of two DBRs ($d = 20\ \mu m$). (b) Emission pumped 3× above lasing threshold. The spectrum is plotted on a log scale to emphasize the contrast between the laser lines and the fluorescent background. Spatial patterns of the cell laser output above threshold ($E_p = 2$ nJ) superimposed on a DIC image of the cell (black and white channel). Scale bar: 5 µm. (c) Schematic of the bacteria laser and the measurement setup. 465 nm pump light is reflected and focused into a microcavity filled with GFP-expressing bacteria (cavity length, $d = 18\ \mu m$). Laser emission from the structure is collected through the same lens, transmitted though the dichroic mirror, and analyzed by a camera or spectrometer. (d) Output spectrum of bacteria laser well above threshold ($E_p = 650$ nJ). Emission pattern of bacteria laser pumped with an expanded excitation beam covering the entire field of view (pump beam diameter = 150 µm) pumped 3× above threshold. The inset is a zoom-in of the marked area in the center. Scale bar: 50 µm. (From Gather, M.C. and Yun, S.H., *Nat. Photonics*, 5(7), 406–410, 2011 and *Opt. Lett.*, 36(16), 3299–3301, 2011. With permission.[83,84])

light (from an optical parametric oscillator). Threshold behavior and discrete peaks in the emission spectrum indicated that lasing was indeed occurring, as shown in Figure 6.18. Without mirrors in place, no lasing behavior was observed.

Stimulated emission (lasing) from biological organisms and molecules is very intriguing. Bioinspired optofluidic lasers have also been demonstrated using luciferin as the gain medium.[85,86] Many different applications have been postulated. Optofluidic biolasers have been suggested as a highly sensitive way to measure changes in biological molecules.[85] Work in this direction may lead to laser light generated within living organisms, improved speed, and through put for analytical techniques, and improved imaging capabilities.[83] GFP lasers in cells may enhance the performance (sensitivity) of fluorescence-based biosensors.[87] Biolasers may also lead to implantation and integration into biological organisms for things such as sensing and targeted therapeutics.[84] Others have postulated that biological lasers may enable self-healing lasers, as well as bridge future hybrid optical and electronic devices, blurring the line between the biotic and abiotic.[88] There are many hurdles still to be overcome. First, green light does not travel far within tissue, potentially limiting their *in vivo* applicability. Second, these lasers currently require a significant number of external components to function (mirrors, spaces, optical pump, etc.). Despite this, it looks like a very intriguing path to follow.

6.6 Opsins

Opsins are the light-sensitive component of the photoreceptor systems (rods and cones) found in the retina of the human eye as well as those of other animals.[89] They are transmembrane proteins meaning that they reside in and span the thickness of the photoreceptor cell membrane in the outer segment of the rods and cones. The opsin proteins belong to a family of G-protein-coupled receptors and are covalently bound to a pigment, a retinal-based chromophore. Together they form the rhodopsin photoreceptor, whose structure can be seen in Figure 6.19. Incoming photons at the opsin's absorption wavelength trigger isomerization and conformational change within the rhodopsin, which triggers a subsequent molecular signaling cascade. Within the vertebrate retina, these pigment-bound opsins perform a crucial part of the image-forming process as well as nonimage-forming functions such as controlling circadian rhythm and pupillary reflex.[89] Opsins are not only found in retinas; evidence suggests that opsins are found in the skin of vertebrates and invertebrates, such as cephalopod, lizard, and even human skin.[92] More than a thousand animal opsins have been identified so far.[93] Animal opsins are classified as Type II opsins, whereas similar light-sensitive 7 transmembrane proteins found in prokaryotes (microbes) are classified as Type I. These microbial opsins have proven to be very interesting. The microbial opsins act as ion channels, ion pumps, light sensors, and more where they help carry out metabolic processes such as control proton gradients, maintain membrane potential, control ionic transport, and phototaxis.[92,93] There are multiple Type I opsins (such as those seen in Figure 6.19), for example, bacteriorhodopsin, which pumps protons out of the cytoplasm into the extracellular medium, and halorhodopsin, which pumps chloride

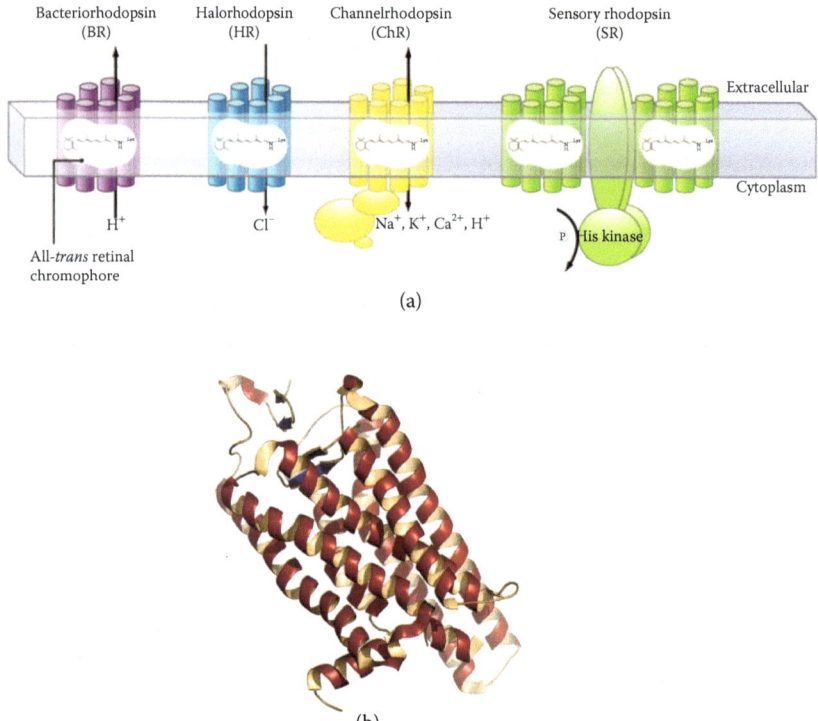

Figure 6.19 (a) Bacteriorhodopsins (and proteorhodopsins) pump protons from the cytoplasm to the extracellular medium, and halorhodopsins pump chloride into the cytoplasm; all three hyperpolarize the cell. Sensoryrhodopsins lack transmembrane ion transport in the presence of the His kinase transducer protein Htr; and algal channelrhodopsins conduct cations across the membrane in both directions but always along the electrochemical gradient of the transported ions. In sensoryrhodpsins and channelrhodopsins, proton translocation within the protein is linked to efficient photocycle progression, but these protons are not necessarily exchanged between the intra- and extracellular spaces. (b) Chemical structure representation of rhodopsin (visual purple) light perception protein showing the 7 transmembrane helices.

(*Continued*)

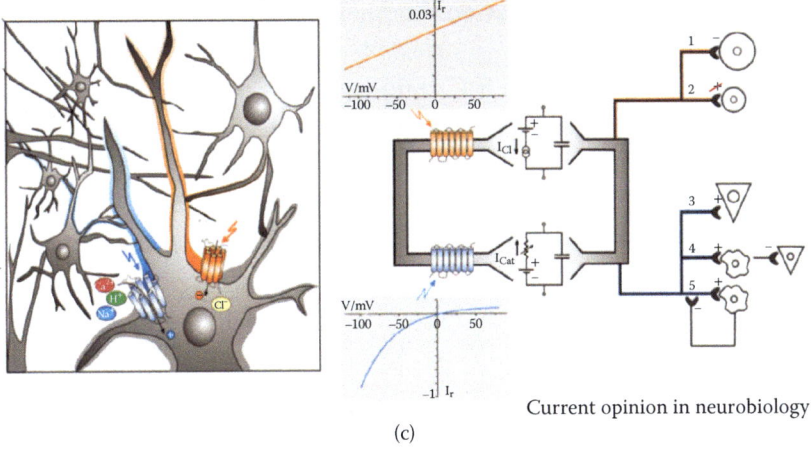

Figure 6.19 (*Continued*) (c) Optical control of neuronal circuits. Inset: Pictorial view of communicating nerve cells that are schematically depicted on the right side. The central one expresses either ChR2 (blue) or NpHR (orange). Blue light depolarizes the cell due to the cation flow through ChR2 (ICat), whereas hyperpolarization is caused by chloride pumping by NpHR (ICl). Representative current–voltage relationships for NpHR (top diagram with orange line) and for ChR2 (bottom diagram with blue line) are shown. Hyperpolarization may trigger OFF responses in retinal ganglions (1), or silence the NpHR expressing neuron (2), depolarization in a presynaptic cell can lead to the activation via synaptic transmission (3), or address feed-forward (4), or feedback inhibition (5). (From Bamann, C. et al., *Curr. Opin. Neurobiol.*, 20(5), 610–616, 2010 and Zhang, F. et al., *Cell*, 147(7), 1446–1457, 2011. With permission.[90,91])

into the cytoplasm. One very exciting use for opsins is as gene-based therapeutics for congenital disorders of the retina such as color blindness and for use as part of a retinal prosthetics approach.[94,95]

The function of the microbial opsins and other similar proteins has also inspired a revolution in the way we explore cellular behavior. In particular, they are having a tremendous impact on our ability to understand, for example, neuronal organization, function, and connectivity.[91,96] In the emerging area of optogenetics, the photoinduced-functionality of the microbial opsins has inspired a method of inducing specific actions within a targeted cell type. By delivering opsin genes into specific cells, and thereby introducing light-sensitive functionality to the cell, we can probe and control the cells' behavior. For example, channelrhodopsin is an opsin found in the unicellular algae *Chlamydomonas reinhardtii*.[90] This channelrhodopsin is sensitive to 470-nm light and can act as an inwardly rectifying cation channel. Halorhodpsin is an opsin found in the archaeum *Natronomonas pharaonas* that is sensitive to 580-nm light and acts as a Cl pump. When these two opsins are transfected into a mammalian cell, as shown in Figure 6.19c, light at 470 nm applied to the area can be used to depolarize the cell turning the cell "on," whereas light at 580 nm can hyperpolarize the cell turning it "off." There are several aspects of microbial-specific opsins that make them ideal for this application.[91] First, both the light-sensitive components and the final effector components of the protein are encoded within the same gene. These single-component systems are relatively compact and well suited for targeted delivery. Second, the microbial opsins have nearly no dark activity (meaning that activity will only be triggered when light of the specific wavelength is applied) and the response time once light is applied is very fast. The reversible structural changes that rhodopsin undergoes can happen as fast as tens of milliseconds. It is this rapid timescale and the high spatial resolution of the technique (by targeting delivery to specific cell types or specific cellular spaces), which makes these opsins a very powerful tool. Optogenetics is being used to examine and affect the behavior of neurons and neuronal circuitry, for example, by triggering or inhibiting neuronal signaling, both *in vitro* and *in vivo* in freely moving animals. This is a powerful tool to tease apart the complex, and intricate nature of neural (and non-neural) systems. Optogenetics is not limited to the use of microbial opsins or neuroscience, either.[97] Scientists in this field are also attempting to optimize, alter, or identify opsins with differing wavelength sensitivities and functionalities such as specific ion selectivity.[98–100] Other opsins, such as animal opsins, are being used (for example) to examine intracellular signaling. Chimeric rhodopsins, hybrids from multiple different microorganisms, are being created with specifically desired properties, such as tracking (fluorescence), kinetics, and light responsivity. The use of the optogenetics tool kit has also begun to expand beyond basic neuroscience, looking toward other areas such as therapeutics (for epilepsy or Parkinson's, for example), prosthetics, regenerative medicine, and stem cell research.[98]

References

1. M. D. Shawkey, N. I. Morehouse, and P. Vukusic, *J. R. Soc. Interface* **6 Suppl 2**, S221–231 (2009).
2. M. Kumar, *React. Funct. Polym.* **46** (1), 1–27 (2000).
3. H. Sashiwa and S.-i. Aiba, *Prog. Polym. Sci.* **29** (9), 887–908 (2004).
4. L. Upadhyaya, J. Singh, V. Agarwal, and R. P. Tewari, *Carbohydr. Polym.* **91** (1), 452–466 (2013).
5. L. H. Chen, T. Li, C. C. Chan, R. Menon, P. Balamurali, M. Shaillender, B. Neu, X. M. Ang, P. Zu, W. C. Wong, and K. C. Leong, *Sens. Actuators, B* **169** (0), 167–172 (2012).
6. C. Li Han, A. Xiu Min, C. Chi Chiu, M. Shaillender, B. Neu, W. Wei Chang, Z. Peng, and L. Kam Chew, *IEEE J. Select. Topics Quantum Electron.* **18** (4), 1464 (2012).
7. L. H. Chen, C. C. Chan, W. Yuan, S. K. Goh, and J. Sun, *Sens. Actuators, A* **163** (1), 42–47 (2010).
8. M. Iftekhar Shams, M. Nogi, L. A. Berglund, and H. Yano, *Soft Matter* **8** (5), 1369–1373 (2012).
9. P. Vukusic, J. R. Sambles, C. R. Lawrence, and R. J. Wootton, *Proc. R. Soc. B* **266** (1427), 1403–1411 (1999).
10. O. Hakimi, D. P. Knight, F. Vollrath, and P. Vadgama, *Compos., B* **38** (3), 324–337 (2007).
11. C. L. Craig and G. D. Bernard, *Ecology* **71** (2), 616–623 (1990).
12. C. L. Craig, G. D. Bernard, and J. A. Coddington, *Evolution* **48** (2), 287–296 (1994).
13. D. J. Little and D. M. Kane, *Opt. Lett.* **36** (20), 4098–4100 (2011).
14. D. J. Little and D. M. Kane, *Opt. Express* **19** (20), 19182–19189 (2011).
15. *Eco-friendly Optics: Spider Silk's Hidden Talents Brought to Light for Applications in Biosensors, Lasers, Microchips,* http://www.osa.org/en-us/about_osa/newsroom/newsreleases/2012/eco-friendly_optics_spider_silks_hidden/. (Angela Stark, The Optical Society, 2012).
16. B. D. Lawrence, J. K. Marchant, M. A. Pindrus, F. G. Omenetto, and D. L. Kaplan, *Biomaterials* **30** (7), 1299–1308 (2009).
17. http://www.wired.com/wiredscience/2009/09/spider-silk/.
18. F. G. Omenetto and D. L. KapLan, *Nat. Photonics* **2** (11), 641–643 (2008).
19. H. Tao, D. L. Kaplan, and F. G. Omenetto, *Adv. Mater.* **24** (21), 2824–2837 (2012).
20. B. D. Lawrence, M. Cronin-Golomb, I. Georgakoudi, D. L. Kaplan, and F. G. Omenetto, *Biomacromolecules* **9** (4), 1214–1220 (2008).
21. K. Tsioris, H. Tao, M. K. Liu, J. A. Hopwood, D. L. Kaplan, R. D. Averitt, and F. G. Omenetto, *Adv. Mater.* **23** (17), 2015–2019 (2011).
22. H. Tao, J. J. Amsden, A. C. Strikwerda, K. B. Fan, D. L. Kaplan, X. Zhang, R. D. Averitt, and F. G. Omenetto, *Adv. Mater.* **22** (32), 3527 (2010).
23. H. Tao, A. C. Strikwerda, M. K. Liu, J. P. Mondia, E. Ekmekci, K. B. Fan, D. L. Kaplan, W. J. Padilla, X. Zhang, R. D. Averitt, and F. G. Omenetto, *Appl. Phys. Lett.* **97** (26) (2010).

24. D. M. Lin, H. Tao, J. Trevino, J. P. Mondia, D. L. Kaplan, F. G. Omenetto, and L. Dal Negro, *Adv. Mater.* **24** (45), 6088 (2012).
25. S. Toffanin, S. Kim, S. Cavallini, M. Natali, V. Benfenati, J. J. Amsden, D. L. Kaplan, R. Zamboni, M. Muccini, and F. G. Omenetto, *Appl. Phys. Lett.* **101** (9) (2012).
26. H. Tao, J. M. Kainerstorfer, S. M. Siebert, E. M. Pritchard, A. Sassaroli, B. J. B. Panilaitis, M. A. Brenckle, J. J. Amsden, J. Levitt, S. Fantini, D. L. Kaplan, and F. G. Omenetto, *Proc. Natl. Acad. Sci. U.S.A.* **109** (48), 19584–19589 (2012).
27. H. Tao, M. A. Brenckle, M. M. Yang, J. D. Zhang, M. K. Liu, S. M. Siebert, R. D. Averitt, M. S. Mannoor, M. C. McAlpine, J. A. Rogers, D. L. Kaplan, and F. G. Omenetto, *Adv. Mater.* **24** (8), 1067–1072 (2012).
28. M. Cronin-Golomb, A. R. Murphy, J. P. Mondia, D. L. Kaplan, and F. G. Omenetto, *J. Polym. Sci. Pt. B-Polym. Phys.* **50** (4), 257–262 (2012).
29. R. Capelli, J. J. Amsden, G. Generali, S. Toffanin, V. Benfenati, M. Muccini, D. L. Kaplan, F. G. Omenetto, and R. Zamboni, *Org. Electron.* **12** (7), 1146–1151 (2011).
30. J. Zhang, E. Pritchard, X. Hu, T. Valentin, B. Panilaitis, F. G. Omenetto, and D. L. Kaplan, *Proc. Natl. Acad. Sci. U.S.A.* **109** (30), 11981–11986 (2012).
31. D. H. Kim, N. S. Lu, Y. G. Huang, and J. A. Rogers, *MRS Bull.* **37** (3), 226–235 (2012).
32. S. W. Hwang, D. H. Kim, H. Tao, T. I. Kim, S. Kim, K. J. Yu, B. Panilaitis, J. W. Jeong, J. K. Song, F. G. Omenetto, and J. A. Rogers, *Adv. Funct. Mater.* **23** (33), 4087–4093 (2013).
33. S. W. Hwang, G. Park, H. Cheng, J. K. Song, S. K. Kang, L. Yin, J. H. Kim, F. G. Omenetto, Y. G. Huang, K. M. Lee, and J. A. Rogers, *Adv. Mater.* **26** (13), 1992–2000 (2014).
34. T. I. Kim, J. G. McCall, Y. H. Jung, X. Huang, E. R. Siuda, Y. H. Li, J. Z. Song, Y. M. Song, H. A. Pao, R. H. Kim, C. F. Lu, S. D. Lee, I. S. Song, G. Shin, R. Al-Hasani, S. Kim, M. P. Tan, Y. G. Huang, F. G. Omenetto, J. A. Rogers, and M. R. Bruchas, *Science* **340** (6129), 211–216 (2013).
35. S. Kim, A. N. Mitropoulos, J. D. Spitzberg, H. Tao, D. L. Kaplan, and F. G. Omenetto, *Nat. Photonics* **6** (12), 817–822 (2012).
36. R.-H. Jin and J.-J. Yuan, in Learning from Biosilica: Nanostructured Silicas and Their Coatings on Substrates by Programmable Approaches, Advances in Biomimetics, edited by Marko Cavrak. (2011).
37. A. Polini, S. Pagliara, A. Camposeo, R. Cingolani, X. Wang, H. C. Schröder, W. E. G. Müller, and D. Pisignano, *Sci. Rep.* **2** (2012).
38. J. E. N. Dolatabadi and M. de la Guardia, *Trac-Trends Anal. Chem.* **30** (9), 1538–1548 (2011).
39. M. A. Ferrara, P. Dardano, L. De Stefano, I. Rea, G. Coppola, I. Rendina, R. Congestri, A. Antonucci, M. De Stefano, and E. De Tommasi, *PloS One* **9** (7) (2014).
40. M. P. Andrews, A. Hajiaboli, J. Hiltz, T. Gonzalez, G. Singh and R. B. Lennox, in *Photonic and Phononic Properties of Engineered Nanostructures*, edited by A. Adibi, S. Y. Lin and A. Scherer (2011), Vol. 7946.

41. R. Gordon, D. Losic, M. A. Tiffany, S. S. Nagy, and F. A. S. Sterrenburg, *Trends Biotechnol.* **27** (2), 116–127 (2009).
42. T. Fuhrmann, S. Landwehr, M. El Rharbi-Kucki, and M. Sumper, *Appl. Phys. B-Lasers Opt.* **78** (3–4), 257–260 (2004).
43. L. De Stefano, P. Maddalena, L. Moretti, I. Rea, I. Rendina, E. De Tommasi, V. Mocella, and M. De Stefano, *Superlattices Microstruct.* **46** (1–2), 84–89 (2009).
44. L. De Stefano, I. Rea, I. Rendina, M. De Stefano, and L. Moretti, *Opt. Express* **15** (26), 18082–18088 (2007).
45. L. De Stefano, L. Rotiroti, M. De Stefano, A. Lamberti, S. Lettieri, A. Setaro, and P. Maddalena, *Biosens. Bioelectron.* **24** (6), 1580–1584 (2009).
46. N. Kroger and N. Poulsen, in *Annual Review of Genetics* (Annual Reviews, Palo Alto, 2008), Vol. 42, pp. 83–107.
47. V. C. Sundar, A. D. Yablon, J. L. Grazul, M. Ilan, and J. Aizenberg, *Nature* **424** (6951), 899–900 (2003).
48. J. Aizenberg, J. C. Weaver, M. S. Thanawala, V. C. Sundar, D. E. Morse, and P. Fratzl, *Science* **309** (5732), 275–278 (2005).
49. http://www.plosbiology.org/article/info%3Adoi%2F10.1371%2Fimage.pbio.v12.i02, Image licensed under creative commons 3.0, http://creativecommons.org/licenses/by/3.0/.
50. M. McFall-Ngai, *Curr. Biol.* **18** (22), R1043–R1044 (2008).
51. W. J. Crookes, L.-L. Ding, Q. L. Huang, J. R. Kimbell, J. Horwitz, and M. J. McFall-Ngai, *Science* **303** (5655), 235–238 (2004).
52. D. G. DeMartini, A. Ghoshal, E. Pandolfi, A. T. Weaver, M. Baum, and D. E. Morse, *J. Exp. Biol.* **216** (19), 3733–3741 (2013).
53. M. Izumi, A. M. Sweeney, D. DeMartini, J. C. Weaver, M. L. Powers, A. Tao, T. V. Silvas, R. M. Kramer, W. J. Crookes-Goodson, L. M. Mathger, R. R. Naik, R. T. Hanlon, and D. E. Morse, *J. R. Soc. Interface* **7** (44), 549–560 (2010).
54. R. M. Kramer, W. J. Crookes-Goodson, and R. R. Naik, *Nat. Mater.* **6** (7), 533–538 (2007).
55. G. Qin, P. B. Dennis, Y. Zhang, X. Hu, J. E. Bressner, Z. Sun, W. J. Crookes-Goodson, R. R. Naik, F. G. Omenetto, and D. L. Kaplan, *J. Polym. Sci. Pt. B-Polym. Phys.* **51** (4), 254–264 (2013).
56. L. Phan, W. G. Walkup, D. D. Ordinario, E. Karshalev, J.-M. Jocson, A. M. Burke, and A. A. Gorodetsky, *Adv. Mater.* **25** (39), 5621 (2013).
57. A. R. Tao, D. G. DeMartini, M. Izumi, A. M. Sweeney, A. L. Holt, and D. E. Morse, *Biomaterials* **31** (5), 793–801 (2010).
58. http://www.raytheon.com/technology_today/2012_i1/infrared.html and in Technology Today - highlighting Raytheon's technology, 2012: Taking our Cue from Nature: Bio-inspired Shutters and Apertures for Infrared Imaging Applications, http://www.icb.ucsb.edu/research/photonic-and-electronic-materials/bio-inspired-electrically-switchable-polymer-based.
59. D. D. Ordinario, L. Phan, W. G. Walkup, J.-M. Jocson, E. Karshalev, N. Huesken, and A. A. Gorodetsky, *Nat. Chem.* **6** (7), 597–603 (2014).
60. E. A. Widder, *Science* **328** (5979), 704–708 (2010).

61. E. A. Widder and B. Falls, *IEEE J. Sel. Top. Quantum Electron.* **20** (2), 10 (2014).
62. K. J. Osborn, S. H. D. Haddock, F. Pleijel, L. P. Madin, and G. W. Rouse, *Science* **325** (5943), 964–964 (2009).
63. *"PanellusStipticusAug12 2009"* by Ylem—Own work. Licensed under Public domain via Wikimedia Commons, http://commons.wikimedia.org/wiki/File:PanellusStipticusAug12_2009.jpg-mediaviewer/File:PanellusStipticusAug12_2009.jpg.
64. S. Hosseinkhani, *Cell. Mol. Life Sci.* **68** (7), 1167–1182 (2011).
65. J. M. Claes, H. C. Ho, and J. Mallefet, *J. Exp. Biol.* **215** (10), 1691–1699 (2012).
66. H. Ghiradella and J. T. Schmidt, *Integr. Comp. Biol.* **44** (3), 203–212 (2004).
67. M. Gronquist, F. C. Schroeder, H. Ghiradella, D. Hill, E. M. McCoy, J. Meinwald, and T. Eisner, *Chemoecol.* **16** (1), 39–43 (2006).
68. M. J. McFall-Ngai, *Ann. Rev. Ecol. Syst.* **30**, 235–256 (1999).
69. M. Zimmer, *Chem. Soc. Rev.* **38** (10), 2823–2832 (2009).
70. C. M. M. a. S. H. D. Haddock and J. F. Case, *"The Bioluminescence Web Page,"*. http://lifesci.ucsb.edu/Bbiolum/.
71. "Aequorin 1EJ3" by Yikrazuul—Own work by uploader; PMID 10830969. Licensed under Creative Commons Attribution-Share Alike 3.0 via Wikimedia Commons, http://upload.wikimedia.org/wikipedia/commons/archive/6/6b/20090721180920%21Aequorin_1EJ3.png.
72. R. Y. Tsien, *Angewandte Chemie-International Ed.* **48** (31), 5612–5626 (2009).
73. J. M. Perkel, *Science* **339** (6117), 350–352 (2013).
74. M. Kobayashi, D. Kikuchi, and H. Okamura, *PloS One* **4** (7) (2009).
75. P. Sengupta, S. Van Engelenburg, and J. Lippincott-Schwartz, *Dev. Cell* **23** (6), 1092–1102 (2012).
76. E. Sasaki, H. Suemizu, A. Shimada, K. Hanazawa, R. Oiwa, M. Kamioka, I. Tomioka, Y. Sotomaru, R. Hirakawa, T. Eto, S. Shiozawa, T. Maeda, M. Ito, R. Ito, C. Kito, C. Yagihashi, K. Kawai, H. Miyoshi, Y. Tanioka, N. Tamaoki, S. Habu, H. Okano, and T. Nomura, *Nature* **459** (7246), 523–U550 (2009).
77. C. Combs, National Geographic, http://news.nationalgeographic.com/news/2009/05/photogalleries/glowing-animal-pictures/, 2009.
78. X. J. Yin, H. S. Lee, X. F. Yu, E. Choi, B. C. Koo, M. S. Kwon, Y. S. Lee, S. J. Cho, G. Z. Jin, L. H. Kim, H. D. Shin, T. Kim, N. H. Kim, and I. K. Kong, *Biol. Reprod.* **78** (3), 425–431 (2008).
79. K. Philipkoski, in *Wired.com* (2002).
80. Glowingpets.com.
81. http://www.glofish.com/.
82. J. W. Lichtman, J. Livet, and J. R. Sanes, *Nat. Rev. Neurosci.* **9** (6), 417–422 (2008).
83. M. C. Gather and S. H. Yun, *Nat. Photonics* **5** (7), 406–410 (2011).
84. M. C. Gather and S. H. Yun, *Opt. Lett.* **36** (16), 3299–3301 (2011).
85. X. Fan and S.-H. Yun, *Nat. Methods* **11** (2), 141–147 (2014).

86. X. Wu, Q. Chen, Y. Sun, and X. Fan, *Appl. Phys. Lett.* **102** (20) (2013).
87. A. Jonas, M. Aas, Y. Karadag, S. Manioglu, S. Anand, D. McGloin, H. Bayraktar, and A. Kiraz, *Lab Chip* **14** (16), 3093–3100 (2014).
88. J. Cartwright, Science Magazine, ScienceNOW, 2011.
89. B. Nickle and P. R. Robinson, *Cell. Mol. Life Sci.* **64** (22), 2917–2932 (2007).
90. C. Bamann, G. Nagel, and E. Bamberg, *Curr. Opin. Neurobiol.* **20** (5), 610–616 (2010).
91. F. Zhang, J. Vierock, O. Yizhar, L. E. Fenno, S. Tsunoda, A. Kianianmomeni, M. Prigge, A. Berndt, J. Cushman, J. Polle, J. Magnuson, P. Hegemann, and K. Deisseroth, *Cell* **147** (7), 1446–1457 (2011).
92. L. M. Mathger, S. B. Roberts and R. T. Hanlon, Biology letters **6** (5), 600-603 (2010).
93. A. Terakita, *Genome Biol.* **6** (3) (2005).
94. S. Nirenberg and C. Pandarinath, *Proc. Natl. Acad. Sci. U.S.A.* **109** (37), 15012–15017 (2012).
95. M. Neitz, J. Neitz, K. Mancuso, J. Kuchenbecker, and T. B. Connor, *Faseb J.* **26** (2012).
96. O. Yizhar, L. Fenno, F. Zhang, P. Hegemann, and K. Diesseroth, *Cold Spring Harbor Protoc.* **2011** (3), top102 (2011).
97. D. Tischer and O. D. Weiner, *Nat. Rev. Mol. Cell Biol.* **15** (8), 551–558 (2014).
98. M. Oheim, M. van 't Hoff, A. Feltz, A. Zamaleeva, J. M. Mallet, and M. Collot, *Biochim. Biophys. Acta-Mol. Cell Res.* **843** (10), 2284–2306 (2014).
99. M. L. Rein and J. M. Deussing, *Mol. Genet. Genomics* **287** (2), 95–109 (2012).
100. A. S. Chuong, M. L. Miri, V. Busskamp, G. A. C. Matthews, L. C. Acker, A. T. Sorensen, A. Young, N. C. Klapoetke, M. A. Henninger, S. B. Kodandaramaiah, M. Ogawa, S. B. Ramanlal, R. C. Bandler, B. D. Allen, C. R. Forest, B. Y. Chow, X. Han, Y. Lin, K. M. Tye, B. Roska, J. A. Cardin, and E. S. Boyden, *Nature Neurosci.* **17** (8), 1123–1129 (2014).

7
SENSORS

7.1 Introduction

All creatures interact with the world in some manner. Our senses enable us to capture vast amounts of information to which we are constantly exposed. These complex, sophisticated systems help us take in information about our environment. Along with the signal processing units in our peripheral and central nervous system, we manage the data that drive our behavior, growth, and survival as individuals and as a species. Humans have significantly more than five senses. Beyond sight, taste, smell, pressure (touch), and sound, we have senses of time, pain, proprioception, temperature, hunger, and more. Animals have many variations on the senses we share, and some have sensing capabilities humans do not. Their senses operate at different wavelengths, different sensitivities, and via different structures with different transduction principles. Squid, spiders, and many insects can sense the polarization of light, as has been discussed earlier in Chapters 4 and 5.[1,2] Some birds, insects, and mammals are able to sense magnetic field directions via specific biological magnetoreceptors, which are used for navigation.[3,4] In each organism, evolutionary pressure has driven our bodies to prioritize certain sensing modalities (certain types of information capture) over others.

Sensors are our technological attempt to mimic senses and expand our ability to monitor and adjust to our environment. Man-made sensors have become critical and pervasive in modern human life with an ever-increasing number and variety, particularly in the last decades. We monitor our buildings and roadways for structural safety, we monitor our population for security, and we monitor ourselves for health and well-being. Sensors are everywhere from our gardens to our cars; our cell phones to our bodies; and the market for sensors continues to grow. In the United States alone, sales of sensors are in the tens of billions of dollars.[5] One of the fastest growing segments seems to be personal health monitoring.[6] But the market is increasing in almost every space. Globally, the sensor market is estimated to reach nearly $100 billion by 2016, with biological and chemical sensors expected to see the highest growth.[7]

This increase in use, application, and market for sensors pushes us to constantly improve existing paradigms and architectures, in addition to identifying completely new sensor technologies. Many improvements have occurred to enable modern sensors that are small, inexpensive, and robust. The advent of the smartphone, for example, combined with rapid advancements in the performance capabilities of sensor technologies, has led to a wider variety of sensor systems either included in—or sold as add-ons to—our existing devices. Along with their ubiquity comes a host of widely varying system requirements. As in most modern device applications, the requirements for the sensor not only include its sensing performance (sensitivity, noise, error, etc.) but also additional requirements, such as size, weight, power consumption, and flexibility.

Unfortunately, the path to improving sensor performance is not always easy or straightforward. The single ideal sensor, one with near-zero size, weight, and power requirements with perfect performance parameters and a sensing mechanism that causes absolutely no effect on the object it is measuring, is impossible to attain.[8] No one sensor fits all applications. Different applications have varying performance requirements where they can accept compromises between performance characteristics. Changes in the sensor design have a profound effect on its performance. There are many trade-offs that must be made and these performance parameters are often not independent variables. One cannot simply reduce the size of the sensor, for example, and expect the performance to remain the same, let alone improve because the behavior of materials changes below certain length scales. Between competing requirements, any given application will prioritize their metrics differently. As the sensor size is reduced, some parameters become significantly more influential than others.

Sensor development faces challenges in fabrication, sensitivity, noise, integration, and more and novel ideas, novel structures, novel materials, and novel approaches are required to move the field forward. In our desire for fresh ideas and insights, we look to nature to inspire solutions to some of our technical challenges. The biological sensory capabilities of insects, for example, are very different from human sensory organs, qualitatively and quantitatively. We can find examples of mechanoreceptors, photoreceptors, chemoreceptors, and thermoreceptors, to name a few. They see in different wavelengths, hear in different audio bands, and sense movement in ways we do not. Biological sensors offer a host of potential benefits from their design, simplicity, robustness, as well as their efficiency of fabrication and function. We can learn lessons from nature in integration, redundancy, and fault tolerance. Integration in nature far exceeds that of human-made systems. We are not yet able to make completely integrated single-material–based sensor systems. Animals and plants often demonstrate duplicative, overlapping, and distributed sensor capabilities: sensors with different mechanisms, partially overlapping ranges for more reliable information, and system robustness. Different processes make it difficult to compare biological and classical miniature sensors. It is not straightforward to determine which sensor is "better" in all cases. This would require a "perfect metric," and it can be difficult to quantify features such as robustness.[8]

There have been many developments and approaches in bioinspired photonic sensors in a variety of sensing modalities: thermal, chemical sensing, pressure sensing, ion, and pH sensing, for example.[9] In this chapter, we focus on two general examples: inspiration from the sensor itself and inspiration from a natural photonic structure that we adapt. We will also focus on two specific areas of optical sensing: thermal (infrared) and gas/vapor sensing.

7.2 Infrared Sensing

Infrared (IR) radiation covers a very broad span of wavelength ranges, extending out from ~1 to 300 µm. When discussing IR radiation, we often split the large IR band into several smaller discrete bands: near IR (NIR) = <2 µm, short-wave IR (SWIR) = 2–3 µm, mid-wave IR (MWIR) = 3–5 µm, long-wave IR (LWIR) = 8–12 µm, very long-wave IR (VLWIR) = 12–25 µm, and far IR (FIR) >25 µm. The mid- and long-wave bands are interesting because those are the wavelengths of body and environmental heat emission, combined with a clear atmospheric transmission window at these bands. Sensors capable of capturing information about IR radiation at these wavelengths can be used, for example, in imagers to visualize the heat emitting from these bodies or emissions from more coherent IR light sources, such as otherwise nonvisible lasers. Near IR is interesting because, among other reasons, it can transmit through biological tissue more deeply than other wavelengths. For a long time, IR imaging and thermal detection have been used primarily for military applications, in part because of materials requirements, the difficulty of fabrication, and high cost. However, it has begun to expand beyond the exclusively military market into more commercial markets. This technology extends our human vision beyond the visible wavelength ranges, enhances our ability to see in the dark, and enables navigation through blinding dust or smoke. It enables reconnaissance and targeting applications. It also allows us to detect and identify chemicals to some extent, as most chemical species have spectral signatures in the IR due to fundamental absorption processes associated with vibration states of the molecules.[10] More recently, as new lower cost, more robust sensors have been developed, the market for thermal imaging is growing and broadening. Application of the technology has extended into medical imaging, astronomy, security, firefighting, building inspection (to see heat lost), surveillance and industrial imaging, and pollution monitoring.[11] It is being used in night vision option on automobiles.[12]

Most of the effective IR sensors today rely on photon capture/detection. Mercury cadmium telluride (HgCdTe) is one example of a material system commonly used as a photodetector for IR sensors. However, HgCdTe is expensive, and fabrication is a significant challenge as the material is brittle, and processing can result in toxic by-products. In addition many of these technologies, such as the HgCdTe photodetectors, require expensive cryogenic cooling to eliminate noise and enhance performance. There is a push to identify approaches for uncooled, lower cost IR imaging sensors to enable larger format, multicolor sensors capable of operating at higher, often ambient, temperatures. Some examples of uncooled approaches are resistive bolometry, pyrolelectric sensing, thermoelectric sensing, and microcantilever (thermomechanical) bimorphs. The excellent performance of existing cooled bolometers [35–200 mK noise equivalent temperature difference (NETD, see Box 7.1)] comes at the expense of complexity of design and is still below the record values of nature.[11,13]

BOX 7.1 NOISE EQUIVALENT TEMPERATURE DIFFERENCE

The ultimate sensitivity of a thermal detector can be described as it is NETD. This is the temperature difference where the signal can no longer be distinguished from the background noise, which is a constant challenge in sensor operation. For a signal to be detectable, the temperature or temperature difference must be higher than the NETD.

As this technology develops, we attempt to optimize existing capabilities such as power density, stability, responsiveness, sensitivity to thermal variation, resolution, and bandwidth in a system that avoids process and design complexity. We also are looking to design in additional capabilities, such as conformal designs, TV rate scanning speeds, broadband operation, 4Pi field of view, and visual readout with the unaided human eye.[13] We want to combine multiple functionalities in the same system: color, polarization, gain, extended wavelengths, and so on.[10] We look for innovative designs that overcome the limitations of the existing technologies and the thermal responsiveness of natural structures are attracting significant interest.

There are many animals and insects that have the ability to sense IR radiation at some level, more specifically heat. We humans, for example, have a distributed system of heat receptors in our epidermis and dermis.[14] These heat receptors are unspecialized free nerve endings that detect changes in temperature over the steady state. That is, we can detect changes in temperature, but not exact absolute temperature. We also have thermal sensing capability in other areas, such as our gastrointestinal tract that will cause specific sensory and reflex response when we detect warm or cold stimulus. For example, our stomach contracts reflexively when our abdomen senses cold internally, but relaxes when we sense warmth. However, our human thermoreceptors are not specific; they provide only a relative detection capability. Other animals, such as bats, snakes, and beetles, have developed specific specialized organs that give them the ability to detect changes in thermal energy better than humans. They combine this input with other sensory systems (such as visible imaging, chemical detection, or mechanical information) for an effective strategy to feed, hunt, mate, and survive.

7.2.1 Snakes

Certain snakes, such as those shown in Figure 7.1, have the ability to use IR detection to help target prey and avoid predation: crotaline snakes (Crotalinae subfamily), such as rattlesnakes, cottonmouths,

Figure 7.1 Examples of snakes capable of directional sensing infrared (IR) light via facial pit organs: (a) Rattlesnake, (b) Cottonmouth, (c) Mexican Rattler *Crotalus basiliscus*, (d) Palm viper *Bothriechis schlegelii*, (e) Green pit viper *Trimeresurus albolabris*, (f) Big eyed pit viper *Cryptelytrops macrops*.

and copperheads; and some boid snakes (Boidae family) such as the python.[14] Members of both families have pit organs that enable them to image the radiant heat emitted by warm-blooded prey. The organs have spectral sensitivities between 3–5 μm and 8–12 μm coinciding with the IR emission of their targeted prey and the relatively clear atmospheric windows. Both are able to detect stimuli warmer or colder than the background temperature independent of the ambient temperature change or the snakes own body temperature. Response in the pit organs trigger and guide the snakes strike toward its prey. It is speculated that they may be able to detect IR much like an eye, with high spatial accuracy; however, this is unclear and recent studies imply that it is less spatially specific although still highly sensitive.[15,16]

BOX 7.2 LIGHT TRAVELING IN THE ATMOSPHERE

Electromagnetic radiation (light) traveling through a medium (air, water, glass, for example) can be absorbed or scattered by the medium. The absorption and scattering gradually weakens (attenuates) the radiation. This attenuation describes how far the light can be transmitted within the medium before it has too low an intensity to be detected (extinction). Transmission is described by the Beers-Lambert law, and can generally be said to drop exponentially with distance. In air, light attenuation will depend on the air quality, humidity, weather, and wavelength. Visible wavelengths travel well on a clear day, but fog, smoke, dust, or air pollution are all particulates in the atmosphere and the scattering and absorption of the light off of these particulates result in reduced visibility. When light of a given range of wavelengths is able to be transmitted through the atmosphere relatively freely with little scattering or absorption, that is, the atmosphere is relatively transparent, the band of wavelengths is called an atmospheric window. These windows, as well as the amount of attenuation you can expect, and hence, the distance that the light can travel, change with atmospheric conditions.

These IR detectors come in the form of pit organs; small cavities located on the face (facial pits) or around the mouth (labial pits) of the snake.[14] Crotaline snakes have two facial pits, one on either side of the face between the eye and the nostril, hence their alternate name: pit vipers. The red arrow in Figure 7.2 shows the facial pit on the head of the Pacific rattlesnake *Crotalus o. oreganus* Holbrook 1840. The rattlesnake pit organ is highly sensitive and able to detect changes in temperature smaller than 0.003°C. Pit vipers also have additional thermoreceptors inside their mouth, in the oral mucosa, which helps them aim when their mouth is open and their fangs are extended.

The pit cavities are open cavities, approximately 1–5 mm in diameter. Inside the facial pit cavity in the crotaline snakes is a thin membrane (15 μm) that is separated from the bottom of the cavity. An illustration of the architecture and a cross section of the rattlesnake facial pit organ can be seen in Figure 7.3. Within this membrane lies a high concentration of nerve fibers (6000–7000 myelinated fibers), which branch and bend and terminate near the outer membrane surface. The terminated nerve endings are sensitive IR receptors and as a result the membrane is extremely sensitive to IR radiation. The nerve bundles are surrounded by mitochondria and are protected and fed via a dense bed of capillaries that supply blood for cooling, energy,

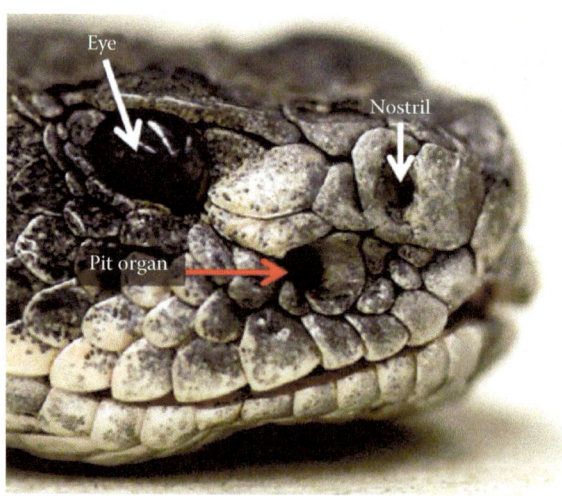

Figure 7.2 The head of a western diamondback rattlesnake *Crotalus atrox* Baird and Girard 1853 showing the location of the facial pit organ. (From Bakken, G.S. and Krochmal, A.R., *J. Exp. Biol.*, 210(16), 2801–2810, 2007.[16])

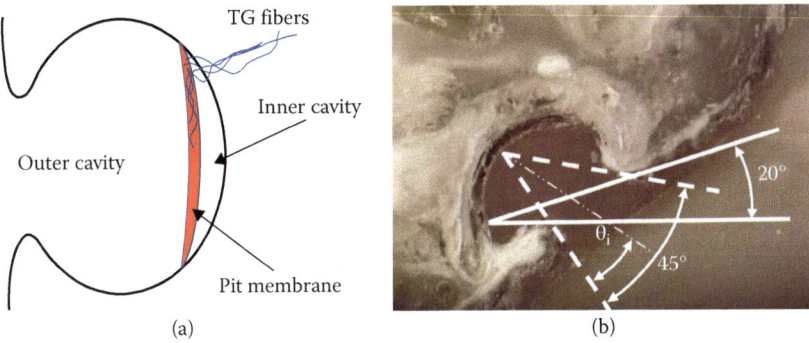

Figure 7.3 (a) Illustration of the pit organ structure showing innervated pit membrane and air gap structure. (b) Cross section of the pit organ of a Pacific rattlesnake *Crotalus o. oreganus* Holbrook 1840. The anterior air chamber was filled with red acrylic before the entire head was infiltrated and embedded. The angular aperture θ_i varies from 23° (including angle 45°) laterally to 10° (including angle 20°) when looking ahead and to the contralateral side. (From Bakken, G.S. and Krochmal, A.R., *J. Exp. Biol.*, 210(16), 2801–2810, 2007 and Gracheva, E.O. et al., *Nature*, 464(7291), 1006–U1066, 2010. With permission.[16,19])

and oxygenation. The nerve system seems to feed into the visual portions of the snake brain, potentially integrating the IR and visual sensory inputs.

In addition to the sensitive membrane, the physical shape and structure of the pit organ (its morphology) play a strong role in its performance. The facial pit organ is slightly asymmetric. The IR sensitive membrane is quite close to the surface of the epithelium (only 2–30 µm below the surface) and so less heat is lost before reaching the thermoreceptors. The membrane is also thermally isolated by an air gap on both sides, which significantly reduces heat loss to its surroundings. The ultrastructure of the surface of the pit organ scatters and reflects electromagnetic radiation at certain wavelengths. The epidermal surface of the pit wall (and other surfaces) is covered with arrays of small pores, called micropits, on the order of 300 nm wide and separated by an average of approximately 800 nm. It is speculated that these micropits may act as spectral filters or antireflective coatings scattering light at other wavelengths, and increase the absorption of IR light. The ultrastructure of the inner wall of the pit viper pit organs also contains close-packed domed structures, which may trap reflected light.

The boid snakes, in contrast, have a larger number of labial pit organs (up to 13) on their upper and lower jaws. However, these pit organs are not nearly as sensitive as the pit vipers, with a comparatively poorer sensitivity of around 0.03°C. These pit organs do not have the suspended membrane seen in the crotaline snakes. Instead, they are shallow depressions and cavities with an innervated floor, with an order of magnitude smaller number of nerve fibers. The shape of these cavities varies with their location (e.g., those facing forward vs. those facing to the side) perhaps providing specialized directionality.

7.2.2 Bats

Vampire bats also have pit organs and a similar sensory system as snakes.[14,17] The common vampire bat (*Desmodus rotundus*) feeds on blood and as one author put it: "evolved specialized systems to suit their sanguinary lifestyle."[18] The bats are able to detect hot spots on their warm-blooded prey, potentially to aid in location of suitable feeding sites. They have three IR-sensing facial pit organs, as well as nostrils, in their nasal structure, called the nose leaf.[14] The IR organ is semicircular section separated into three areas: an apical pit (toward the top) and two lateral pits (on either side of the mid-plane of the animal) as shown in Figure 7.4. Each pit is approximately 1 mm wide and 1 mm deep and faces a different direction. The apical pit faces upward and forward, and the lateral pits face outward. The nose leaf is very flexible, and the bat moves it around constantly when looking for prey. The bat can also change the aperture (the size of the opening) of the pit organ. These factors may enable the bat

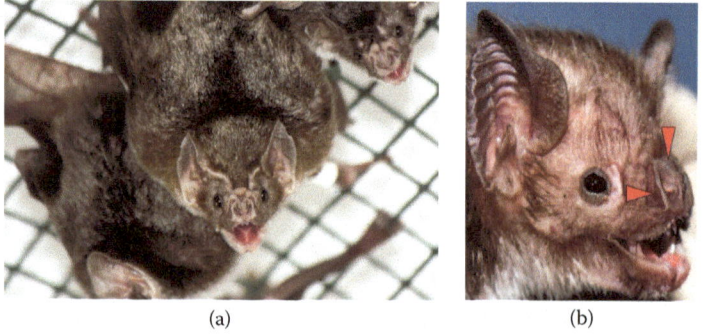

Figure 7.4 (a) Vampire bat (*Desmodus rotundus*). (b) Red arrowheads mark pit organs surrounding the vampire bat's nose. (From Gracheva, E.O. et al., *Nature*, 476(7358), 88, 2011.[18])

to provide directionality and some measure of focus to the sensor. Much like the boid snake pits, the floor of these pit organs is densely packed with tissue and nerves. Thermography also suggests that there is some amount of thermal isolation, perhaps from connective tissue, as the nose leaf area is 9°C cooler than the surrounding tissue. Also like the boid snakes, the pits are not especially sensitive, with some study suggesting that the bat must be within ~16 cm for detection, implying that the bat uses some other means to locate its prey, and potentially using the thermal detection to localize where to feed.

Compared with the bright and iridescent structures of butterflies and beetles, the IR-sensing systems of bats and snakes seem to have received somewhat less study. Recent work seems to focus on understanding the quality of imaging capable by the vipers.[15,16,18] Researchers from the University of California, San Francisco, and colleagues, have also been looking to understand the molecular origin of exactly how the nerves within the pit organs sense heat and how it differs from the photochemical process seen in eyes.[17,19]

7.2.3 Beetles

There are quite a few examples of insects that can sense heat. The *Triatoma infestans*, a blood-sucking bug, may have a thermoreceptor in its antenna, which allows it to seek blood for food.[14] Several examples of Jewel Beetle (family Buprestidae) and the *Acanthocriemus nigricans* ash beetle of Australia are a few of the pyrophilic (fire loving) insects.[20] When it comes to the beetle studied the most and explicitly identified as inspiration in thermal sensors, the *Melanophila acuminata* "fire-beetle" (Figure 7.5) is the clear winner. Combined with olfactory receptors on their antennae that can sense smoke, their thermal receptors enable *M. acuminata* to sense forest fires up to approximately a hundred miles away.[14] This capability is vital to the survival of the fire-beetle. The beetles lay their eggs under the bark of freshly killed conifers, just after the fire was extinguished, and the wood has reached temperatures that will not harm the eggs. The beetle's larvae are wood-boring, remain in the dead trees for a year, and feed on the bark. But the larvae cannot survive the defensive mechanisms of live trees, and so the fiery death of the tree eliminates the chemicals harmful to the larvae. Forest fires in the beetle's native

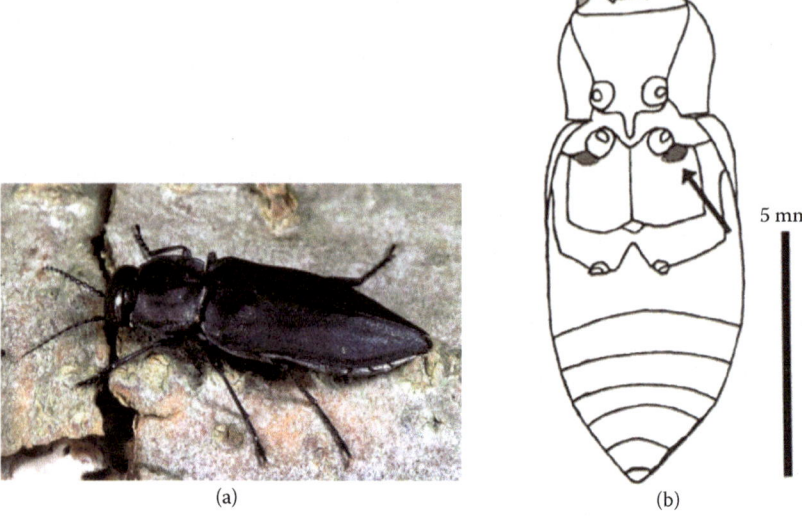

Figure 7.5 (a) *Melanophila acuminata* "fire-beetle." (b) Illustration of *M. acuminata* with arrows indicating the position of IR organs and receptors (legs and antenna not drawn). (Courtesy of Schmitz, H., *IEEE Sens. J.*, 12(2), 281–288, 2012.[20])

habitat happen approximately every 50–200 years, and so it is possible that the long-distance sensitivity was an evolutionary adaptation to the somewhat rare occurrence.

The temperature of forest fires (700°F–2100°F [435°C–1150°C]) corresponds to a wavelength of approximate 2–4 μm.[14] The forest fire IR wavelengths, and the peak sensitivity of the *M. acuminata* receptor (~3 μm), coincide with the atmospheric window at wavelengths of 3–5 μm (For discussion of atmospheric window, see Box 7.2). It does not seem to take much to stimulate the nerve response. Similarly architected mechanoreceptors are sensitive to deformations as small as a nanometer. It is estimated that less than 5 mW/cm² of light at 3 μm wavelength could cause enough deformation in the photomechanical receptor to result in neuronal activity. These photomechanical sensors are fast, too, responding on the order of under 10 μs.

The fire-beetle's two thermoreceptor pit organs are located midway down the thorax, one on each side (Figure 7.5b). The ultrastructure of the pit organ and the receptors inside has been investigated using, for example, scanning electron microscopy (SEM), tunneling electron microscopy (TEM), and focused ion beam (FIB) milling.[14,20]

Inside each pit are approximately 70–100 hemispherical domes closely packed together. Each 12–15 µm diameter hemisphere is a single IR receptor, called a sensilla. SEM image of the dome-shaped sensilla as well as an illustration of the interior structure can be seen in Figure 7.6. The topmost layer of each hemisphere is a hard exocuticular shell that is reinforced with chitin. This hard outer layer (marked in the figure) seems to be covered or partially secreted by a wax gland and is believed to protect the sensilla from dirt, smoke, and desiccation.[14] A small depression sits at the apex of each hemisphere, and a portion of the outer shell extends down into the internal core of the structure. Just below the outer shell is a softer mesocuticular layer, followed by a second, hard layered (or lamellar) structure that has been described as "lenslike." The inner core of the sensilla is a spongy mesocuticle, and this core acts as a pressure chamber.[21] The open spaces in the spongy core as well as the interstitial spaces in the rest of the structure are filled with fluid. Fluid transport between reservoirs is facilitated by micro- and nanofluidic channels in the lamellar region.[20–22] A single mechanosensitive nerve fiber in each sensilla comes up from below with the tip of the sensory dendrite anchored in the core.

There is a continuing debate on the details of how the sensilla receptor functions.[20] It is generally believed that these sensilla evolved from the more common hair mechanoreceptors. Given that, the most

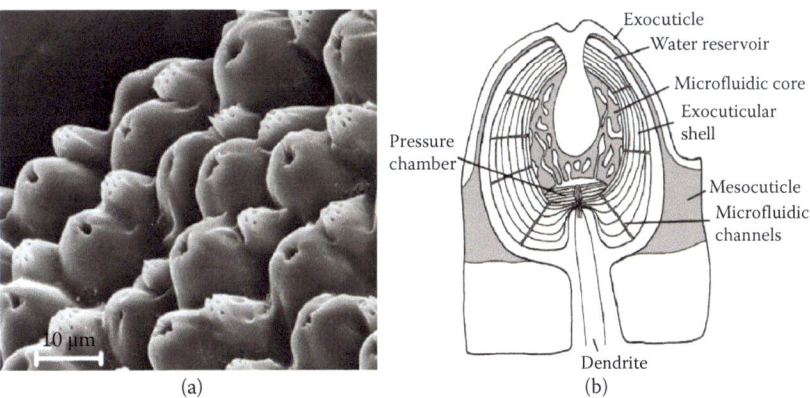

Figure 7.6 (a) SEM of the interior surface of the IR sensing organ of a *Melanophila* beetle. Domed IR sensilla are intermingled with porous wax glands. (b) Illustration of the interior structure of an IR-sensillum of *M. acuminata*. (From Schmitz, H. et al., *IEEE Sens. J.*, 12(2), 281–288, 2012 and Kahl, T. et al., In A. Lakhtakia and R. J. MartinPalma [eds.], *Bioinspiration, Biomimetics, and Bioreplication 2012*, Bellingham, Spie-Int Soc Optical Engineering, 2012. With permission.[20,21])

likely explanation is that the sensilla acts as a photomechanical receptor. IR light hits the sensilla and is absorbed by the protein and chitin in the cuticle and transfers the heat to the interior of the sensilla. In addition, some of that heat travels down the extension of the cuticle toward the core, acting like a waveguide and bringing heat to the water in the core. The outermost structure is stiff, and as the water heats it expands, building pressure within the chamber and deforming the dendrite, which causes a neuronal response. In addition to providing a significant area for thermal conduction between the structure and fluid, the complex microfluidic optical structure seems to have an interesting sophisticated additional function: potentially providing the entire system with a mechanism to eliminate variation in the sensor performance by changes in ambient temperature. The small channels between liquid reservoirs enable something of a pressure release valve for long slow changes, to compensate for slow changes in pressure.

7.2.4 Thermal Sensors Inspired by the Fire-Beetle

There are several examples of beetle inspired thermal sensors in the literature.[13,14,20–23] Two groups in particular stand out with some interesting approaches. Researchers from Georgia Tech describe what they call a thermopneumatic IR imager.[13] Noting that the beetle sensilla registers the thermal expansion of gas or liquid in its core, and wanting to create a sensor that has a small size and high sensitivity, a sensor array was designed inspired by the beetle's mechanism of deformation of the membrane. Microarrays (64 × 64, 4096 pixels) of 80 μm diameter, 100 μm deep cavities were fabricated. Deposited over these cavities was an ultrathin (~50 nm thick) polymeric membrane, the properties of which had to balance the competing needs of robustness and gas impermeability, and reflectivity and flexibility. The films were grown layer by layer with a layer of gold nanoparticles for reflectivity. Individual sensors in the array were separated and isolated from each other by 15 μm open channels. When IR light (heat) hits the sensor, it warmed the gas inside that expanded. The deflection of the sensors was measured optically, interferometrically. This type of sensor is called a Golay cell. Figure 7.7 shows an illustration of the Golay cell operation and the deformation of the membrane versus temperature. The data show that the deformation (in nanometers) is nearly

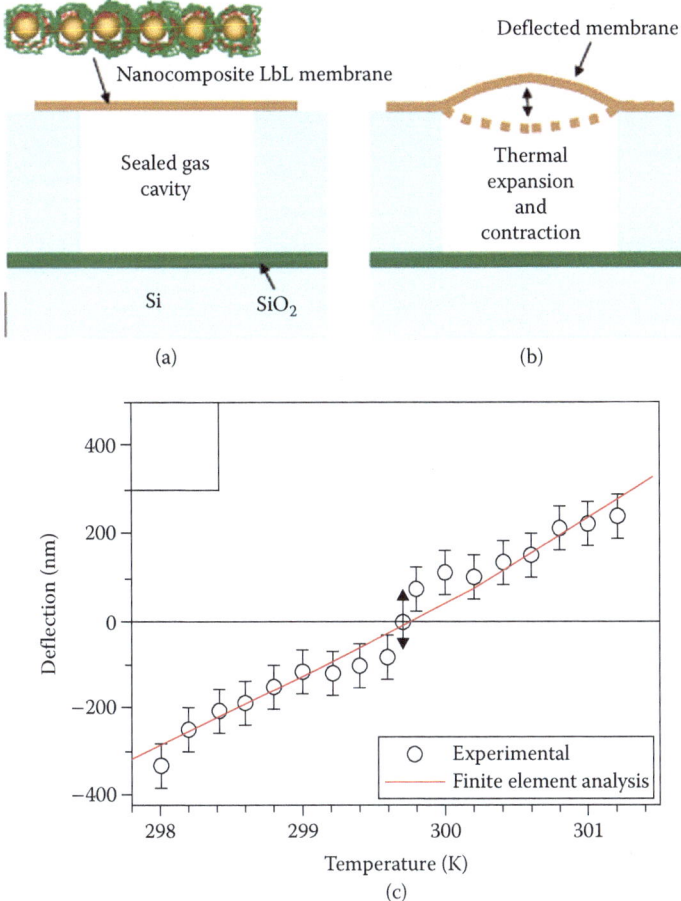

Figure 7.7 Illustration of the photothermal microcavity sealed with a freely suspended nanomembrane: (a) at room temperature and (b) in deformed states. (c) Graph of deflection vs. temperature for the freely suspended nanomembrane as measured from optical fringes. (From Jiang, C. et al., *Chem. Mater.*, 18(11), 2632–2634, 2006. With permission.[54])

linear with temperature above and below ambient, and that there is a large deformation for a small temperature differential. When the temperature reaches ambient, or room temperature, the data shows a nonlinearity, as the deformation changes direction. Sensitivity of the sensor on either side of room temperature is approximately 0.12 nm mK^{-1} with a speed approaching 60 ms, outperforming other related membrane and cantilever approaches.

When the membranes were cooled down past a critical temperature, buckling was seen at the trench edges (Figure 7.8), which was described as "wormlike," as well as out of plane buckling of the sealed

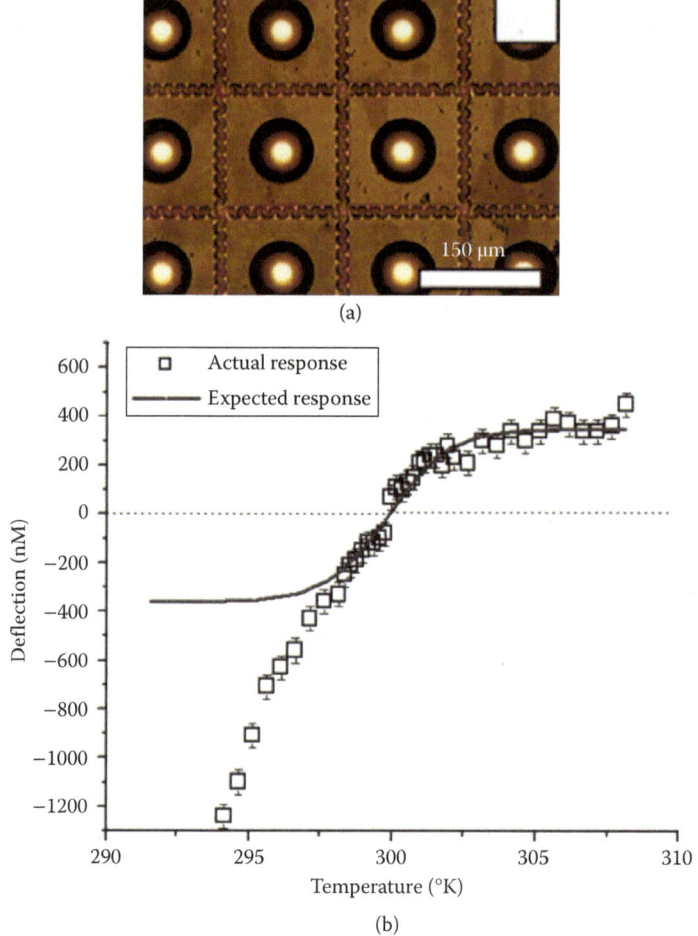

Figure 7.8 (a) Image of the nanomembrane array at below room temperature. (b) Relatively large-scale interferometer results, black squares. The blue line indicates the theoretical behavior of thermal-pneumatic sensors. There is a large increase in the sensitivity below room temperature that indicates the onset of out-of-plane buckling. (From Jiang, C. et al., *Chem. Mater.*, 18(11), 2632–2634, 2006 and McConney, M.E. et al., *Adv. Funct. Mater.*, 19(16), 2527–2544, 2009. With permission.[13,54])

sensors.[24] With the sensor cooled, the buckled state could be used to amplify the response to heating as the warmed sensor membrane unbuckles. The cooled sensor showed a tremendous, asymmetric response with sensitivity of 356 nm K^{-1} at around room temperature, which is extremely large. The cooled sensor was able to resolve a temperature difference as small as 10 mK. This is comparable with uncooled IR sensors and is comparable with other cooled (but more challenging to fabricate) compound semiconductor-based sensors.[24,25]

SENSORS

Researchers out of the Center of Advanced European Studies and Research in Bonn, Germany have also been exploring uncooled bio-inspired IR sensors.[22] Inspiration was taken from the structure of the beetle sensilla, including a liquid microfluidic compensation component. An illustration of the beetle structure and the intended microelectromechanical sensor design can be seen in Figure 7.9. The primary capacitive sensor consists of a fluid-filled cavity with an IR transparent window on the bottom and a flexible membrane above. A gold electrode is fabricated on the flexible membrane. Above this membrane,

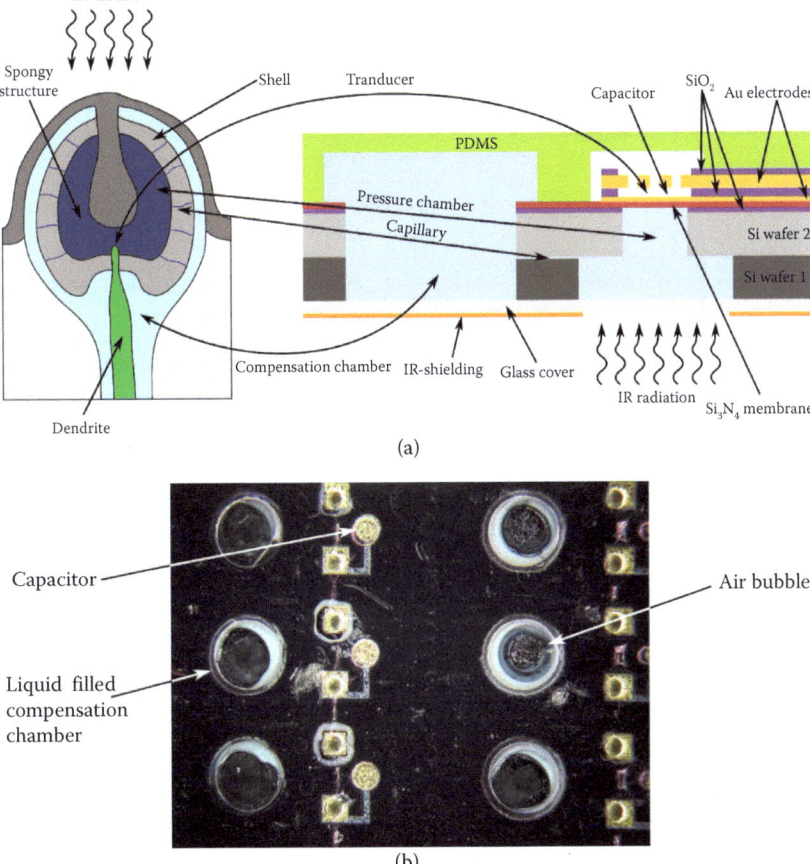

Figure 7.9 (a) Illustration of the biological sensor and the technological sensor. Left: Simplified cross section of the beetle's sensillum. Right: Cross section of the technological IR-sensor (not to scale). (b) Completely assembled sensors, including polydimethylsiloxane cover and filling with water. The micro-channel between the compensation chamber and the pressure chamber are not visible. (From Siebke, G. et al., In R.J. MartinPalma and A. Lakhtakia [eds.], *Bioinspiration, Biomimetics, and Bioreplication 2013*, Bellingham, Spie-Int Soc Optical Engineering, 2013.[22])

but separated by a gap is a second electrode. The gap-separated electrodes act as a capacitor, the capacitance of which will change as the gap distance or shape of the electrodes changes. When the IR radiation enters through the window, the fluid in the cavity is heated up which deforms the membrane. Next to the sensor, and connected by a meandering microfluidic channel, sits a pressure compensation chamber. This multicomponent fluidic system would allow fluid to move between the two chambers, equalizing the pressure for slow changes in ambient temperature, much as is postulated for the beetle system. The entire system was fabricated on two silicon wafers (top and bottom halves of the structure) and wafer bonded together. A top-down view of the finished sensors and compensation chambers can also be seen in Figure 7.9. The sensor performance has not yet been reported.

Fabricating complex multicomponent systems can be especially challenging, and special process adaptations must be explored to cleverly overcome difficulties. Two particular aspects of the fabrication were especially challenging: the fluid filling of the structure and the fabrication of the capacitor. The challenge with filling the structure with fluid is that bubbles often formed within the cavities and could not be eliminated.[22] A patented approach was developed to overcome this challenge. The researchers completely assemble the sensor except for the glass cover, to which glue that could cure in liquid was applied. Both parts were then placed in a chamber with a container of the liquid to be filling the structure, and the chamber is evacuated to below the vapor pressure of the liquid, causing the liquid to evaporate. The vaporized liquid then clings to the exposed structures relatively uniformly in a very thin layer. This is called wetting. Once the thin layer was down, assuming it was complete, fluid loaded into the structure flowed much more easily across the surface, and bubbles did not form. The two pieces were then submerged in the liquid, brought together, and cured.

The second challenge involved fabrication of the capacitor. This was challenging for two reasons. First, the spacing between electrodes was very small, roughly 300 nm, while being relatively large in area, ~1 mm². Second, one of the electrodes needed to be deformable. A typical MEMS process for something like this would involve depositing a sacrificial layer (SiO_2) between the electrode layers, and then wet etching away the sacrificial layer and cleaning. Unfortunately, this led to the two electrodes sticking to each other as the cleaning fluids dried and capillary forces pulled to deformable electrode up, destroying the sensor. Two new approaches for fabrication were developed, which can be seen in Figure 7.10. In the first approach, an additional amorphous-silicon (α-Si) support structure was added between the

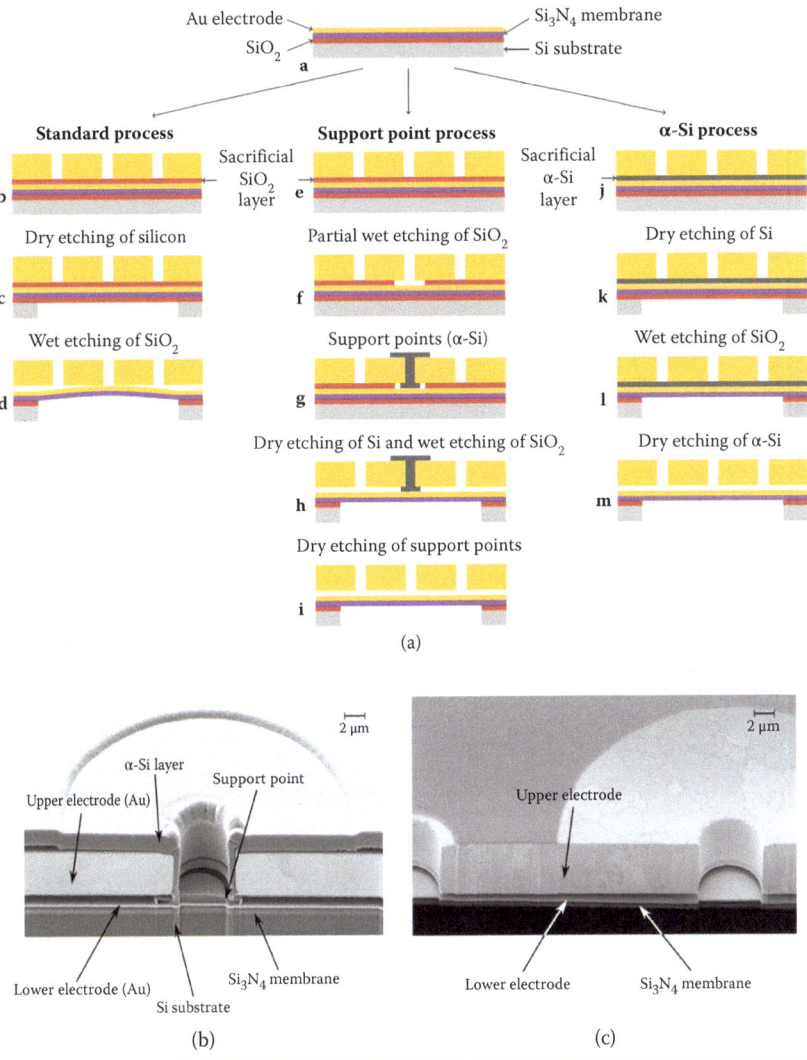

Figure 7.10 (a) Simplified process flow for the fabrication of the capacitor (not to scale). a: Common starting point. b–d: Standard process. e–i: Support point process. j–m: Amorphous Si (α-Si) process. (b, c) SEM images of the capacitor cross section (via focused ion beam [FIB]) showing steps in the support point process. (b) After etching a part of the cleared space is filled with α-Si forming a support point. (c) The sacrificial SiO_2 layer and the support points are removed and the membrane is fixed only at its border. (From Siebke, G. et al., In R.J. MartinPalma and A. Lakhtakia [eds.], *Bioinspiration, Biomimetics, and Bioreplication 2013*, Bellingham, Spie-Int Soc Optical Engineering, 2013.[22])

electrodes. Once the wet-etch and cleaning were completed, this α-Si support could be dry-etched away, which eliminated the capillary forcing of the electrodes together. SEM images of this process can be seen in Figure 7.10. However, this approach added complexity to the fabrication process that was less than ideal. In the second approach, the entire former SiO_2 sacrificial layer was replaced with α-Si. Inductively coupled plasma etching of the new sacrificial layer removed the α-Si and a gap was formed.

With this type of fluid-based sensor, there are additional potential paths for improvement.[20,21] Early attempts described above, as well as by others, mention water as the liquid filler. Alternate liquids are another potential way to increase the pressure inside the chamber, and therefore, increase the maximum membrane deflection. The liquid would have to have the optical product of density and heat capacity, as well as the capability to absorb IR light. Additional inclusion of nanoparticles or plastic mesh in the liquid may facilitate absorption.

7.2.5 Butterflies

There are some butterflies, such as those in Figure 7.11, that use their wings as thermoregulators, maintaining a specific range of body temperature (30°C–40°C).[14] Some use wing position to control the amount of exposure to the sun. The chitin of which the wing is made is a good absorber of solar radiation, and butterflies and other insects use this to their advantage. Some butterflies hold their wings in specific positions, open or closed to specific angles, to facilitate absorption. Some butterflies supplement this position-dependent basking behavior with additional optical structures designed to enhance or reduce solar absorption. The *Pachliopta aristolochiae* and the *Troides plateni*, for example, have strong black wing markings. The dark pigment-based markings enhance heat absorption in the wing, which is then transferred to the body by convection of the warmed air and by conduction by wing veins. Conversely, the *Pieris brassicae* has white wings. Rather than bask with their wings wide open, as the *T. plateni* might, these butterflies hold their wings in an intermediate position, partially, but not entirely closed. It is speculated that they use the reflective white surfaces of their forewings to reflect warm sunlight onto their dark thorax, to help them maintain their body temperature.

Figure 7.11 Butterflies that utilize wing position and color structures for thermoregulation—*Pachliopta aristolochiae*, *Pieris brassicae*, and *Troides Magellanus*.

The nanostructures found in black and white butterflies, which may play a role in their thermoregulation were discussed in Chapter 3. *Troides* butterflies also have multiple simple thermoreceptors, which enable them to monitor temperature and warming rate, and avoid overheating. One sits along the anal vein in the wings, consisting of terminal branching dendrites beside the nerve, and additional receptors exist on their antennae.

7.2.6 Thermal Expansion and Optical Sensor Structures

Inspired by the optical structures on butterflies, and their role in thermal regulation, as well as thermoexpansion properties of the chitin from which these structures are made, two groups have developed interesting approaches to uncooled IR sensors. In one approach, a team from the University of California (UC), Berkeley used the IR-triggered expansion of chitin within a conventional bimorphic cantilever sensor.[23] In a bimorphic cantilever sensor, a cantilever is made out of two materials with differing coefficients of thermal expansion (CTE), meaning that for a given thermal increase, each will expand differently. When the two are bonded together, the structure bends from the strain of the CTE mismatch. By measuring the deflection, we can determine the temperature. Often these two materials are a metal and a ceramic. However, in the UC Berkeley study, they fabricated a bimorphic sensor using ceramic and chitin. Although actual device measurements have not yet been completed, initial fabrication and modeling of the proposed devices shows notable potential performance.

The design of the sensor seen in Figure 7.12 consists of two bimorphic beams anchored to the substrate and connected to a flat pad in the center. The beams are only attached to the substrate at the anchor points, and the beams and pad floats above the base substrate. The bending of the beams with temperature will move the pad up and down, creating an optically (interferometrically) measurable change (this change can also be measured electrically, but requires additional system complexity.) Because the legs are symmetric, the pad movement is nearly perfectly vertical.

Chitin is an advantageous material to use here, as it absorbs IR light at mid-(3–5 μm) and long-(8–10 μm) IR wavelengths. Chitin is also relatively rugged, resistant to chemical and radiation exposure, and has a higher relative CTE than other standard bimorph materials. This implies that it will be able to stand up to most processing techniques, will have a strong operational lifetime and higher potential sensitivity compared with other materials. In addition, this type of sensor is less complex to fabricate than others, and amenable to wafer-scale fabrication, and so is scalable to large, relatively inexpensive arrays.

Fabrication began with a p-type silicon wafer substrate. A 200-nm-thick highly doped poly-Si etch-stop layer was deposited on the substrate using a chemical vapor deposition (CVD) technique, followed (on both sides of the substrate) by a 1.5 μm sacrificial oxide layer, SiO_2. The top layer of the SiO_2 was then patterned using reactive ion etching (RIE). The anchors are patterned and etched into this sacrificial layer. The 400 nm poly-Si structural layer was then deposited also using CVD and patterned to form the legs and the pad.

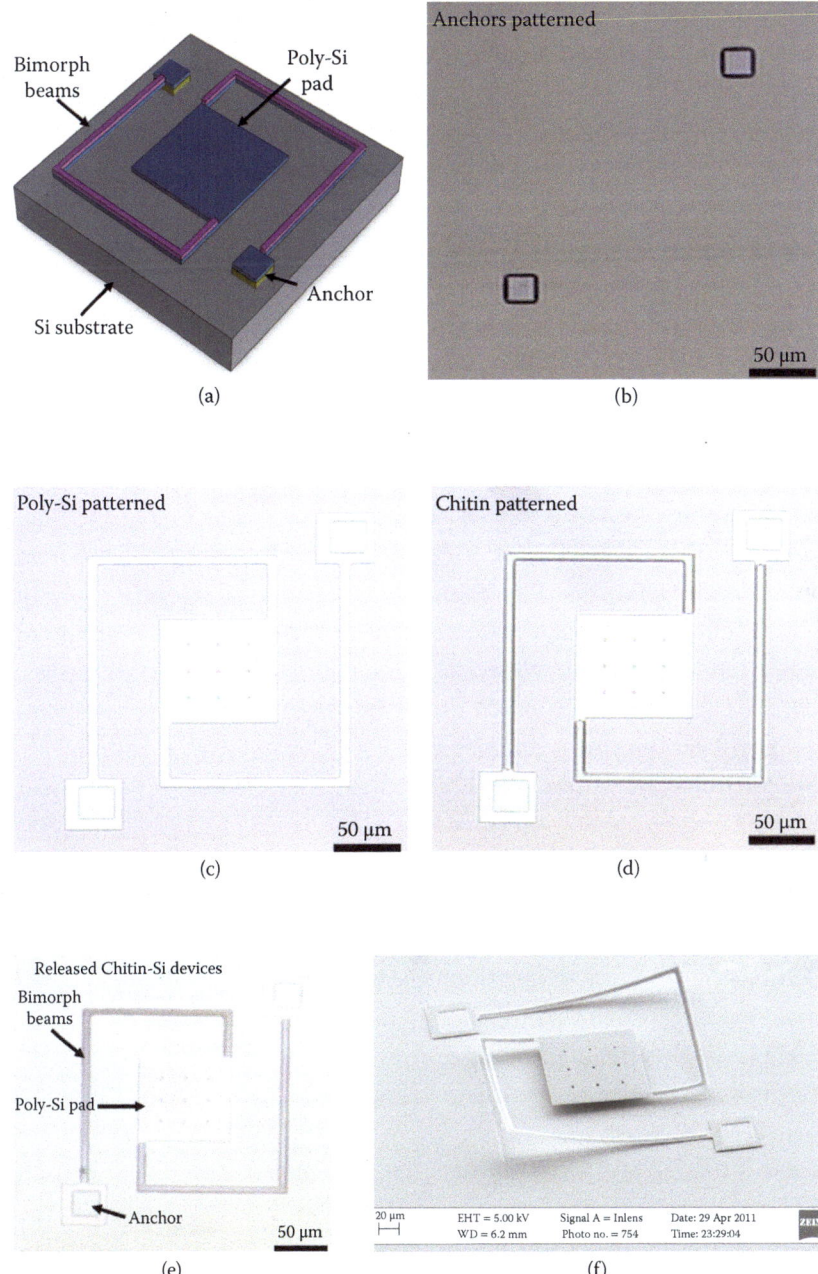

Figure 7.12 (a) Illustration of the bimorph sensor design. (b)–(e) SEM images illustrating various points during the bimorph device fabrication process. (f) SEM image of the chitin–polysilicon bimorph IR sensor. (From Zhang, N. et al., *Bioinspired, Uncooled Chitin Photomechanical Sensor for Thermal Infrared Sensing*, New York, IEEE, 2011.[23])

A layer of chitosan, a slightly more fabrication friendly form of chitin, in solution was spun-coat onto the wafer and then converted to chitin. The chitin layer was then protected using a poly(methyl methacrylate) layer, lithographically patterned, and then oxygen plasma etched. Finally hydrogen fluoride (HF) vapor was used to etch away the sacrificial layer, releasing the sensor structure from the substrate, except at the anchor points. Figure 7.12 also shows SEM images of the process described above as well as an image of the completed structure, where the beams can be seen bending due to residual stress in the films.

Modeling of the sensor performance was done using finite element analysis comparing the chitin/poly-Si structure to an equivalent Au/Si structure. Comparison of the deflection response for the two devices versus input IR power density can be seen in Figure 7.13. On the scale of the graph, the Au/Si device (shown by the red line) presented a relatively flat response. However, the chitin/poly-Si device (shown by the blue line) showed a significantly increased linear deflection response, demonstrating an approximately 50× improvement in sensitivity compared with the Au/Si cantilever. The team also modeled the chitin/poly-Si device for varying design parameters, such as thickness ratio of the chitin/poly-Si layer. The finite element analysis (FEA) model indicates that there is an optimum thickness ratio of ~2, where the chitin layer is approximately twice the thickness of the poly-Si. The chitin thickness can have a big impact on the device sensitivity. The thin layer studied to compare with the modeled Au/Si device demonstrated a sensitivity of ~0.8 nm/(pWμm^2). (For reference, the Au/Si device demonstrated a sensitivity of only 0.017 nm/[pWμm^2].) The thicker film at the optimum ratio achieved the best sensitivity of the modeled device at 2.56 nm/(pWμm^2), a significant increase.

GE's Global Research Center took an entirely different approach to butterfly-inspired IR sensor design, while also leveraging the thermal expansion of chitin.[11] *Morpho* butterfly wings, as described in Chapter 3, have complex multiscale three-dimensional photonic structures on their wing scales that are responsible for much of their startling iridescent color. Working with *M. sulkowskyi* butterflies, it was demonstrated that the butterfly wing itself could be used as an

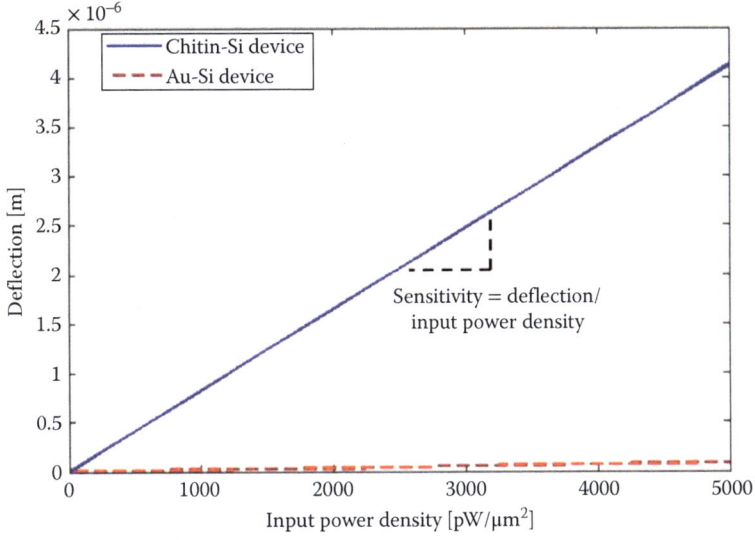

Figure 7.13 Graph of the deflection response versus. input power density for the chitosan–polysilicon biomorph sensor compared with the performance of a similarly fabricated Au-silicon biomorph sensor. (From Zhang, N. et al., *Bioinspired, Uncooled Chitin Photomechanical Sensor for Thermal Infrared Sensing*, New York, IEEE, 2011.[23])

IR sensor. The photonic crystal scales on the *Morpho* wing are made up primarily of chitin, which is secreted from epidermal cells during development. On the wing, micron-sized scales are anchored to surface of the wing membrane. The top surfaces of the *Morpho* scales have complex nanostructures, long ridges with cross-bars (lamellae) and microribs, and when light hits the scale, multilayer interference (from the horizontal lamellae) and diffraction (from the vertical ridges), it results in beautiful iridescent coloring. A picture of the *Morpho* butterfly and a TEM image of a cross section of the photonic nanostructure can be seen in Figure 7.14. The reflectance spectra of the wing, the complex interference, and diffraction pattern that we see, are created by light transmission, reflection, and refraction though the nanostructure. As the chitin in the wing scales absorbs IR radiation, it expands, changing the nanoarchitecture of the structure, as well as causing a small reduction in the index of refraction (n). This results in a measurable change in the reflection spectrum. The approach by GE enhanced the thermal absorption of the chitin nanostructure by functionalizing the surface of the *Morpho* wing nanostructures with

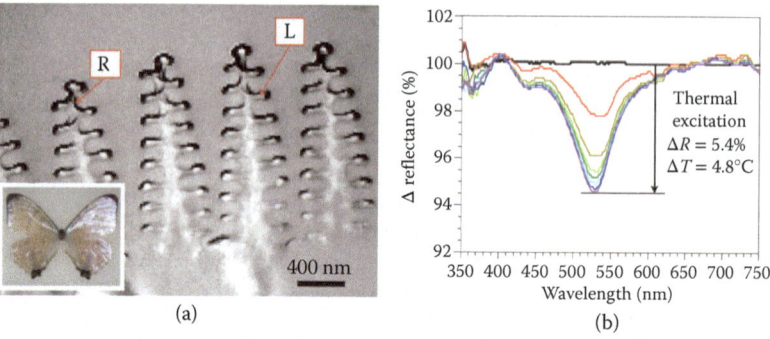

Figure 7.14 (a) TEM image of a cross section of the *Morpho* nanostructure. Inset: *Morpho sulkowskyi* butterfly, as used in the study. (b) $\Delta R(\lambda)$ spectra of the *Morpho* nanostructure at different temperatures (from heating with MWIR radiation). Signal intensity is proportional to the amount of MWIR radiation applied to the nanostructure, with a maximum signal change of $\Delta R = 5.4\%$ at the steady-state response due to a temperature increase $\Delta T = 4.8°C$. (From Pris, A.D. et al., *Nat. Photonics*, 6(8), 195–200, 2012.[55])

single-wall carbon nanotubes (SWCNTs). The SWCNTs increase the MWIR absorption cross section, increasing the thermal response in the structure. The addition of the nanotubes reduced the reflectance slightly (~15%) overall, but did not change the location of the primary reference peak at ~498 nm.

To evaluate the effect of thermal absorption on the *Morpho* reflectance spectrum, spectra were taken at different effective temperatures, and compared with the spectrum of the unheated *Morpho* sample. The differential spectral measurement for a given wavelength, $\Delta R(\lambda) = 100\%(R(\lambda)/R_o(\lambda))$, where $R(\lambda)$ is the reflectance measurement at temperature and $R_o(\lambda)$ is the baseline reflectance, eliminates any common features between the two spectra. Changes in the reflectance spectra that are caused by the temperature differential are emphasized. This differential spectrum can then be used to calculate the temperature sensitivity (TS) of the *Morpho* reflectors:

$$\text{TS} = N_{rms} (\Delta T/\Delta R(\lambda))$$

where N_{rms} is the root mean square noise and ΔT is the temperature differential. Thermal effects on the reflectivity of *Morpho* structures were explored using an actual MWIR source and a heating stage. The heating stage allowed the interrogation over large temperature range, up to $\Delta T = 30°C$. The spectral features were reproducible for repeated temperature cycles, and were consistent with spectra seen with an actual MWIR source. The data, seen in Figure 7.14, show the differential reflectance versus wavelength for varying temperature. The data show a large peak between 450 and 650 nm with a full width at half maximum (FWHM) value of ~50 nm, with a maximum $\Delta R(\lambda = 450–460$ nm$)$ achieved for a $\Delta T = 4.8°C$. The TS was calculated to be 18–62 mK, which is as good or better than commercial bolometers. Modeling and further analysis was done to evaluate the source of the change in reflectance, potentially from a heat-related reduction in index of refraction, or thermally induced movement in the support structure. Evidence indicated that thermal changes in the nanostructure were the dominant cause of the change in reflectance, although there was a small change resulting from the change in RI.

There are several potential benefits to this approach. The materials are environmentally friendly, and sensitive to changes in temperature. In addition, these temperature changes can be read optically in visible wavelengths (through a spectrophotometer), essentially transducing the IR signal to visible. The structure seems to have a relatively fast response time, as well, from 35 to 40 Hz. Finally, the material itself provides an additional potential system benefit. Chitin has a low thermal mass (a measure of how well a material stores and releases heat), and the structure is therefore responsive (fast) and is able to recover from the added heat quickly. This is in contrast to existing metal and semiconductor materials, which require a cooler or large metal heat sink. This significantly reduces system complexity, as well as enabling potential new system architectures, such as the transmission mode seen in Figure 7.15. The IR light hits the back of the *Morpho* nanostructure. The observer is located at the front of the structure (which is illuminated with white light) where they can then capture the modified visible light reflection. Finally, because the visible and IR optical properties are due to the material and architecture of the nanostructure, the materials and structure may be optimized to enhance the visible response. Doping the structures, such as the SWCNT functionalization described here, to alter the CTE or absorption cross section, provides yet an additional avenue of investigation. One final benefit of this approach is the potential ability to eliminate individually addressed pixels. That is, if one could fabricate these structures over a large scale, each nanostructure is in effect

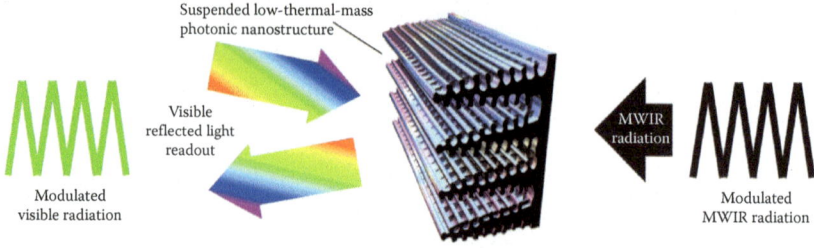

Figure 7.15 Illustration of experimental design showing the free-standing *Morpho* nanostructure, its excitation from behind with MWIR radiation and readout at the front of the visible response. R, ridge; L, lamella; CR, crossrib. (From Pris, A.D. et al., *Nat. Photonics*, 6(8), 195–200, 2012.[55])

a pixel. If an IR image were focused onto the area of the butterfly sensor, each individual ridge or lamella would respond to variations in the intensity of the IR light slightly differently, creating a direct map of the incoming image. The effective pixel size could be much smaller than current approaches, which have pixels on the order of micrometers.

7.3 Gas and Vapor Sensors

Another area in sensing where bioinspired systems are exhibiting some interesting potential is in the area of gas and vapor sensing. Examples of gases and vapors important for detection and monitoring include carbon monoxide, radon, toxic industrial chemicals, and volatiles from breath. These sensors have a potential to identify specific gas or vapor molecules at very low quantities, identify concentrations of multiple gases or mixtures, and identify toxic levels of specific chemicals in the presence of other similar materials. A host of applications use chemical vapor or gas sensors, and are looking for ways to improve their performance.[26,27] Environmental monitoring of waste products from a variety of industrial processes; monitoring home, workplace, agriculture, and homeland safety and security; and health diagnostics and monitoring of performance metrics and disease biomarkers are all areas pushing for advancements in chemical sensor technology.

Gas and vapor sensors operate on two main principles: either direct spectroscopic measurements or indirect measurements that involve a sensing material. In direct spectroscopic measurements, gas or vapor molecules are probed with electric or magnetic fields or with different types of ionization sources. The gas or vapor molecules will then exhibit a uniquely or nearly uniquely identifiable spectroscopic or ionization signature depending on the mechanism of interaction between the probe and the molecules. Examples of such sensors include UV sensors for monitoring of ozone absorption, Raman sensors for monitoring of nitrogen (N_2) gas, IR and near-IR sensors for detection of numerous dissolved gases. In contrast to sensing based on intrinsic spectroscopic properties of the analyte (the material we are trying to sense), indirect sensing uses a responsive sensing material. This approach expands the range of detected species, improves sensor performance (e.g., analyte detection limits), and is more straightforwardly adaptable for miniaturization. These attractive features can be offset by some limitations of indirect sensors, for example, insufficient selectivity, poisoning, poor long-term stability, and slow response and recovery times. Nevertheless, indirect sensors constitute the most active research area in developing sensing approaches that cannot be addressed with direct sensing.

The benchmark, but bench-sized, gas chromatograph–mass spectrometer is the standard for high sensitivity, and high accuracy lab-based chemical identification. However, these are far too large and expensive for anything other than lab work, in addition to requiring special handling to maintain its calibration and sensitivity. Many applications are moving toward small, inexpensive, robust, low-power, unobtrusive form factors, requiring little direct interaction from the user, and few additional regularly needed components (such as reagents or test strips).[26] Each of these applications places different emphasis on which requirements are the highest priority. However, in general a short list of properties is typically targeted for the focus of sensor improvement: sensitivity (how small a sample can be detected), specificity (how selectively and accurately it can be captured), and response time (how fast the identification takes place). In addition, properties such as low false alarm rate, reversibility (meaning that the sensor can "reset" in some manner to provide a continuous functionality without "using up" the sensor), and sensor systems that can accurately identify multiple potential targets simultaneously, are also of interest.

Sensor development is big business, and thousands of researchers give presentations on new chem/biosensor technologies and techniques each year, only a subset of which depends on photonics. In biology, there are few examples of animals or plants that can naturally sense changes in chemical composition within a timescale of interest to most of the sensor community. The scales of the freshwater fish *Channa punctatus* have been reported to change over time based on the presence of heavy metals and other pollutants in their habitat.[28] The feather coloration of some birds is also very sensitive to food source and environmental contaminants.[29] However, these are not chemical sensors in the way we are seeking high-performance technologies.

In development, many researchers are working on approaches that mimic the olfactory senses of animals, as they seem to possess the sensitivity, selectivity, and reusability we require.[30,31-33] Of course, we also still use trained animals for a wide variety of sensing tasks. The other path that many researchers are exploring in the area of bioinspired photonic chemical sensors is directed to the biological structural color photonic structures themselves.[9,26] As we have discussed in earlier chapters, the optical properties of these photonic crystals are extremely sensitive to their architecture. By designing a system that targets specific changes to the structure under sensing conditions we can design sensor materials that change color in the presence of target materials (colorimetric sensor).

One type of bioinspired photonic crystal of gas sensor under investigation is the relatively simple one-, two-, and three-dimensional crystal. One-dimensional multilayers or three-dimensional colloidal photonic crystals in opal or inverse opal structures, as well as bioinspired liquid crystal structures are examples.[9,26,34-37] The structural color-based coloration of several of these biological structures, such as the structures in beetle elytra and bird feathers, has been shown to be sensitive to humidity and a humidity sensor based on the nanostructure in the Hercules beetle elytra (Figure 7.16) has been demonstrated.[38-43]

For sensing of significantly more complex molecules, sensing in the presence of highly similar molecules or sensing of multiple molecules and mixtures, several groups have investigated colloidal structures, as well as the even more complex *Morpho* butterfly photonic structures.[35,44-47] It is not possible to reliably fabricate structures as complex as the *Morpho* nanostructure yet, and the gas sensing in the above

Figure 7.16 (a) Photograph of a Hercules beetle. The exoskeleton of the Hercules beetle changes color from (b) khaki-green in a dry atmosphere to (c) black in high humidity level. (From Kim, J.H. et al., *2010 IEEE Sensors*, IEEE, New York, 2010.[42])

examples was done on portions of an actual *Morpho* wing, although the long-term hope is to investigate fabrication of potentially optimizable structures in the future. Another example of using actual biological structures as backbones for chemical sensing is the use of diatoms.[48–50]

7.3.1 Diatoms

The porous silica (SiO_2) nanostructures in the diatom skeleton (frustule), once divested of its organic matter, make them very attractive from a nanotechnology perspective. Diatom frustules are beautifully micro- and nanostructured, as shown in Figure 7.17. The pores create a three-dimensional structure with an open architecture and a significant amount of surface area, characteristics which are attractive in a gas sensor, as it provides space for the target analyte to adsorb and interact with the sensor.[48] The diatom structure can be functionalized, either during growth of the living diatom or to the frustule after processing, with additional molecules that bind to specific targets.[51] The fact that diatoms are plentiful, and "fabrication" of additional frustule material is easy to scale, implies a low potential cost, which would also be a significant benefit.

Aulacoseira diatom structures have demonstrated the ability to (among other things) sense nitric oxide (NO) gas.[48] *Aulacoseira* diatoms have an elongated cylindrical shape, and are ~10 μm long. The interior of the cylinder is hollow, with a circular opening capping each end, as seen in Figure 7.17. Arrays of fine nanoscale holes pierce the cylinder along its length. Approaches are being explored to convert the SiO_2 of the diatom structure to materials more amenable to large-scale processing and integration into systems, such as silicon, without losing fidelity of the frustule nanostructure. Being able to achieve these nanostructures in silicon is highly desirable as that allows them to be compatible with large-scale commercial fabrication processes, such as wafer-scale processing. However, many existing potential processes are extremely high temperature and most often ruined the resultant replica nanostructure, and thus new techniques must be developed.

First, the SiO_2 frustules were reacted with magnesium gas (Mg) at 650°C to change the SiO_2 into a magnesia/silicon (MgO/Si) composite structure. Hydrochloric acid then removed the magnesia leaving an entirely silicon version of the original SiO_2 frustule. A silicon frustule was then placed on a silicon nitride substrate and connected by platinum electrodes to gold pads. A bias voltage was placed across the diatom circuit (10 mV at 100 Hz). Baseline impedance across the circuit is measured in the presence of flowing argon gas. Then NO was added to the argon gas stream and the change in impedance was measured for differing amounts of NO. Figure 7.18 shows the silicon frustule attached to the platinum electrodes. A graph of the ratio of differential impedance (argon gas/NO vs. argon gas alone) shows how the ratio changes with time with the introduction of NO concentrations. The diatom frustule showed a relatively rapid response to the presence of the NO gas, and a clear signal at just 1 part per million (ppm) NO. In this first approach, the diatom acted as a functionalizable sensor in a circuit, and the adsorption of the NO is what changed the electrical properties of the circuit.

The photoluminescent properties of diatoms are another potential avenue of exploration as a sensing mechanism.[50] The photoluminescence

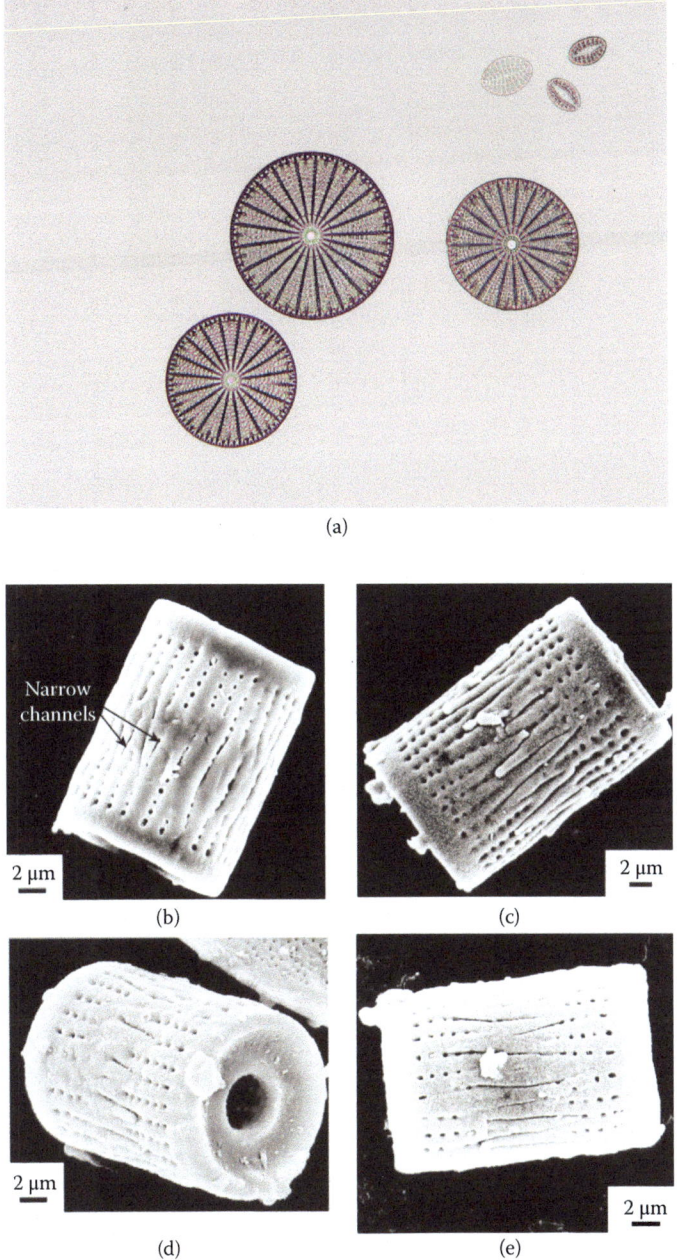

Figure 7.17 (a) The round diatom *Arachnoidiscus* at magnification 100×. (b–e) Secondary electron images of (b) an *Aulacoseira* diatom frustule, (c) a MgO/Si composite replica after reaction of an *Aulacoseira* frustule with Mg(g) at 650°C for 2.5 h, (d) a silicon-bearing replica produced by selective dissolution of magnesia from a MgO/Si replica in an HCl solution, and (e) a silicon replica after the HCl treatment and an additional treatment in a HF solution. (From Bao, Z. et al., *Nature*, 446(7132), 172–175, 2007.[48])

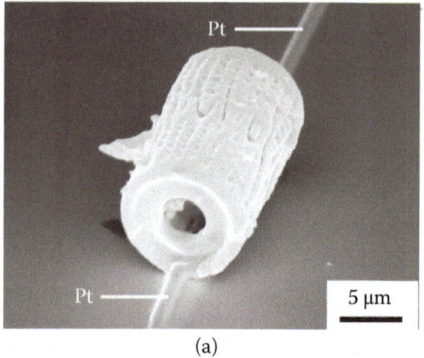

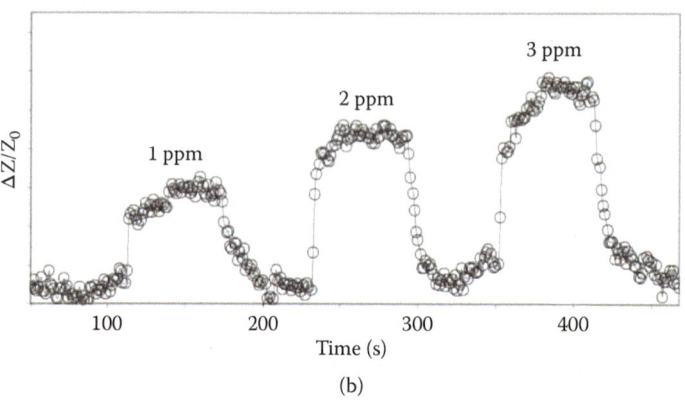

Figure 7.18 (a) Secondary electron image of a microporous silicon frustule replica with attached electrode. (b) Graph of the electrical response to NO (g) vs. time for the single silicon frustule sensor. ΔZ = impedance change upon exposure to NO (g), and Z_o = sensor impedance in pure flowing argon. (From Bao, Z. et al., *Nature*, 446(7132), 172–175, 2007.[48])

of the diatoms *Thalassiosira rotula* Meunier, 1910 and *Coscinodiscus wailesii* was used to explore their response to nitrogen dioxide (NO_2) gas. Figure 7.19 shows SEM images of the two diatom species, as well as close-ups of their multiscale pore structures. Measurement of the (room temperature, continuous wave) photoluminescent spectra (photoluminescence vs. wavelength) of the diatoms in an artificial dry air environment and in the presence of varying concentrations of NO_2 gas show a relatively broad peak (600 nm FWHM) centered near ~548 nm. The general trend of the spectra with increasing NO_2 concentration shows the peak height decreasing until the intensity is effectively quenched (eliminated) at a critical concentration. Plotting the ratio of the differential photoluminescent yield (integrated spectrum over the emission wavelength range) to the baseline with no

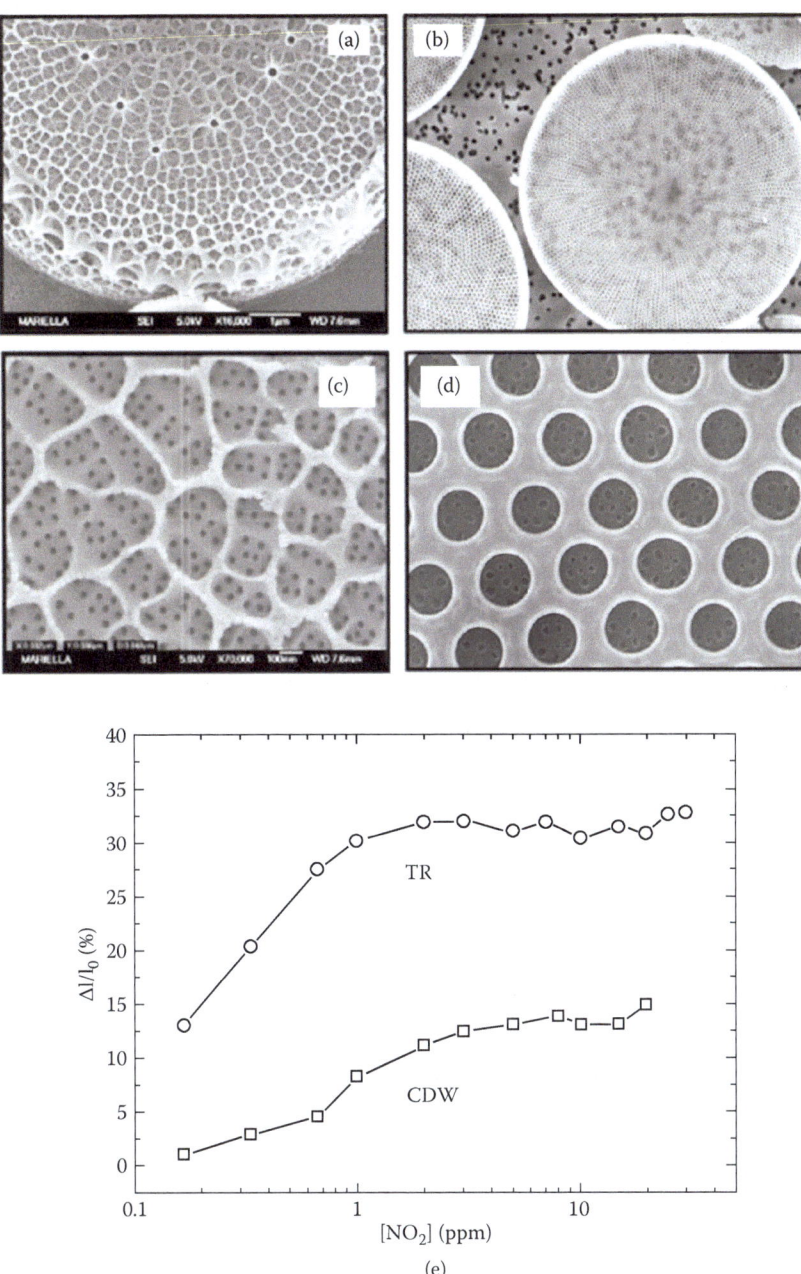

Figure 7.19 SEM images of (a and b) *Thalassiosira Rotula* and (c and d) *Coscinodiscus Wailesii* frustules, deposited on silicon substrates. (e) The relative variations of measured differential photoluminescence intensity (yield) as a function of NO_2 concentration for TR (*Thalassiosira Rotula*, circles) and CDW (*Coscinodiscus Wailesii*, squares). (From Lettieri, S. et al., *Adv. Funct. Mater.*, 18(8), 1257–1264, 2008.[50])

NO_2 ($\Delta I/I_0$) versus the NO_2 concentration shows an identifiable yield at extremely low concentrations. Data shown in Figure 7.19 suggest a detection limit of around 50 parts per billion. These structures are, however, likely limited to low-concentration sensing applications, because the photoluminescence is quenched at high NO_2 concentrations (10 ppm for the *C. wailesii*, for example). Further examination of the system seems to imply that the NO_2 chemically bonds to the surface of the diatom. The photoluminescence seems to be related to the oxygen vacancies of –OH surface terminations. The chemical adsorption potentially reduces the density of radiative states, resulting in the change in the height of the spectral peak but not its shape or wavelength, and eventually quenching the photoluminescence.

7.3.2 Butterfly Wings as Sensors

Much like the clever use of the inherent materials' properties and natural structurally dependent reflective properties of the *Morpho* butterfly as a thermal sensor, teams of researchers have also been investigating the *Morpho* and similar structures for chemical sensing. Except by evolutionary processes, the butterfly coloration does not rapidly change in response to chemicals in its environment as it flies from place to place, certainly not in any color or intensity visible to the naked human eye. And yet it has been demonstrated that by measuring the reflectance spectra as a function of wavelength of white light as it reflects off of a butterfly wing small changes in the spectra can be identified in the presence of a variety of vapors.

Researchers from GE's Global Research Center looked at the wing structures of the *Morpho sulkowskyi* and wondered if those structures could be used as sensors.[44] By illuminating the structures with white light, the reflectance spectrum (reflectance as a function of wavelength) of the wing, was measured. When the structures interacted with different vapors, the reflectance spectrum changed, and that change could be detected using a spectrophotometer. In fact, when analyzing the data, it was determined that several closely related vapors could be identified. The *M. sulkowskyi* butterfly is highly iridescent, as shown in the insert to Figure 7.20, but without the intense blue coloration associated with its other *Morpho* counterparts. It does not have pigment to absorb stray reflections and so is highly reflective at visible wavelengths while having a denser ridge and more regular lamellar structure than other *Morphos*. The reflectance spectra taken across the wings can be seen in Figure 7.20, and seems to be highly reproducible. There is a significant peak in reflectance between approximately 440 and 530 nm.

To investigate the wings response to different vapors, a small (5 mm × 5 mm) portion of a wing was used and illuminated with white light. Using a fiber optic spectrograph with a CCD detector the differential reflectance spectra $\Delta R(\lambda) = 100\%(R(\lambda)/R_o(\lambda) - 1)$ were monitored, where $R(\lambda)$ is the reflectance measured when the structure is exposed to the target vapor, and $R_o(\lambda)$ is the reflectance measured when the structure is exposed to a dry N_2 carrier gas. The wing structures were then exposed to several similar solvent vapors (water, methanol, and ethanol), which have similar polarity and indices of refraction. Although the overall spectral features remained alike, the reflectance spectra showed differences with each solvent, as well as varying with concentration of the vapor, particularly at the longer wavelengths, as can be seen in Figure 7.21. To tease out the details of this change in reflectance, they turned to a technique called principal component analysis (PCA, see Box 7.3), a method of multivariate analysis. This allows them to map in a form of multidimensional space the differences in spectra for the three vapors, where each vapor can be clearly differentiated from the others. The smallest concentration detected for

Figure 7.20 Characterization of reflectivity of scales of an intact *Morpho* butterfly demonstrating the stability of the reflectivity peak (100 μm step size). Inset, image of *Morpho* butterfly used in the study (wingspan = 87 mm) with arrow indicating direction and scan position. (From Pris, A.D. et al., *Nat. Photonics*, 6(8), 195–200, 2012.[55])

BOX 7.3   PRINCIPAL COMPONENT ANALYSIS

PCA is a statistical method of analyzing complex data sets. It is a technique that is used in neuroscience, image analysis, and meteorology, for example, where the data sets are large, the variables not independent, and not easily analyzed by other means. It takes high-dimensional data and represents it in a lower-dimensional form, where the data are now represented by principal components, which are weighted sums of the original variables.[52] By plotting the principal components against each other relationships between the variables that might otherwise be hidden may emerge.

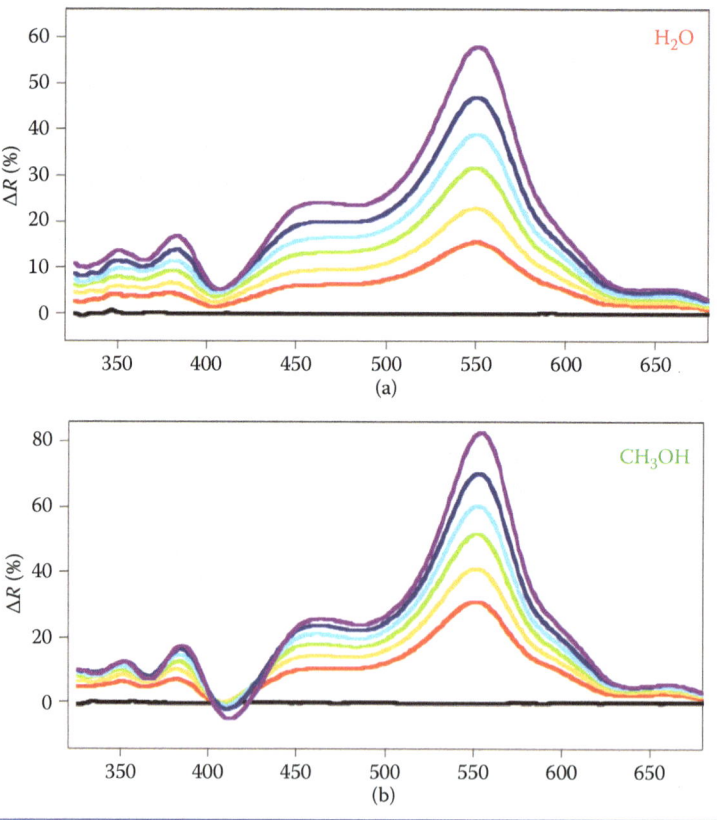

Figure 7.21   Graphs of ΔR vs. wavelength for sensor exposure to (a) water, (b) methanol

(Continued)

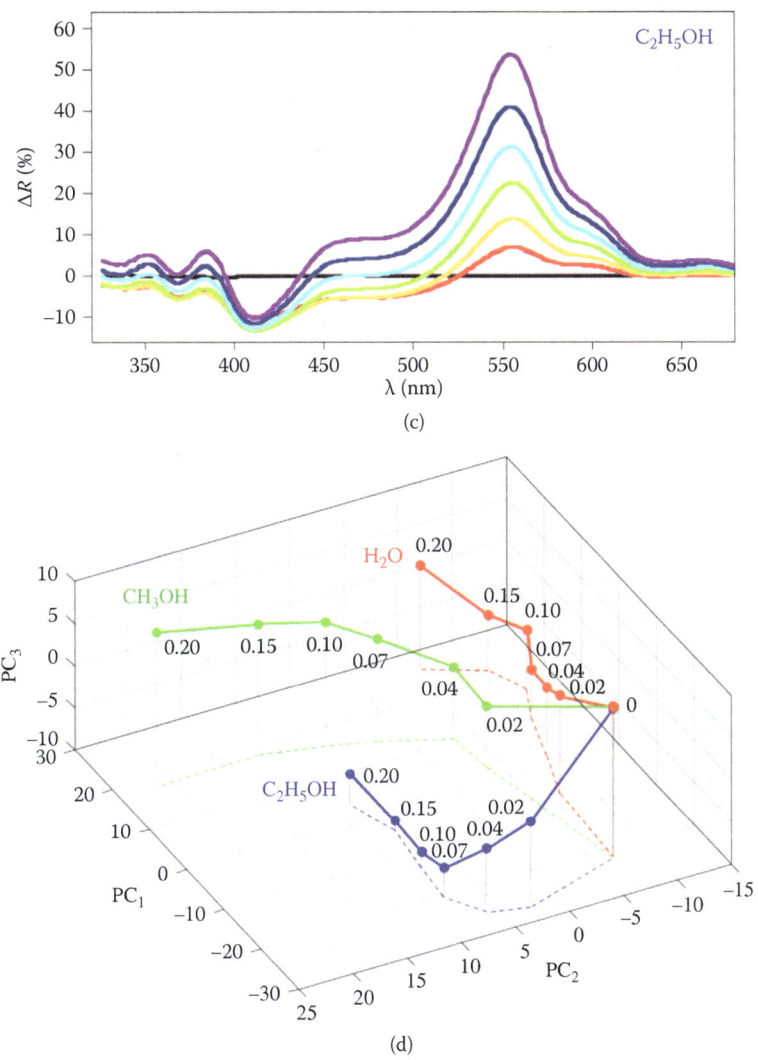

Figure 7.21 (*Continued*) (c) ethanol vapors of increasing partial pressure ranging from 0 to 0.2 P_0. Vapor concentrations were 0 (black line), 0.02 (red), 0.04 (yellow), 0.07 (green), 0.1 (light blue), 0.15 (dark blue), and 0.2 (purple) P_0. (d) Principal component analysis results of the ΔR spectra showing discrimination between the three vapors after mean-centering the spectra. (Scores plot of the first three principal components (PCs) where PC1 = 71.7%, PC2 = 18.8%, and PC3 = 6.9%, capturing 97.4% of the total variance in the spectra.) (From Potyrailo, R.A. et al., *Nat. Photonics*, 1(2), 123–128, 2007.[44])

these vapors was 1–2 ppm. Selectivity or the ability to confidently identify the presence of a specific target vapor is a major challenge, particularly at low concentrations. The presence of additional vapor molecules, humidity, and molecules of similar structure (confuser molecules) often results in false sensor readings (positive and negative). The ability to

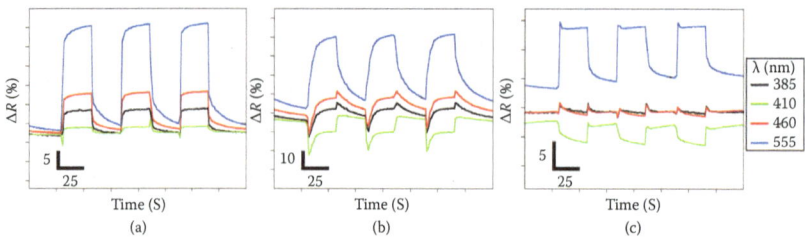

Figure 7.22 Graphs showing the dynamic response (ΔR vs. time) of scales of *M. sulkowskyi* to different vapors at 385, 410, 460, and 555 nm. (a) Water, (b) methanol, and (c) *trans*-1,2-DCE vapors. Shown are replicate ($n = 3$) responses to 0.07 P0 of vapors. Numbers in the bottom left of the panels are numeric values for *x* and *y* axes. (From Potyrailo, R.A. et al., *Nat. Photonics*, 1(2), 123–128, 2007.[44])

differentiate three separate vapors from each other on the same graph indicates that these sensors have the potential to enable us to identify and quantify target vapors out of a range of others.

To test whether the wing structures would enable the unique identification of even more closely related vapors, wing samples were exposed to the three isomers of dichloroethylene (DCE): the 1,1-DCE, the *trans*-1,2-DCE, and the *cis*-1,2-DCE. The indices of refraction of the three vapors were again very similar (1.425, 1.449, and 1.449, respectively). PCA of the resulting differential reflectance spectra shows that, although the sensitivity to these vapors was lower than the initial alcohol and water vapors, the changes in reflectance spectra enabled the differentiation of the three isomers. In addition, the dynamic response can be seen in Figure 7.22 for water, methanol, and *trans*-1,2-DCE vapors. The dynamic response shows clearly different behavior for the three vapors, which may be due to interaction of the gas with specific portions of the nanoarchitecture (ridges, lamellae, microribs, etc.). Next steps would be to look at mixtures of vapors to see if the individual vapors can be discriminated and quantified.

Since the first publication of demonstration of vapor sensing on a butterfly wing in 2007, other groups interested in their potential for bioinspired sensor applications have investigated butterfly wing structures.[45–47] Researchers from the Center for Natural Sciences in Budapest Hungary looked to a different butterfly species.[45] *Polyommatus icarus*, *Polyommatus bellargus*, and *Polyommatus coridon* have a very different structure than that of *M. sulkowskyi*. Instead of the Christmas treelike structure of the *Morpho*, the structure of these species has been described as a "pepper-pot" structure. *P. icarus* and *P. bellargus* can be seen in Figure 7.23, along with SEM images of the wing nanostructures. As in the SEM images, the general structure is a series of multiple porous layers supporting long ribs and suspended cross-rib structures.[53] Seen in the images from above, it seems as if the open spaces between the ribs are filled with overlapping layers of nanopores.

To explore the sensing capabilities of these structures, a small portion of a butterfly wing was placed in a temperature-controlled airtight aluminum container with a quartz glass window.[45] The temperature controller, that is, a thermoelectric cooler, can change the temperature of the wing structure during sensing. Using the spectrum seen in the presence of synthetic air (a mixture of 80% nitrogen [N_2] and 20% oxygen [O_2]) as a baseline, measurements of the change in reflectance were taken at different temperatures (17°C, 19°C, 24°C), to explore not only the chemical sensitivity but also the temperature sensitivity of the response. A UV/VIS/NIR light source illuminated the wing segment, and the reflectance was collected through a fiber optic spectrophotometer at the angle of maximum reflection. The dynamic response was measured by flowing the synthetic air across the structure, followed by a synthetic air/vapor mixture with the vapor to be sensed at a specific concentration. This was cycled at 10 seconds of air/vapor, followed by 50 seconds of synthetic air to clean and "reset" the structure and system. The vapor concentration was increased with sequential cycles. The change in reflection was captured as a function of time and wavelength, as seen in Figure 7.23. As these measurements are made relative to the synthetic air spectrum (designated as 100%), data capture both increases and decreases in reflectance.

Detection of ethanol, acetone, and water vapor were investigated in wing portions of the three butterfly species. The three species were chosen because the *P. icarus* and *P. bellargus* have very strong features in their reflectance spectra, narrow peaks at 380 and 435 nm, respectively, that reflect the bright blue iridescence. *P. coridon* has a similar structure but lacks the strong peaks seen in the other two. By comparing the response in these three butterflies, the role of shape of spectral features and structure related to vapor sensitivity could be further examined. The dynamic response of the *P. Icarus* butterfly

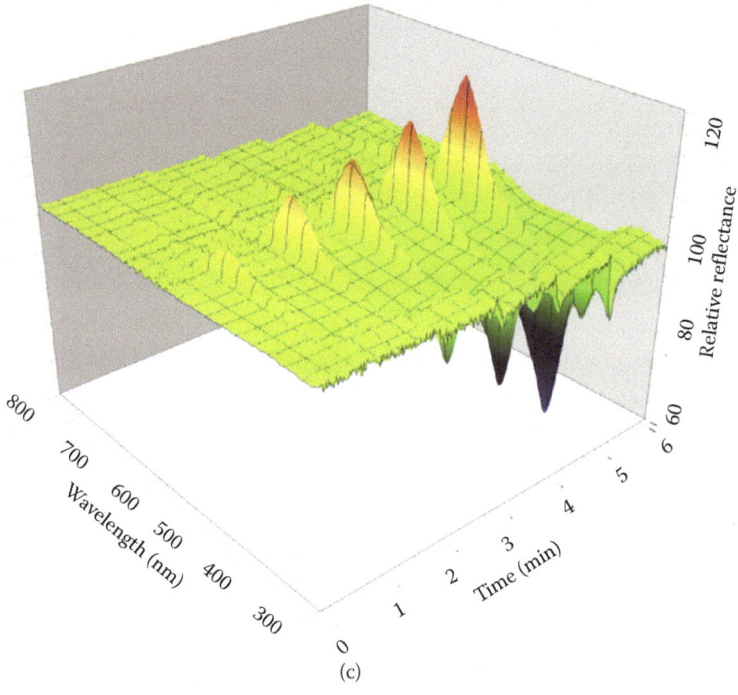

Figure 7.23 (a) *Polyommatus icarus* image and SEM image of the wing scale nanostructures. (b) *Polyommatus bellargus* optical and SEM image. (c) Graph of the relative optical reflectance vs. wavelength and time for *Polyommatus icarus* wing structure sensor. The wing in synthetic air was used as reference (100%). Increasing concentrations of ethanol vapor flow were applied for 10 seconds followed by 50 seconds synthetic air flow for cleaning the cell. (From Kertesz, K. et al., *Appl. Surface Sci.*, 281, 49–53, 2013.[45])

wing to ethanol can be seen in Figure 7.24. The reflectance spectra were integrated over a 30-nm band around the maximum increase in reflectivity relative to the baseline (480–510 nm), displayed as the solid black lines, as well as a 30-nm band around the maximum decrease in reflectivity (300–330 nm), displayed as a dotted line. The data clearly show five separate response peaks corresponding to varying concentrations, and the response increases roughly linearly with increasing concentration. The normalized integral is graphed as a function of time at three different temperatures and the response

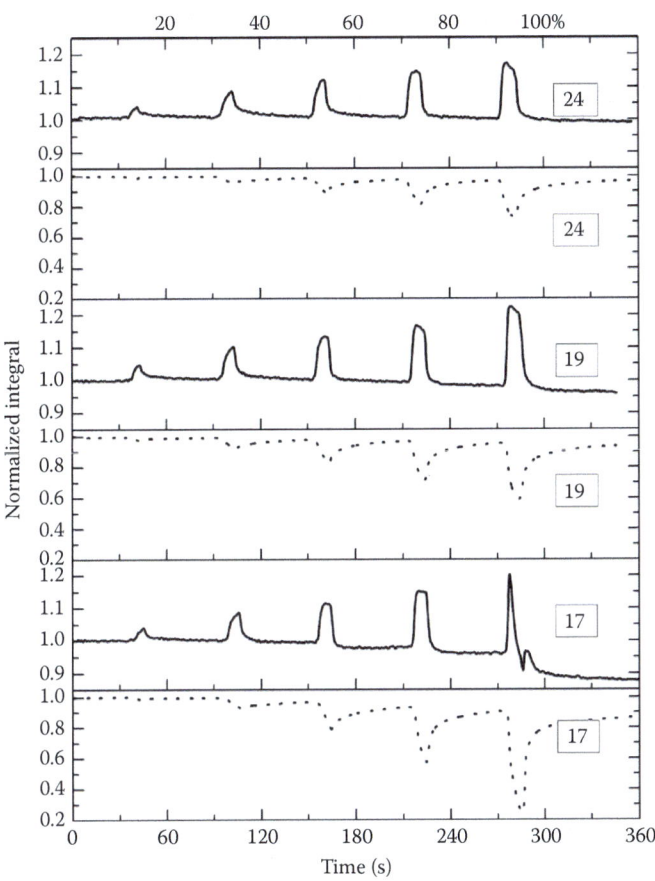

Figure 7.24 Graph of the dynamic response tracking the change in maxima (480–510 nm, continuous line) and minima (300–330 nm, dotted line) (integrated response signal variation vs. time). During the measurements the *Polyommatus icarus* wing scale was kept at 17°C, 19°C respective 24°C. The axes of the three pairs of graphs are the same. Horizontal upper axis indicates the ethanol vapor concentrations for each response. (From Kertesz, K. et al., *Appl. Surface Sci.*, 281, 49–53, 2013.[45])

increases with reduced temperature. The linear response with concentration as well as the increasing response at lower temperatures is a trend that is consistent for the other vapors (water and acetone) as well as the other butterfly species examined. *P. coridon* lacking the strong reflectance peak in the baseline spectrum shows the weakest response.

Understanding the subtleties of how the butterfly wing structure and material properties interact with the vapor to drive the reflection response is very important if we are to move past using the wing themselves and creating optimized sensors. Again, the open architecture of this structure provides a significant amount of available surface area to facilitate interaction between the structure and an introduced vapor. The photonic response of the structure is easily perturbed by the minute changes induced by this interaction. When vapor passes through the wing structure, the vapor settles into the pores and crevices of the nanostructure through a process called capillary condensation. This effect is seen in other nanostructures being investigated as vapor sensors, such as porous Si.[44] The vapor fills the pores in the structure and condenses into liquid. The reflective properties are exquisitely sensitive to changes in the structure, including the changing of spacing, index of refraction, and so on, which may be caused by the presence of condensed vapor. The geometry of the multiscale and nonuniform pores in *Polyommatus*, for example, plays a significant role in the dynamics of the adsorption and desorption process of the vapor.[45] In the case of the *Polyommatus*, some hysteresis was seen at high concentrations, which was attributed at least in part due to nonuniformities of the nanopores, which empty and fill at different rates because of varying equilibrium vapor pressures in the pores, as well as potential chemisorption due to available surface bonds.

The material properties of the wing also play a significant role in its vapor sensing functionality. The presence or lack of specific hydrophobic or hydrophilic areas on the scales of the *Morpho* was investigated by staining and no specific hydrophobic domains were found.[44] No particular elemental compositional difference was found between any parts of the nanostructure, such as ridges, lamellae, and so on. It was noted that the response of the *Morpho* wing to a polar solvent (the alcohols) was different to its response to a nonpolar solvent (DCE). Most recently it has been discovered that the *Morpho* butterfly wing nanostructures actually have a small, but not insignificant polarity gradient from the tops of the "tree" to the base.[52] It is postulated that this polarity gradient has no critical role in the butterfly's survival, but rather a benign side effect of the material and process of scale development. The epicuticle surface layer of the nanostructures, seen in SEM images in Figure 7.25, is a composite of lipids, proteins, and polyphenol molecules and its composition and role in adsorption was examined. The tops of the nanostructures have a higher surface polarity, whereas the base is less polar. Vapors of different polarities are preferentially adsorbed onto specific areas of the ridges.

In attempting to understand all of the crucial details involved in the vapor/structure interaction, a host of experimental and modeling techniques must be used. For example, in the experiments to understand the polarity gradient on the *M. sulkowskyi*, staining of the ridges using polarity-sensitive dyes along with spatially resolved optical characterization experiments, such as fluorescence microscopy, to map the spatial polarity distribution; time-of-flight secondary ion mass spectrometry (ToF-SIMS); solvent-extraction high-performance liquid chromatography ToF-MS to understand the epicuticular composition; plasma treatment to reduce any initial surface chemical gradient; exposure of the structures to vapors with particular differentiable characteristics; and more were all used.[52]

Modeling and simulation of variations on these structures, simplified Bragg stacks and idealized *Morpho* structures, also illuminate details and begin to point toward potential structures for real-world use.[44,46,47,52] Optical models of simplified and idealized versions of these structures have shown that solvent film thickness plays a strong role in the change in optical response.[44] Vapor adsorption was modeled as a uniform conformal coating of the nanostructure by liquid solvent of varying thickness. Figure 7.25 shows the $\Delta R(\lambda)$ response for an idealized *Morpho* structure. The top graph shows the simple reflectance measurements versus wavelength. Both spectra show that the model structure results in two primary spectral features, around 400 and 520 nm. As the film thickness of the adsorbed vapor increases from 0 to 3 nm, the reflectance at the shorter wavelengths decreases, and the reflectance at the longer wavelength peak increases as well as moving slightly to longer wavelengths. This is essentially what was seen in the experimental study. By examining variations in the model structure, we can begin to understand how the architecture impacts the vapor response. One can change the thickness or existence of vertical microribs between the lamellae or change the variations in width of the lamellar stack. In the case of the microribs, the spectra look very similar to the original model structure, with a

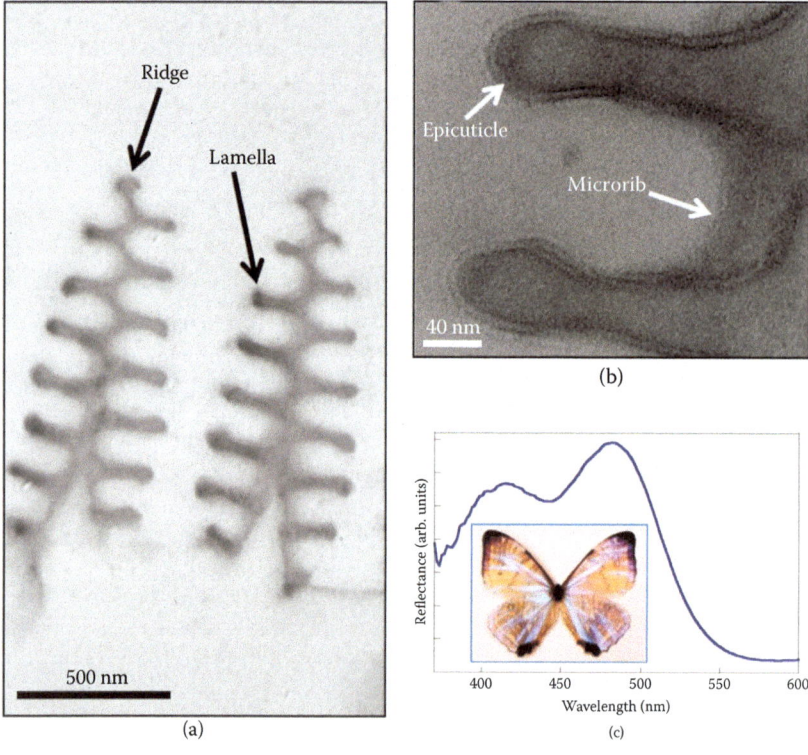

Figure 7.25 (a) Ridge nanostructures of *M. sulkowskyi* with lamellae. (b) Image of the conformal epicuticle on the lamellae. An out-of-plane microrib is also visible. (c) Reflectance spectrum of an *M. sulkowskyi* butterfly. (Inset: Reflected light image.) (b) Spectral response $\Delta R(\Lambda)$ to five tested vapors: water, propanol, butanol, methanol, and ethanol. Concentrations of vapors were 0.02 P/P_0, where P = vapor partial pressure and P_0 = saturated vapor pressure.

(*Continued*)

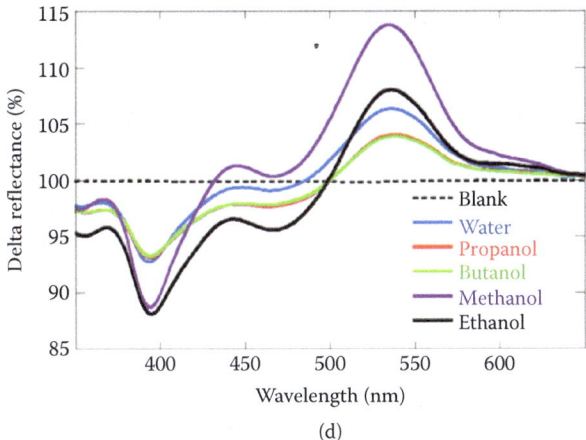

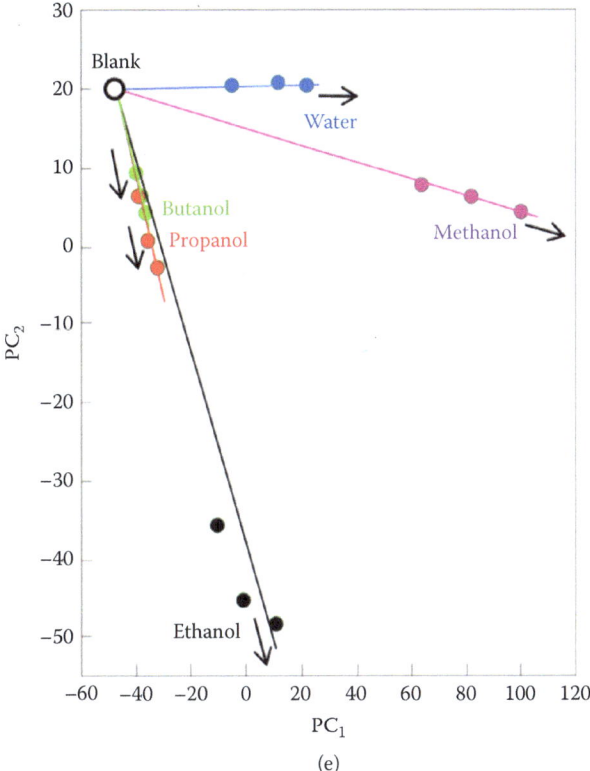

Figure 7.25 (*Continued*) (d) PCA scores plot of spectral response $\Delta R(\Lambda)$ to the five tested vapors. Concentrations of tested vapors were 0.02, 0.04, and 0.06 P/P_0. Lines in (d) are $n \times d$ (vapor index of refraction × adsorbed layer thickness) fits to the scores data for individual vapors. (From Potyrailo, R.A. et al., *Proc. Natl. Acad. Sci. USA*, 110(39), 15567–15572, 2013.[52])

slight decrease in spectral response overall. When the model system has a uniform width to the lamellar stack, the response, however, changes. The double peak is lost, and only a single peak is seen in the reflectance spectra.

When examining the effect of the polarity gradient, we can begin to see how different sections of the nanostructure can affect the reflectance spectrum.[52] By splitting the vertical structure into three sections and examining how the response changed for three different vapors, researchers begin to understand how physisorption in those specific regions may affect response. Figure 7.26 shows the reflectance spectra if the vapor results in a uniform coating of condensed liquid covering the entire structure. When analyzed by PCA the data plots on a single curve relative to the optical thickness of the coating, $n \times d$, where n is the index of refraction and d is the thickness. The reflectance spectra, when the condensed liquid coats only one of the three sections of nanostructure, show distinct differences, reflected in the widely separated PCA data, which show clearly delineated responses. The result is that vapors with different molecular properties will selectively settle on different portions of the vertical structure leading to both the sensitivity and selectivity of the sensor response.

What is exciting about these results is that they offer the promise of selectivity as well as sensitivity in a "reusable" or "resettable" vapor sensor with a relatively simple colorimetric readout. These structures were responding to relatively low vapor concentrations, the smallest at single digit parts per million, while also providing a differentiable responses for each vapor. It will be interesting to see how broad-spectrum and at the same time how unique these responses are. In current sensor systems, in order to sense and identify more than one vapor target, arrays of sensors functionalized with specific binding molecules must be created. It adds system complexity, cost, and potential error. The more unique the vapor response, the greater number of vapors can be differentiated, and the challenge becomes building the spectral library. Optimizing these structures with specialized multimaterial photonic architectures, and tailored surface properties and functionalization could offer a significantly improved sensor performance compared with other nanostructure approaches. Taking material and structural lessons from these biological and bioinspired sensor systems may have a significant future impact on our ability to sense and identify target vapors in the presence of other confusing vapors in real-world operation, which is an ultimate challenge for vapor sensors.

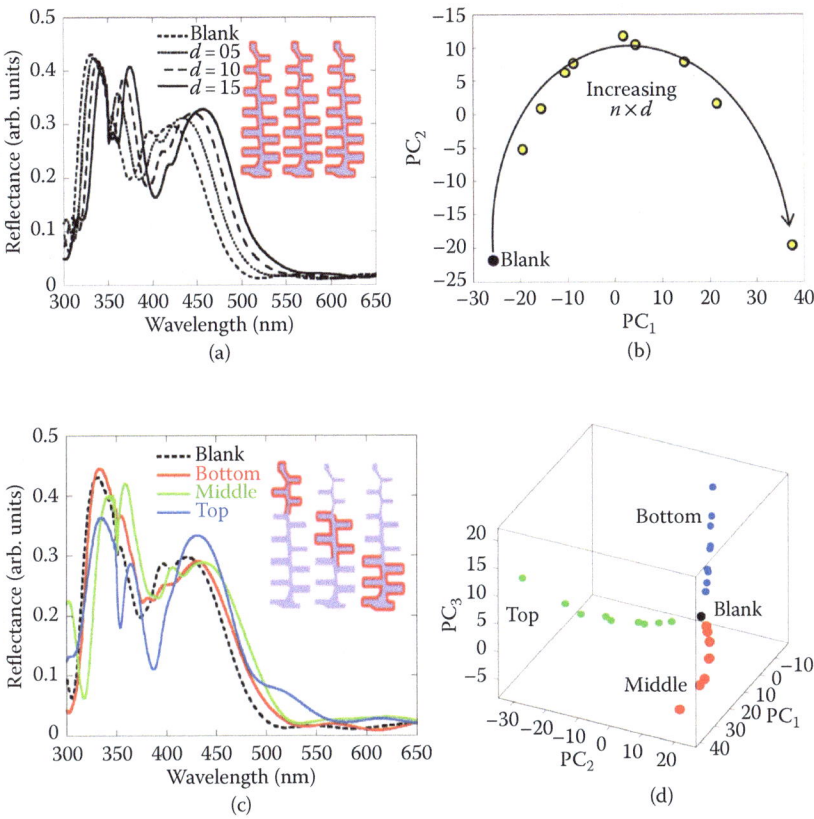

Figure 7.26 (a) Simulated reflectance spectra of uniform vapor adsorption onto the *Morpho* nanostructure. Vapor refractive index $n = 1.5$; vapor concentrations related to condensed liquid layers of thickness $d = 5$, 10, and 15 nm. (b) PCA scores plot of spectral response of the *Morpho* nanostructure on the uniform coverage $n \times d$ with adsorbed vapors. (c) Reflectance spectra of "gradient" vapor adsorption onto three regions of the *Morpho* nanostructure covering only the top, middle, and bottom regions of the nanostructure, respectively. Vapor refractive index $n = 1.5$; $d = 15$ nm. Insets in (a) and (c) show the modeled distribution of vapors. (d) PCA scores plot of spectral response of the *Morpho* nanostructure on the three-region coverage $n \times d$ with adsorbed vapors.

(Continued)

314 BIOINSPIRED PHOTONICS

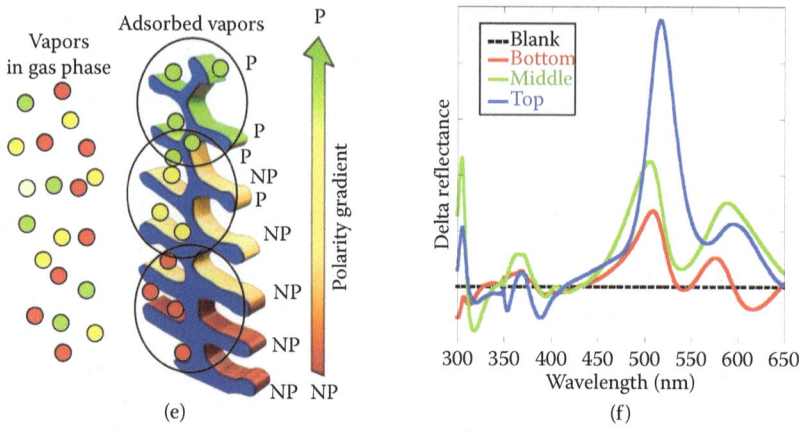

Figure 7.26 (*Continued*)   (e) Illustration of the gradient of the surface polarity of the ridges that runs from the most-polar ridge tops down to the less-polar ridge bottoms. The polarity gradient facilitates preferential adsorption of vapors of different polarity onto the respective regions of the nanostructured ridge. P, polar; NP, nonpolar. (f) The adsorption of vapors at certain locations along the height of the ridge corresponds to the respective regions in simulated $\Delta R(\lambda)$ spectra. (From Potyrailo, R.A. et al., *Proc. Natl. Acad. Sci. USA*, 110(39), 15567–15572, 2013.[52])

References

1. M. Dacke, T. A. Doan, and D. C. O'Carroll, *J. Exp. Biol.* **204** (14), 2481–2490 (2001).
2. L. M. Mathger and R. T. Hanlon, *Biol. Lett.* **2** (4), 494–496 (2006).
3. N. Lambert, Y. N. Chen, Y. C. Cheng, C. M. Li, G. Y. Chen, and F. Nori, *Nat. Phys.* **9** (1), 10–18 (2013).
4. M. Winklhofer, *Science* **336** (6084), 991–992 (2012).
5. http://www.photonics.com/Article.aspx?AID=53739.
6. http://www.biospectrumasia.com/biospectrum/opinion/194407/personal-health-monitoring-anytime-.UnVsoZGEzwI.
7. http://www.bccresearch.com/market-research/instrumentation-and-sensors/sensors-technologies-markets-ias006d.html.
8. M. Motamed, J. Yan, and IEEE, A Review of Biological, Biomimetic, and Miniature Force Sensing for Microflight. (IEEE, New York, 2005).
9. H. Wang and K. Q. Zhang, *Sensors* **13** (4), 4192–4213 (2013).
10. S. Krishna, Optoelectronic and Microelectronic Materials and Devices (COMMAD), 2010 Conference on Optoelectronic and Microelectronic Materials and Devices (COMMAD), 14 December 2010.
11. A. D. Pris, Y. Utturkar, C. Surman, W. G. Morris, A. Vert, S. Zalyubovskiy, T. Deng, H. T. Ghiradella, and R. A. Potyrailo, *Nat. Photonics* **6** (3), 195–200 (2012).
12. J. Frank, "*Eyeing the thermal imaging market*," http://spie.org/x25354.xml. (2002).

13. M. E. McConney, K. D. Anderson, L. L. Brott, R. R. Naik, and V. V. Tsukruk, *Adv. Funct. Mater.* **19** (16), 2527–2544 (2009).
14. A. L. Campbell, R. R. Naik, L. Sowards, and M. O. Stone, *Micron* **33** (2), 211–225 (2002).
15. G. S. Bakken, S. E. Colayori, and T. Duong, *J. Exp. Biol.* **215** (15), 2621–2629 (2012).
16. G. S. Bakken and A. R. Krochmal, *J. Exp. Biol.* **210** (16), 2801–2810 (2007).
17. E. O. Gracheva, J. F. Cordero-Morales, J. A. Gonzalez-Carcacia, N. T. Ingolia, C. Manno, C. I. Aranguren, J. S. Weissman, and D. Julius, *Nature* **476** (7358), 88 (2011).
18. A. B. Safer and M. S. Grace, *Behav. Brain Res.* **154** (1), 55–61 (2004).
19. E. O. Gracheva, N. T. Ingolia, Y. M. Kelly, J. F. Cordero-Morales, G. Hollopeter, A. T. Chesler, E. E. Sanchez, J. C. Perez, J. S. Weissman, and D. Julius, *Nature* **464** (7291), 1006–U1066 (2010).
20. H. Schmitz, H. Soltner, and H. Bousack, *IEEE Sens. J.* **12** (2), 281–288 (2012).
21. T. Kahl, N. Li, H. Schmitz, and H. Bousack, in *Bioinspiration, Biomimetics, and Bioreplication 2012*, edited by A. Lakhtakia and R. J. MartinPalma (Spie-Int Soc Optical Engineering, Bellingham, 2012), Vol. 8339.
22. G. Siebke, P. Holik, S. Schmitz, M. Lacher, and S. Steltenkamp, in *Bioinspiration, Biomimetics, and Bioreplication 2013*, edited by R. J. MartinPalma and A. Lakhtakia (Spie-Int Soc Optical Engineering, Bellingham, 2013), Vol. 8686.
23. N. Zhang, J. C. Cheng, C. G. Warren, A. P. Pisano, and IEEE, *Bioinspired, Uncooled Chitin Photomechanical Sensor for Thermal Infrared Sensing.* (IEEE, New York, 2011).
24. F. Niklaus, C. Vieider, and H. Jakobsen, in *Mems/Moems Technologies and Applications III*, edited by J. C. Chiao, X. Chen, Z. Zhou, and X. Li (Spie-Int Soc Optical Engineering, Bellingham, 2008), Vol. 6836.
25. R. Amantea, L. A. Goodman, F. Pantuso, D. J. Sauer, M. Varghese, T. S. Villani, and L. K. White, in *Infrared Technology and Applications XXIV, Pts 1-2*, edited by B. F. Andresen and M. Strojnik (Spie-Int Soc Optical Engineering, Bellingham, 1998), Vol. 3436, pp. 647–659.
26. R. Potyrailo and R. R. Naik, in *Annual Review of Materials Research, Vol 43*, edited by D. R. Clarke (Annual Reviews, Palo Alto, 2013), Vol. 43, pp. 307–334.
27. C. K. Ho, M. T. Itamura, M. Kelley, and R. C. Hughes, Sandia Report: Review of Chemical Sensors for In-Situ Monitoring of Volatile Contaminants SAND2001-0643, 2001.
28. R. Kaur and A. Dua, *Environ. Monit. Assess.* **184** (5), 2729–2740 (2012).
29. T. Eeva, S. Sillanpaa, J. P. Salminen, L. Nikkinen, A. Tuominen, E. Toivonen, K. Pihlaja, and E. Lehikoinen, *EcoHealth* **5** (3), 328–337 (2008).
30. O. Leitch, A. Anderson, K. P. Kirkbride, and C. Lennard, *Forensic Sci. Int.* **232** (1–3), 92–103 (2013).

31. M. Brattoli, G. de Gennaro, V. de Pinto, A. D. Loiotile, S. Lovascio, and M. Penza, *Sensors* **11** (5), 5290–5322 (2011).
32. S. E. Stitzel, M. J. Aernecke and D. R. Walt, in *Annual Review of Biomedical Engineering, Vol 13*, edited by M. L. Yarmush, J. S. Duncan, and M. L. Gray (Annual Reviews, Palo Alto, 2011), Vol. 13, pp. 1–25.
33. W. van der Goes van Naters and J. R. Carlson, *Nature* **444** (7117), 302–307 (2006).
34. C. E. Finlayson and J. J. Baumberg, *Polym. Int.* **62** (10), 1403–1407 (2013).
35. R. A. Potyrailo, Z. B. Ding, M. D. Butts, S. E. Genovese, and T. Deng, *IEEE Sens. J.* **8** (5–6), 815–822 (2008).
36. Z. H. Wang, J. H. Zhang, J. Xie, C. A. Li, Y. F. Li, S. Liang, Z. C. Tian, T. Q. Wang, H. Zhang, H. B. Li, W. Q. Xu, and B. Yang, *Adv. Funct. Mater.* **20** (21), 3784–3790 (2010).
37. T. G. Mackay and A. Lakhtakia, *IEEE Sens. J.* **12** (2), 273–280 (2012).
38. M. Rassart, J. F. Colomer, T. Tabarrant, and J. P. Vigneron, *New J. Phys.* **10**, 14 (2008).
39. M. Rassart, P. Simonis, A. Bay, O. Deparis, and J. P. Vigneron, *Phys. Rev. E* **80** (3), 6 (2009).
40. C. M. Eliason and M. D. Shawkey, *Opt. Express* **18** (20), 21284–21292 (2010).
41. F. Liu, B. Q. Dong, X. H. Liu, Y. M. Zheng, and J. Zi, *Opt. Express* **17** (18), 16183–16191 (2009).
42. J. H. Kim, S. Y. Lee, J. Park, J. Moon, and IEEE, in *2010 IEEE Sensors* (IEEE, New York, 2010), pp. 805–808.
43. J. H. Kim, J. H. Moon, S. Y. Lee, and J. Park, *Appl. Phys. Lett.* **97** (10), 3 (2010).
44. R. A. Potyrailo, H. Ghiradella, A. Vertiatchikh, K. Dovidenko, J. R. Cournoyer, and E. Olson, *Nat. Photonics* **1** (2), 123–128 (2007).
45. K. Kertesz, G. Piszter, E. Jakab, Z. Balint, Z. Vertesy, and L. P. Biro, *Appl. Surface Sci.* **281**, 49–53 (2013).
46. Y. Gao, Q. Xia, G. L. Liao, and T. L. Shi, *J. Bionic Eng.* **8** (3), 323–334 (2011).
47. X. F. Yang, Z. C. Peng, H. B. Zuo, T. L. Shi, and G. L. Liao, *Sens. Actuator A-Phys.* **167** (2), 367–373 (2011).
48. Z. Bao, M. R. Weatherspoon, S. Shian, Y. Cai, P. D. Graham, S. M. Allan, G. Ahmad, M. B. Dickerson, B. C. Church, Z. Kang, H. W. Abernathy, 3rd, C. J. Summers, M. Liu, and K. H. Sandhage, *Nature* **446** (7132), 172–175 (2007).
49. W. R. Yang, P. J. Lopez, and G. Rosengarten, *Analyst* **136** (1), 42–53 (2011).
50. S. Lettieri, A. Setaro, L. De Stefano, M. De Stefano, and P. Maddalena, *Adv. Funct. Mater.* **18** (8), 1257–1264 (2008).
51. Y. Lang, F. del Monte, L. Collins, B. J. Rodriguez, K. Thompson, P. Dockery, D. P. Finn, and A. Pandit, *Nat Commun.* **4** (2013).

52. R. A. Potyrailo, T. A. Starkey, P. Vukusic, H. Ghiradella, M. Vasudev, T. Bunning, R. R. Naik, Z. X. Tang, M. Larsen, T. Deng, S. Zhong, M. Palacios, J. C. Grande, G. Zorn, G. Goddard, and S. Zalubovsky, *Proc. Natl. Acad. Sci. USA* **110** (39), 15567–15572 (2013).
53. Z. Balint, K. Kertesz, G. Piszter, Z. Vertesy, and L. P. Biro, *J. R. Soc. Interface* **9** (73), 1745–1756 (2012).
54. C. Jiang, M. E. McConney, S. Singamaneni, E. Merrick, Y. Chen, J. Zhao, L. Zhang, and V. V. Tsukruk, *Chem. Mater.* **18** (11), 2632–2634 (2006).
55. A. D. Pris, Y. Utturkar, C. Surman, W. G. Morris, A. Vert, S. Zalyubovskiy, T. Deng, H. T. Ghiradella, and R. A. Potyrailo, *Nat. Photonics* **6** (8), 564–564 (2012).

8

ENERGY FROM LIGHT

8.1 Insatiable Appetite for Power and Energy

Modern life in the developed world was made possible by fossil fuels. Coal, natural gas, and oil enabled our Industrial Revolution, our era of rapid technological advances, and our leap into the twenty-first century. We have a seemingly insatiable thirst for energy. Access to these natural resources drives population affluence and political intrigue. The earth currently supports a population of seven billion people, all of whom consume these nonrenewable resources. In the United States and other developed countries, we each of us consume approximately 250 kWh/day.[1] Worldwide, our energy needs in 2001 were ~13.5 TW.[2] Although this may be offset somewhat by advancing technologies with improved and improving efficiency, it is predicted that our energy needs will still continue to increase. Developing nations are beginning to consume in increasing amounts. The world's population is expected to continue to grow over the next several decades before stabilizing at nine billion people by 2050 when our needs will reach an estimated 27 TW or more.[2]

Currently, fossil fuels supply approximately 85%–90% of the global energy consumed.[3] It has been estimated that the available fossil fuels provide an ample energy reserve to meet our needs, particularly if you include some of the newer extraction technologies, such as fracking. Indeed, current resources, from coal, natural gas, shale, tar sands, etc., could potentially meet a 25- to 30-TW global energy consumption rate for several centuries to come. But the acquiring and burning of these fuels is resulting in alarming environmental consequences. "… In the global arena, the problem for the immediate future is not a limitation of fossil fuel reserves but the consequences of its combustion."[3] Mining, drilling, and now fracking all have long histories of toxic waste, spills, and land and water contamination. Increasing carbon levels in the atmosphere over time from the utilization of these fuels is one of the primary attributed causes of climate change. In fact, carbon levels are expected to nearly double by 2050, even taking into account the potential for new, improved technology to reduce the power requirements of consumer products. It is no secret at all that the world is searching for new energy sources.

There are other potential sources of energy that are renewable, or at least carbon neutral, such as nuclear, hydro, wind, tidal, and solar, for example. We need something approaching 10 TW, or a third of our 2050 anticipated consumption, to come from renewable resources to simply maintain current carbon levels. The European Union has set a goal of 20% renewable energy sources by 2020.[4] Several countries, such as Germany and China, have made significant commitment to the investment in and implementation of renewable technologies.[5,6] In the summer of 2014, Germany began to meet a significant amount of its daily power requirements with renewable power, at times over 50%, using photovoltaics. Some sources, such as nuclear and hydroelectric, require significant infrastructure to produce usable power, not to mention public safety concerns. Others, such as solar, are much more scalable and can be used for either large installation or entirely local energy harvesting. We are now in an era of what some have dubbed "Solar Opportunity," where renewable, carbon-free power is undergoing rapid advancement.[2] In this chapter, we will be exploring solar energy harvesting and how lessons from nature may help to solve our future power needs.

8.2 Harvesting Solar Power

At any given moment, a tremendous amount of solar energy hits the earth. In fact, the sun theoretically provides the earth with more than 1×10^5 TW of solar power, many orders of magnitude above what humankind requires in a year.[3] The available amount of sunlight varies with geography, but even with axial tilt, latitude, and atmosphere accounted for the solar intensity still averages at least 100 W/m^2.[1,7] Many paths are being explored to develop technologies capable of harvesting this abundant resource, particularly in the areas of photovoltaics and solar fuels. Nature has been taking advantage of this free energy for billions of years. Plants, algae, and even some bacteria and animals transform sunlight into energy-dense carbohydrates through the chemistry of photosynthesis. Through a variety of technologies, we humans have also been harvesting solar energy, though nothing close to the scale of nature. Photovoltaic devices have only been developed during the last century.[8] As we will discuss, some researchers are taking advantage of the structures described in Chapters 2 and 3 to improve existing designs. Others are looking back toward the natural solar harvesting process for solar-to-electricity production, whereas others are looking at the same process but inspired by the idea of solar fuel production. All of these designs are motivated by the desire for low-cost, high-efficiency, and ubiquitous renewable energy resources. In this chapter, we will focus on the bioinspired and biomimetic approaches and improvements being made at the front end of the process, solar capture.

8.3 Photosynthesis

Inside the leaves of solar harvesting organisms lies a powerhouse, the engine room of life, where the photosynthetic process occurs. These organisms convert a portion of the solar energy they receive into a form of energy that they can use to grow and replicate. Photosynthesis is a complex, multistage process that evolved billions of years ago by which the organism converts sunlight, water, and carbon dioxide to biofuel. Photosynthesis in plants, for example, occurs in large organelles called chloroplasts that reside within the cells of the plant, such as those seen in the cells of *Plagiomnium affine*, Laminazellen, and Rostock in Figure 8.1. A cartoon illustration of the architecture and functional components within the chloroplast can be seen in Figure 8.2. The chloroplast has a double (inner and outer) membrane. Within each chloroplast are stacks of vesicles called thylakoids. The thylakoids are connected to each other by lamellae and are surrounded by a colorless fluid called the stroma. Each thylakoid vesicle consists of a membrane and an inner fluid called the lumen. Most of the components involved in the photosynthetic process reside in the membrane bilayer. The centers of light-harvesting activity in the membrane are the photosystem I (PSI) and photosystem II (PSII) (named in the order they were discovered). These photosystems contain antenna complexes, assemblies of proteins, and pigment molecules bound together.[7] These antenna complexes are composed of large clusters of pigments (>100 molecules) and there are multiple different pigments in each cluster, such

322 BIOINSPIRED PHOTONICS

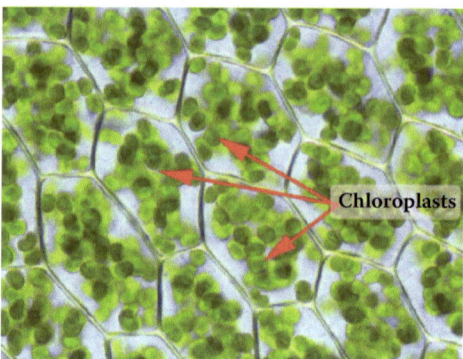

Figure 8.1 Chloroplasts visible in the leaf cells of the many fruited thyme moss *Plagiomnium affine*, Laminazellen, Rostock. (From Wikimedia commons.[9])

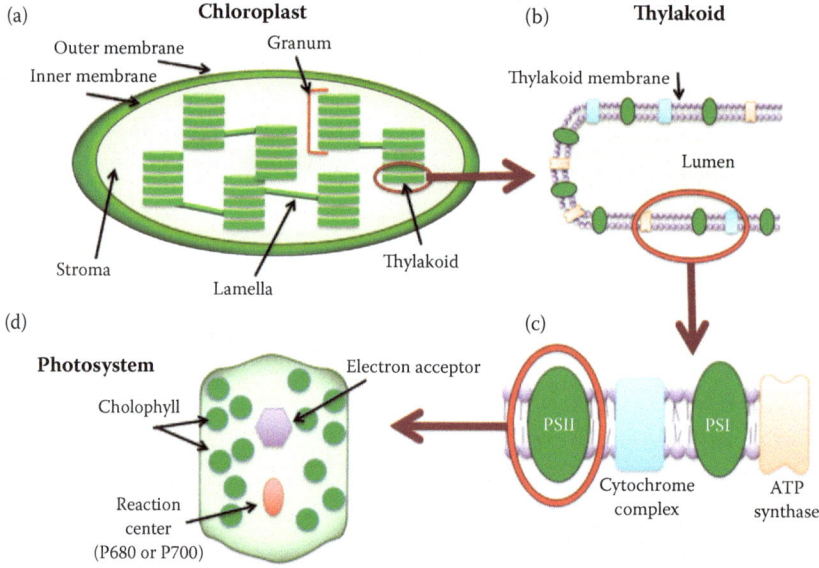

Figure 8.2 Cartoon illustrations of the components of the photosynthetic process. (a) The structure of a chloroplast. The red bracket denotes the stack (called a granum) of individual thylakoid membranes; the red circle highlights a single thylakoid. Arrows denote inner and outer membranes, the stroma (the colorless fluid that surrounds the thylakoids), and the lamella, which connect them. (b) The thylakoid structure, showing the thylakoid membrane and inner portion of the thylakoid, the thylakoid space. (c) Close-up of the thylakoid membrane, showing that the two photosystems (I and II) and other components of the photosynthetic process reside in the thylakoid membrane. (d) The photosystem showing the chlorophyll antenna molecules, the reaction center sensitive to either 680 or 700 nm for photosystem II and I, respectively, and the electron acceptor.

as chlorophyll and carotene. They are bound together with proteins, which provide structural integrity and organization to the complex. The pigments in the antenna complex absorb the incoming (incident) photons, which causes electron excitation. The system then feeds the harvested energy down the electron transfer chain. This pathway may involve multiple intermediate transfers before it reaches the photochemical reaction center. The specific antenna pigments and the structure of the complex vary with organism. They have multiple different pigments to each absorb slightly different wavelengths, which enable the organism to harvest energy from the broadest possible portion of the solar spectrum. Chlorophyll absorbs in the red and blue, for example, which is why chlorophyll-containing organisms look green.

BOX 8.1   LIGHT TO ELECTRONS— THE PHOTOVOLTAIC EFFECT

To transduce light into usable energy, both photosynthesis and photovoltaics work via the photovoltaic effect. When light (e.g., sunlight) is shone on a material (bulk crystal, molecule, etc.), the photons interact with the electrons in the material and an incoming photon will give up its energy to an electron. Electrons in materials can only reside in specific bands of allowed energy (called the electronic band structure) and preferentially fill the bands of lowest energy. In semiconductors, all of the bands up to a certain energy are filled; the highest filled band is called the valence band. Empty but energetically available bands exist, and the band nearest in energy to the valence band, the lowest unfilled band, is called the conduction band. Between the valence and conduction bands in semiconductors is the "bandgap," which is an energy barrier that must be overcome for electrons that are bound to a specific atom in the structure (valence electrons) to be freed to move about the material and conduct. (In metals there is no gap, which is why they conduct electrons well.) Figure 8.3 illustrates both a band structure and the photoelectric/photovoltaic process. The energy the photon donates to the electron ($h\nu$) is dependent on the photon's wavelength.

(*Continued*)

BOX 8.1 (*Continued*) LIGHT TO ELECTRONS—THE PHOTOVOLTAIC EFFECT

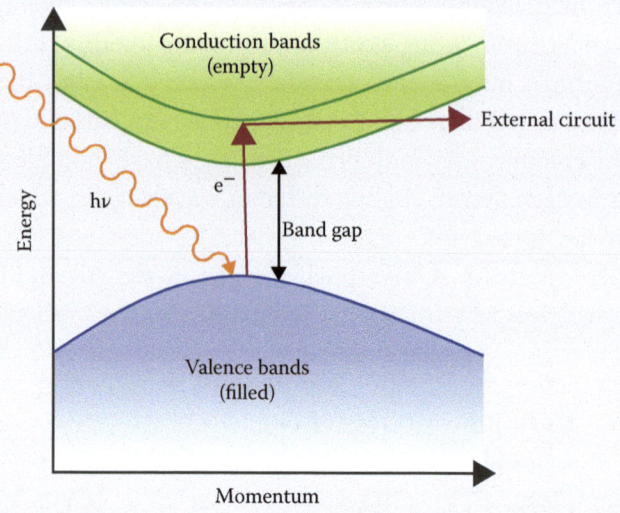

Figure 8.3 Illustration of a semiconductor electronic band structure and the photoelectric/photovoltaic effects. Blue band is the valence band, the highest filled electronic band. The green bands are the empty conduction bands. The energy required to bridge the filled and unfilled bands is called the bandgap. The incoming photon has energy $h\nu$. The red arrow denotes the photoexcitation of the electron (e^-) and its subsequent move to the external circuit in a photovoltaic cell.

When an electron is photoexcited, it can enter a higher energy band, presuming the transferred energy was big enough to overcome the bandgap. Within molecules, electrons are restricted to specific orbitals and the valence and conduction band equivalents go by the names HOMO and LUMO levels (highest occupied molecular orbital and lowest unoccupied molecular orbital, respectively.) When an electron jumps to the conduction band, it leaves behind a hole, which is treated as a positively charged quasiparticle. Together the excited electron and hole are called an exciton, and it is their behavior that drives the remainder of the light to fuel process. Without further perturbation, the electron will relax back into the band it had previously occupied, destroying the exciton. In photovoltaic cells, the electrons are pulled away and fed into an external circuit.

Generally, the equation for photosynthesis can be written as follows:

$$6CO_2 + 6H_2O \xrightarrow{h\nu} C_6H_{12}O_6 + 6O_2$$

That is, plants take in carbon dioxide, water, and sunlight (denoted by the photon energy $h\nu$) and they metabolize it into oxygen and carbohydrates. This chemical equation implies a simple one-step reaction, which is not the case. The photosynthetic process to generate carbohydrate chemical fuels is multistage, driven by two photosystems that work in sequence. A simplified diagram illustrating the photosynthetic process can be seen in Figure 8.4. A 680-nm light is absorbed by the antenna complexes in PSII (the second to be discovered) and excites two electrons. One electron is used to produce protons (H+ ions) from water. The other travels, via the electron transport chain (cytochrome b6f [Cyt b6f] complex and plastocyanin [PC], in part) to PSI. PSI accepts that electron, in addition to photoexcited electrons that resulted from the absorbed 700-nm light, and generates NADPH, a reducing species. NADPH and ATP, created from colocated ATP synthase, are used to convert CO_2 to sugars (e.g., glucose). The overall process is not terribly efficient.[3] It takes the energy from 48 red photons to create a glucose

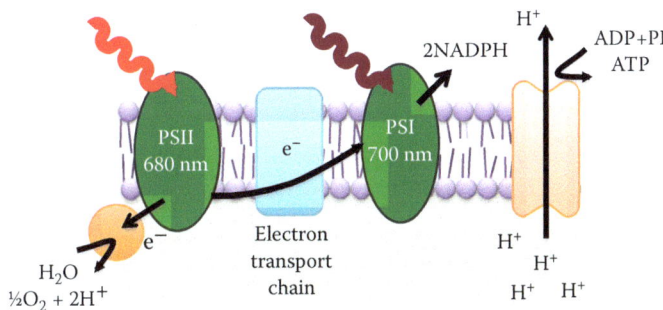

Figure 8.4 Illustration of the light reactions of photosynthesis. 680-nm Light is absorbed by photosystem II (PSII) and two electrons are excited. One electron is used to produce protons (H+ ions) from water. The other travels, via the electron transport chain (cytochrome b6f [Cyt b6f] complex and plastocyanin [PC], in part), to photosystem I (PSI). PSI accepts that electron, in addition to photoexcited electrons due to absorbed 700 nm light, and generates nicotinamide adenine dinucleotide phosphate (NADPH), a reducing species. NADPH and adenosine triphosphate (ATP), created from colocated ATP synthase, are used to convert CO_2 to sugars.

molecule. Much of the energy collected is lost in down-converting absorbed blue photons to the required 680- and 700-nm red photons and to other processes within the plant. The photosystems are also sensitive to light intensity and can sustain damage, so the plant protects itself from photodamage by nonphotochemical quenching of the excited state chlorophylls, returning them to their ground state and dispersing the energy in the form of heat. So, the entire complete energy-forming photosynthetic processes are approximately <2%–4% efficient.[7] However, in some systems, the initial photon absorption process is actually highly efficient. Green sulfur bacteria, for example, survive quite well in low-light conditions where, in these environments, light is a limited resource and the bacteria must make the best use possible of the few photons that reach them. Their quantum efficiency, that is, the efficiency of the photon to electron transduction process, also known as the incident photon to converted electron (IPCE) efficiency, is near 100%.

8.3.1 Quantum Biology

As we have already discussed, the antenna pigments are excited by incoming photons and produce energetic electrons. What happens next is the subject of some debate, and things have gotten very interesting. In most texts, you will find a description of the process that operates in a semiclassical model: the photoexcited electron hops from pigment to pigment, transported along a seemingly random path deeper into the photosynthetic complex until it reaches the reaction center, as seen in Figure 8.5. However, beginning in the mid-2000s, experiments began showing that the truth is more intriguing and complex than we had thought.[10,11] There is evidence that the excitons created could be coherent waves, a hallmark of quantum behavior. Quantum coherence implies that the excitonic waves travel in phase with each other and exist on more than one molecule at a time. With the waves existing in multiple pathways simultaneously, they essentially probe every pathway until they find the most efficient one. Coherence is especially surprising because it was assumed that the inherent "messiness" of a biological system would destroy it. Coherence can be somewhat delicate and easily

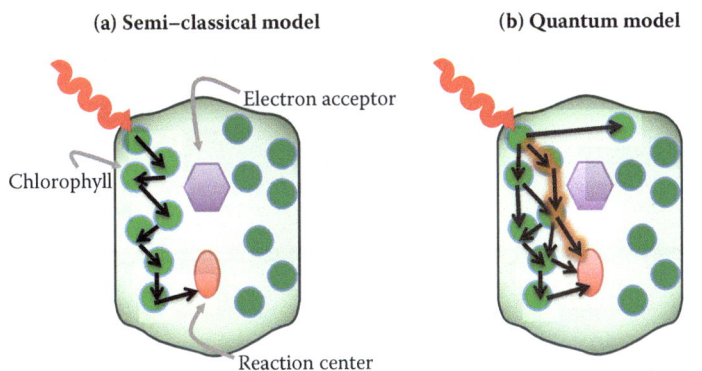

Figure 8.5 Illustration of the electron path to and from the point of photoexcitation to the reaction center. (a) Semiclassical model where the photoexcited electron hops from pigment to pigment, transported along a seemingly random path deeper into the photosynthetic complex until it reaches the reaction center. (b) The quantum biology model where the excitonic waves exist in multiple pathways simultaneously and probe every pathway until they find the most efficient one.

disturbed and destroyed. In fact, there is some thought that perhaps it is that very messiness that makes it possible, that the "environmental noise" may actually free pinned excitons and allow them to continue transport through the system. To examine this phenomenon, researchers looked at green sulfur bacteria, which live deep in the ocean.[12] Their photoexcitation and transport system needs to be highly efficient, as light is a scarce resource at these depths. This is not the only biological system in which we think we see evidence of quantum behavior! Quantum effects are postulated to play a role in avian magnetic navigation, olfactory systems, and more.[10] By studying these effects, we may begin to develop new models for light harvesting.

8.4 Photovoltaics

Photovoltaics as a solar energy–harvesting technology are well known. After decades of research, development, and commercialization, photovoltaics have now generated intense market interest. At some level, all photovoltaics are bioinspired. Some, however, are looking further into how nature accomplishes this task and beginning to explore methods of artificial photosynthesis for creation of energy and biofuels. The synthetic analog to this aspect of nature's technology is

BOX 8.2 PHOTOSYNTHESIS IN ANIMALS

In addition to plants and bacteria, a few examples of photosynthesis in animals and insects have been found. Sea slugs, salamander, and pea aphid, for example, have each been shown to derive some amount of their energy requirements from sunlight.[13–16] In the green sea slugs, *Elysia chlorotica* (in Figure 8.6), each slug baby is born with the ability to produce their own chlorophyll, derived from genes inherited from algae.[13] These sea slugs also eat intertidal algae and store the algal chloroplasts in their own cells. The combination of stored chloroplasts and inherent chlorophyll production enables these slugs to survive if needed without food, living on light. The salamander *Ambystoma maculatum* has intracellular algae that enter the embryo as it develops, captured from their environment, and perhaps also transferred from the mother.[14] There is speculation

Figure 8.6 Photographs of photosynthetic animals: (a) sea slug *Elysia chlorotica*, (b) salamander *Ambystoma maculatum*, and (c) pea aphid *Acyrthosiphon pisum*. (From Wikimedia commons, http://commons.wikimedia.org/wiki/File:Elysia_chlorotica_(1).jpg-mediaviewer/File:Elysia_chlorotica_(1).jpg and The International Aphid Genomics, *PLoS Biol.*, 8(2), 1–24, 2010.[17,18])

BOX 8.2 (*Continued*) PHOTOSYNTHESIS IN ANIMALS

that the intracellular algae provide photosynthetic products to the mitochondria in the cells they inhabit. The pea aphid (*Acyrthosiphon pisum*) can synthesize a different pigment molecule, carotenoid; but it, too, may perhaps use photosynthesis to provide energy to its mitochondria for ATP production.

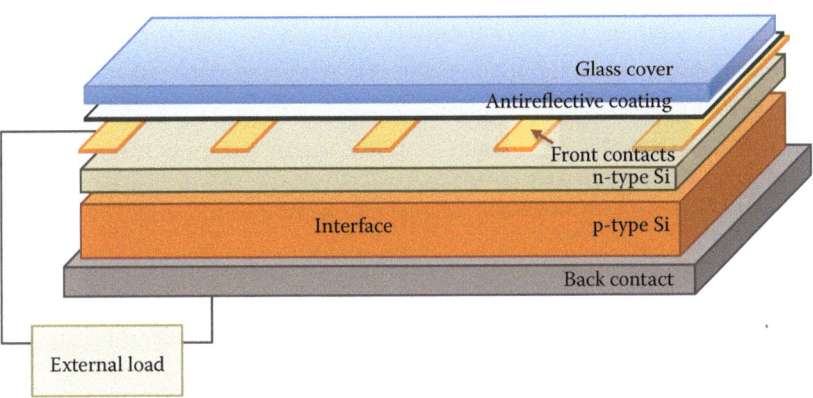

Figure 8.7 Illustration of a simple Si-type solar cell. Bottom layer of the cell is a back contact layer. The active layers of the cell are the p- and n-type layers, where n and p refer to the conduction of the predominant charge carriers in those materials (electrons and holes, respectively). Photoexcited electrons are pulled to the upper contacts and to the external load, creating a current. An antireflective coating Antireflective coatings (ARCs) helps to reduce optical losses (photons that are reflected and not available for use by the system).

the photovoltaic cell, which we use to transduce solar energy from photons into electrical current we can store or use.

The basic single-crystal silicon (Si) solar cell, illustrated in Figure 8.7, is probably the archetypical photovoltaic cell. The Si cells are made of multiple functional layers. At the bottom of the cell is a back contact layer. The active layers of the cell are the p- and n-type layers, where n and p refer to the conduction of the predominant charge carriers in those materials (electrons and holes, respectively). To create the n- and p-type Si, impurities are created in the crystalline Si material by doping with phosphorous (n-type) and boron (p-type), for example. An electric field forms just at the interface of these two layers, called the p-n junction, due to the natural migration of charge carriers between the two materials. The electric field at the junction acts as a diode, allowing electrons to travel from p to n but not in the reverse direction. Photoexcited electrons are pulled to the upper contacts and to the external load creating a current. Electrons are replenished to the system via the back contact. Antireflective coatings (ARCs) help to reduce optical losses (photons that are reflected and not available for use by the system). There are numerous current approaches to the design of solar cells, and the most recent "best of breed" efficiency ranges for four major categories can be found in Figure 8.8.[20] The simplest is probably the single-crystal Si cell described above. The highest efficiency cell demonstrated to date is a four-junction multi-junction compound semiconductor cell with a concentrator, coming in at 44.7%. Other approaches involve amorphous semiconductors, thin films, organics, etc. Each of these technologies is targeted at a specific aspect of the photovoltaic performance. Multijunction cells, for example, are targeting the limits of efficiency. But, for now, these

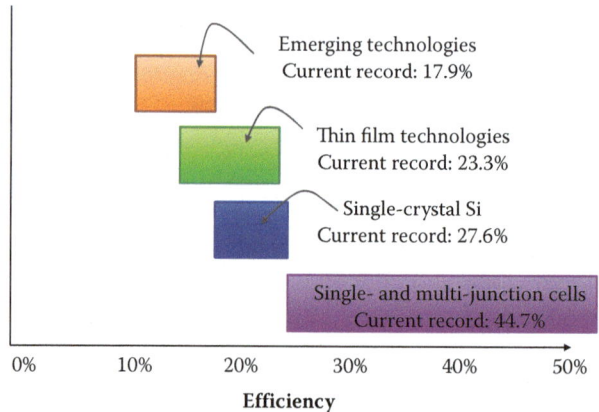

Figure 8.8 Relative ranges of the current best performing photovoltaics. Purple: multijunction solar cells (two, three, and four) and GaAs single-junction cells. Blue: crystalline Si cells. Green: thin-film technologies, such as copper indium gallium selenide (CIGS) cells. Orange: emerging technologies, such as dye-sensitized solar cells (DSSC), perovskites, and organic cells. (Data courtesy of NREL as of August 2014.[19])

cells are very difficult to make for a production-level product, and very costly. Si cells are attractive because Si is abundant, nontoxic, and processing requirements are well known thanks to the integrated circuit industry.[21] Thin-film photovoltaics are targeting applications that require or support low cost, low weight, and flexibility. However, the thin layers are poor light absorbers and defects in the material limit carrier transport.

Photovoltaic technologies have historically been expensive, and cost has become much more important to the adoption of solar technologies in the near term than efficiency.[3] Fossil fuel delivers electricity at a cost of approximately $0.02–0.05\ [\text{kW·hr}]^{-1}$, which includes storage and distribution.[2] To compete with this already entrenched technology, solar must deliver electricity at a low cost/watt. (This also requires other considerations, such as energy storage and other improvements, as well.) This means that the highest performing photovoltaics demonstrated to date, which all have significant fabrication and material costs, are likely priced out of the consumer market. Thin-film technologies are rapidly improving in performance and cost/watt metrics but also require materials such as cadmium, which raise concern for human and environmental health. So, it is important to look at how nature designs the structures that complete this process, to identify areas of inspiration, mimicry, or direct usage of materials and design from biology.

8.5 Antireflective Structures

One cause of low quantum efficiency (IPCE) in photovoltaic cells is optical loss caused by reflectance from the cell surface. Si can reflect as much as 30% of the incident light. Any photons that are reflected are not available to excite electrons and create current. This Fresnel reflection loss occurs due to the abrupt change in refractive index at the air–cell interface. One area of active research is the utilization of bioinspired antireflection structures, such as the "moth-eye" structures, to reduce the amount of light reflecting from the surface.[8,22] As discussed in Chapter 5, these structures are arrays of small cone-shaped pillars found on the surface of the eyes of some moth and butterfly species and the wings of some flying insects, such as the cicada and hawk moth (Figure 8.9). The structures are made from chitin, the same material as the outer layer of the compound eyes

332 BIOINSPIRED PHOTONICS

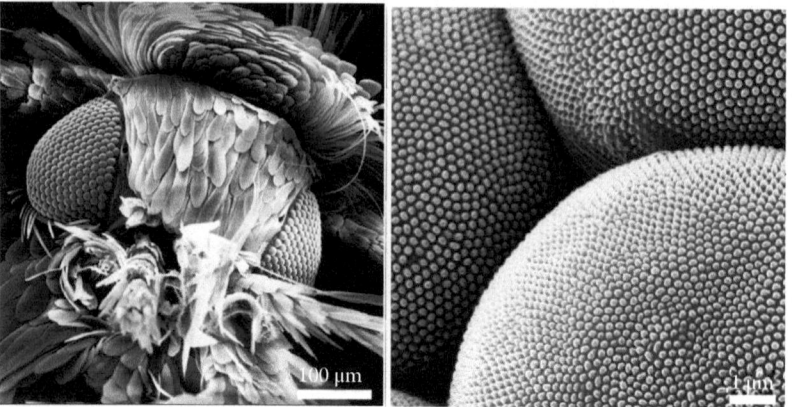

Figure 8.9 (a) Scanning electron microscope (SEM) showing the ventral view of the head and the two compound eyes of the leafminer *Meraria ohridella*. (b) Detailed view of three ommatidia at higher magnification showing the densely packed nipples of parabolic shape on the corneal surface. (From Fischer, S. et al., *Zoomorphology*, 131(1), 37–55, 2012.[23])

and wings. The volume fraction of chitin to air changes with vertical depth into the array as you move to the corneal surface. The moth-eye structures are smaller than the wavelengths of visible light and, therefore, the index of refraction changes gradually from the structure–air interface toward the structure–cornea interface. This graded index greatly reduces the Fresnel reflection over a wide band of wavelengths. (Bandwidth is used to describe the range of wavelengths over which desirable properties, in this case high conversion efficiency, exist). The aspect ratio of the pillars, their height, width, and array density drive the optical properties. Because of the scale of the structures, they provide not only improved solar absorption, but these structures are also often super hydrophobic. Liquids (particularly water) do not stick to the surface, provide the cell with protection from wind and water, and help to shed dirt.

Fabrication of synthetic moth-eye ARCs has been achieved in a variety of ways applied specifically to solar cells; they have been created via templating from actual biological structures or fabricated most often via etching or lithographic techniques. The structures have been made out of Si, glass, TiO_2, and a variety of polymers, for example. Researchers out of Kyung Hee University and Korea Advanced Nano Fab Center in South Korea have created moth-eye arrays in TiO_2 onto gallium arsenide (GaAs) solar cells via a combination of gold nanopatterning and dry etching.[24] TiO_2 was used because it has good optical properties (good optical transmittance and a relatively high refractive index [1.9–2.5]), as well as good mechanical properties (durability and robustness). To create the TiO_2 nanostructures, gold nanoparticles were formed on the surface of a layer of TiO_2 (on top of a GaAs single-junction solar cell) via glancing angle deposition and rapid thermal annealing, as seen in Figure 8.10. These nanoparticles serve as an etch mask and protect the material underneath from the etchant. When the TiO_2 was subsequently dry etched long enough that the gold was fully removed, ~150-nm tall, ~50-nm thick cone-shaped TiO_2 structures remained. The subsequent solar cell with subwavelength structures (TiO_2 SWS) was tested against a comparable cell with no antireflective surface (bare) as well as a comparable GaAs single-junction solar cell with a conventional thin-film TiO_2 layer antireflective surface (single-layer antireflective coating, or TiO_2 SLARC). The graph of IPCE versus wavelength shows that over the entire wavelength region examined in the study, the SWS structures outperform both the bare cell and the SLARC cell.[19] The SWS cell showed, overall, an ~33% improvement in IPCE compared to the bare cell (19.66% vs. 14.74%, respectively) due to the reduced reflectance.

A second type of bioinspired antireflective structure that is being investigated is the light-trapping structural color components on black butterfly wings.[8,22,25] In butterfly wing scales, the black coloration and high solar absorbance are often due to a combination of nanostructure and melanin. Black scales not only contribute to the patterning, shading, and contrast of the butterfly coloration but also seem to play a role in their thermal regulation. The butterfly *Ornithoptera goliath*, for example, has inverse V-shaped ridges on its black wing scales.[26] In some butterflies, such as *Papilio ulysses*, described in Chapter 3, the inverse V-type ridges are coupled with subwavelength holes, effectively trapping light to be absorbed by the melanin.[27] A templating/conversion approach was used to create the *O. goliath* inverse-V structures by turning the chitin in butterfly wing itself into amorphous carbon structure via vacuum sintering, which retained much of the structural fidelity of the natural wing nanostructures.[26] Some have speculated that these black wing structures might be of future use for solar cells; however, their actual application in the area of photovoltaics has not yet been demonstrated.

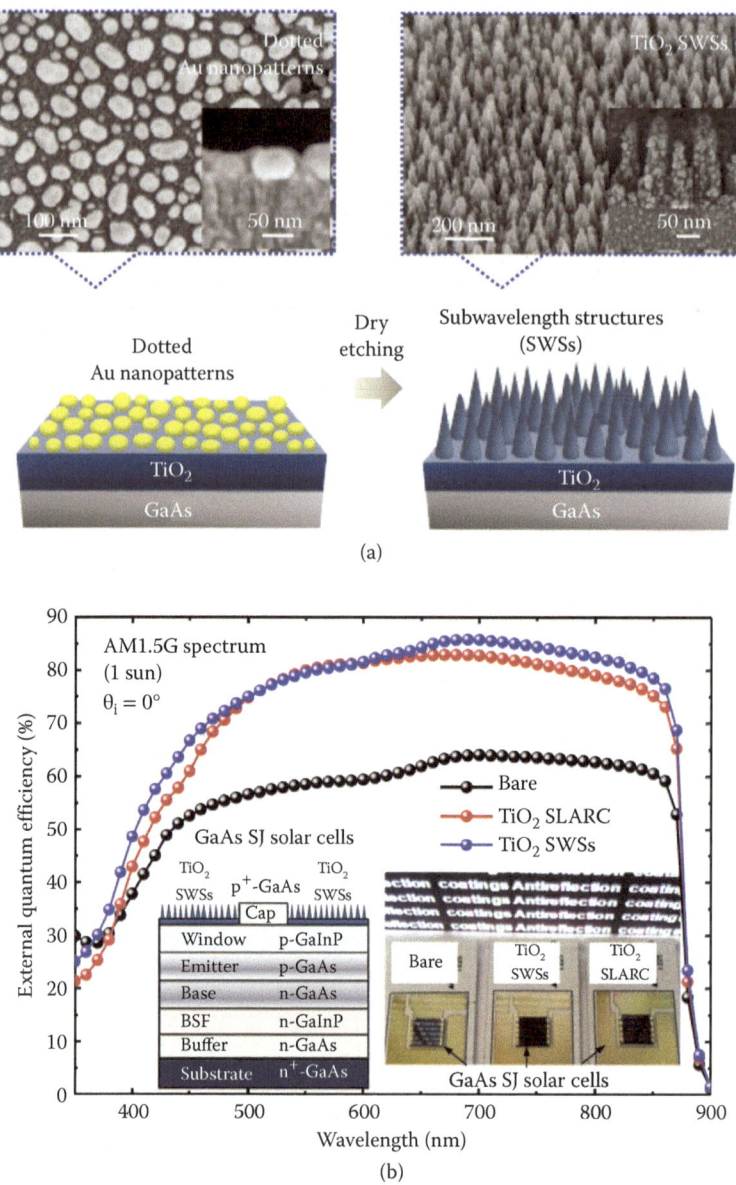

Figure 8.10 (a) Illustration of the fabrication process for cone-like moth-eye or subwavelength structures (SWSs) on the TiO_2 layer/GaAs substrate by dry etching using Au nanopatterns and SEM images of the (left) dotted Au nanopatterns by the thermal dewetting, and (right) etched SWSs on the TiO_2 layer/GaAs substrate. (b) Measured external quantum efficiency (EQE, or incident photon to converted electron) spectra of the III–V GaAs single-junction solar cells with different ARCs of bare surface, TiO_2 layer, and TiO_2 SWSs at $\theta_i = 0°$. In the inset: layer structures (left) of the fabricated III–V GaAs SJ solar cell incorporated with the TiO_2 SWSs as an ARC and photograph (right) of the corresponding assembled solar cells for device measurements. (From Leem, J.W. et al., *Sol. Energy Mater. Sol. Cells*, 127, 43–49, 2014.[24])

8.6 Dye-Sensitized Solar Cells

Another approach to bioinspired solar cell development is the dye-sensitized solar cell (DSSC), also called the Grätzel cell. This type of cell is a composite cell that combines organic and inorganic components, mimicking the multistage photosynthetic photosystem seen in nature. DSSC technology is comparatively young, having been first demonstrated in 1991.[4] As of August 2014, the maximum confirmed reported DSSC research cell efficiency is 11.9%, well below the single-crystalline compound semiconductor approaches.[28] However, these cells, and other organic approaches, may have the advantage of potentially easier fabrication requirements, lightweight systems, flexibility, can be made of abundant materials, and (importantly) lower cost.[4] The general structure for a DSSC is shown in Figure 8.11. The first substrate layer is often glass with a transparent conducting layer (typically an oxide) that operates as a transparent anode. A 5–30 μm thick mesoporous oxide layer on top of the anode (most typically TiO_2) acts as an electronic conductor. Coating the mesoporous TiO_2 is a layer of organic dye, which acts as a "sensitizer"; the dye molecules are covalently bonded to the TiO_2, absorb the incident light, and inject the resulting excited electrons into the TiO_2 conduction band. The

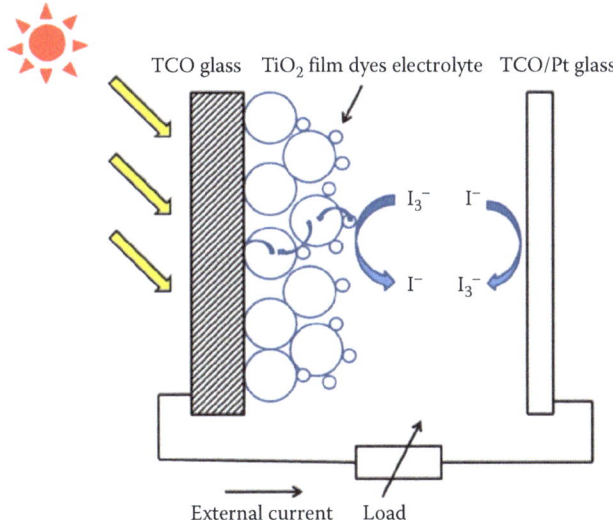

Figure 8.11 Illustration of the general structure of a dye-sensitized solar cell (DSSC). The key component to the structure is the monolayer of charge transfer dye that is covalently bonded to the surface of the mesoporous oxide layer (TiO_2, typically) that enhances light absorption, and the electrolyte that helps regenerate the dye. (From Gong, J. et al., *Renewable Sustainable Energy Rev.*, 16(8), 5848–5860, 2012.[4])

electrons travel to the anode and are utilized at the external load (for example, a battery). An electrolyte layer enables regeneration of the dye electrons and finally a second conductor-coated glass layer coated with a platinum catalyst completes the cell. The porosity and morphology of the TiO_2 layer play a large role in the efficient transfer of electrons, as it affects the amount of surface area available for bonding to dye molecules. The dye and TiO_2 combination is most interesting from a biophotonics point of view, as they mimic the antenna-reaction center structures found in the natural photosynthetic machinery. Development of these cells faces several challenges if they intend to make better-than-incremental improvements in total efficiency. The liquid electrolyte can cause cell failures, the dye performance lifetime is a concern for commercial products, the efficiencies in the longer wavelengths (red and near infrared [IR]) are low, and its ability to absorb light is somewhat poor (extinction coefficient) requiring large area cells. Areas of research for performance improvement in these cells include replacements of the platinum (Pt) catalysts (Pt is a high cost material), investigation of novel transparent conducting oxides, and solid electrolytes, to name a few. Importantly for us, groups are also looking to optimize the structure of the mesoporous anode material, improve the wavelength range over which the dye molecules absorb, and the overall architecture of the cell, to improve the DSSC efficiency.

8.6.1 Biophotonic Crystal Structures

Many biophotonic structures are playing a role in the advancement of photovoltaic systems. As discussed in Chapters 2 and 3, many of the bright colors in nature are due to highly ordered crystalline structures, such as opals, or micro- and nanostructural components in insect integuments. By manipulating the structure of the semiconductor anode layer or by introducing additional nanostructured layers into the cell, researchers may be able to not only increase the surface density of dye-semiconductor bonds but also use the structures to enhance the absorption of specific wavelengths, all of which can improve the conversion efficiency. Currently, many of the photonic crystals being introduced into solar cell structures are modeled off of the structures found in opalline crystals and beetle scales, such as the synthetic structures shown in Figure 8.12.[30] Opal crystals are simply

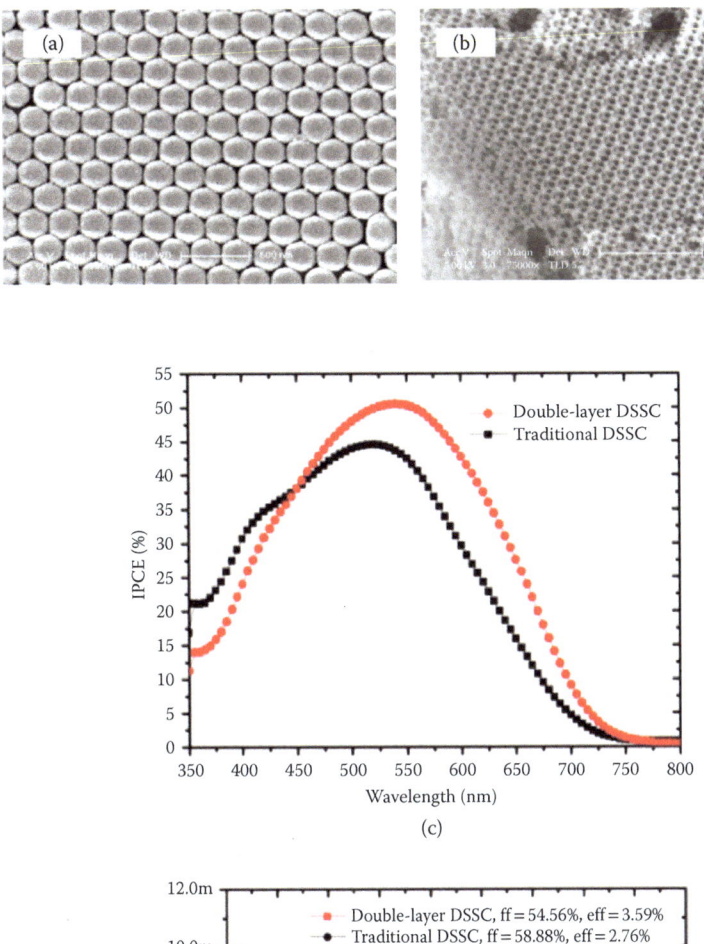
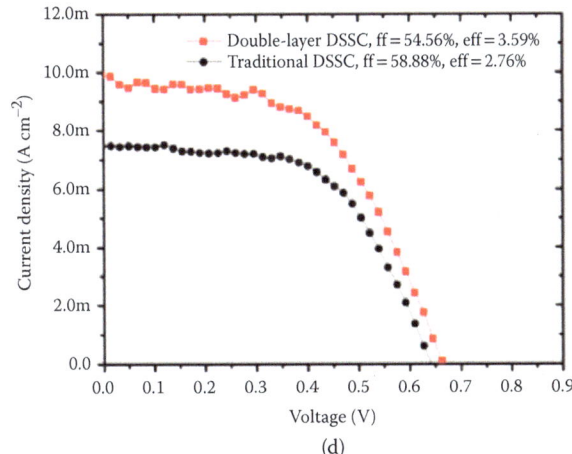

Figure 8.12 (a) SEM of the (111) face of polystyrene opal structure. (b) SEM of the (111) face of TiO$_2$ inverse opal structure. (c–d) Photovoltaic behavior of the optimized double-layer dye-sensitized solar cell (DSSC) and traditional DSSC, (c) incident photon to converted electron spectra, and (d) I–V curves. (From Li, H.R. et al., *Sol. Energy*, 86(11), 3430–3437, 2012.[29])

a regular array of close-packed, layered nanometer-scale colloids or particles. Compared to many of the naturally found biological photonic crystal structures, opalline structures are relatively easy to fabricate via self-assembly. A variant on the opal structures, the inverse opals, is formed by fabricating an opal structure, infilling the interstitial spaces with a chosen material, and then dissolving the original colloids. In the Figure 8.12, the opal structure is fabricated out of polystyrene beads, whereas the inverse opal structure the material used to infiltrate the structure was TiO_2.

Both of the opal and inverse opal structures have been investigated for potential improvement in solar cell performance. One reason is because they can be used to facilitate or block transmission of specific wavelengths and create, essentially, a light trap that concentrates the sunlight and reduces escape losses.[29,31] For light trapped in the structure, scattering within the structure lengthens the optical path and, therefore, increases the absorption and conversion efficiency. Another interesting effect associated with photonic crystals in solar cells is the slow photon effect, where the localized reduction in group velocity also effectively increases the optical path length and enhances the optical absorption.[29,32] Researchers from the Shanghai Jiaotong University, for example, have fabricated a double-layer DSSC in which the double-layer photoelectrode is formed from a fairly standard mesoporous nanocrystalline (NC) TiO_2 layer and an interconnected macroporous inverse opal TiO_2 layer.[29] Graphs of the IPCE versus wavelength and the current density versus wavelength enhancement can be seen in Figure 8.12 comparing the double-layer DSSC to a traditional DSSC. The black squares indicate traditional DSSC data, which have only the mesoporous TiO_2 layer. The red circles denote the cell with the double-layer photonic crystal structure. The photonic crystal structures increase the IPCE to approximately 50% between 500 and 600 nm. Likewise, the current density shows a significant improvement, particularly at lower voltages.

The noncolloidal biological photonic crystal structures that influence coloration in butterflies are being integrated into photovoltaic and photocatalytic architectures, as well, to enhance the wavelength specificity and absorption and ultimately improve efficiency.[8] Porous multilayer structures, for example, localize light of specific wavelengths depending on the layer thickness and spacing. The multilayer can then be fabricated to localize light within the absorption band of the sensitizer dye to enhance the photoabsoption, while remaining transparent at other wavelengths. Two- and three- dimensional photonic crystals enable even further control over the optical properties within the solar cell. Complex multicomponent structures, containing Bragg stacks, NC layers, and mesoporous layers, are being investigated and have shown some improvement over other comparable single-component cells. Researchers at the University of Pennsylvania have demonstrated a complex multicomponent (heterostructured) photoanode in conjunction with a solid-state electrolyte in a DSSC and compared it to the performance of cells constructed with liquid electrolytes and more simply structured anodes.[33] An illustration of the cell architecture can be seen in Figure 8.13 and shows the three components in the heterostructured photoanode. At the bottom of the photoanode structure was a 500-nm-thick layer of organized mesoporous TiO_2 that acted as an interfacial layer (IF layer). The middle layer was a 7- or 10-μm-thick NC TiO_2 layer. The top layer consisted of a 2-μm-thick

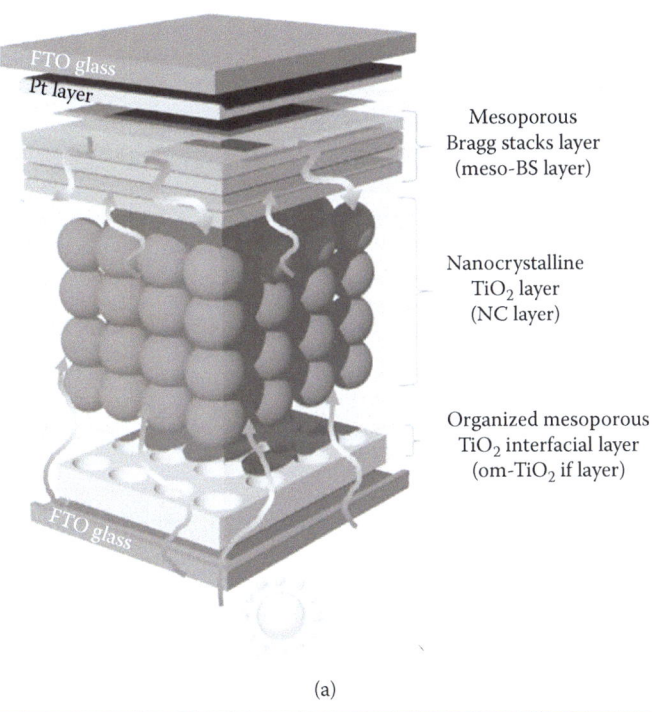

(a)

Figure 8.13 (a) Structure of dye-sensitized solar cells (DSSCs) with the multilayered photoelectrode.

(*Continued*)

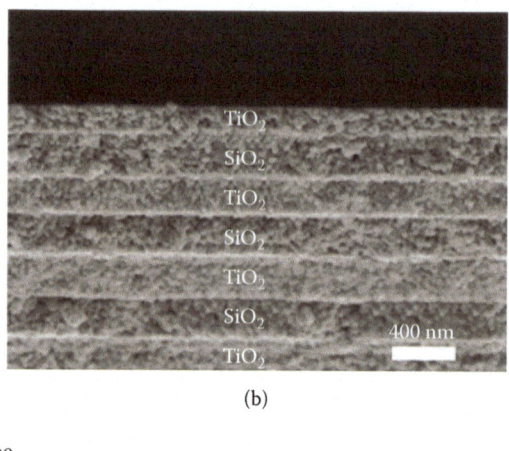

(b)

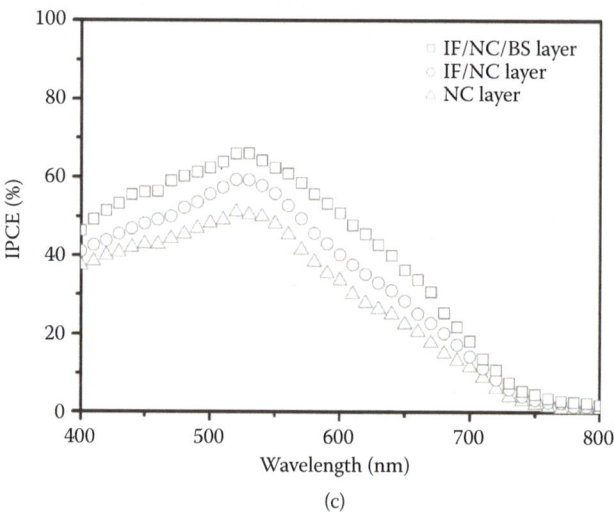

(c)

Figure 8.13 (*Continued*) (b) Cross section of mesoporous Bragg stack layer consisting of organized mesoporous TiO_2 and colloidal SiO_2. (c) Incident photon to converted electron values of the ssDSSCs (solid state dye-sensitized solar cells) assembled with three types of photoelectrode deposited on a fluorine-doped tin oxide glass substrate and solid-state polymerized ionic liquid as a solid electrolyte at 100 mW/cm². The thickness of nanocrystalline layer is approximately 7 μm. (From Park, J.T. et al., *Adv. Funct. Mater.*, 23(17), 2193–2200, 2013.[33])

mesoporous Bragg stack (meso-BS layer) made of alternating layers of mesoporous TiO_2 and colloidal SiO_2. The meso-BS layer (a multilayer reflector) helps to reflect light back into the cell. The mesoporous IF layer improved photovoltaic performance by increasing the visible light transmittance into the cell. It also enhanced electron transfer and created better adhesion to the transparent conducting oxide layer. The heterostructured photoanode improved performance in both a liquid electrolyte and the solid-state electrolyte cell architectures. A graph of IPCE versus wavelength for a solid-state electrolyte cell can also be seen in Figure 8.13, showing clear improvement in efficiency for the three-component photoanode compared to a two-component photoanode, as well as a single-component anode.

Researchers have also investigated bicontinuous gyroid structures such as those found on the wings of butterflies such as *Callophrys rubi* for use in solar cells.[8,34–36] Bicontinuous gyroid structures (such as those seen in Figure 8.14) are interesting because each of the air "lattice" and the chitin lattice in the butterfly wing structures forms a continuous three-dimensional (3D) periodically repeating volume that interpenetrates the other. This structure is attractive for photoanodes in solar cells, because the intimate integration of both the donor and the acceptor material can improve separation and extraction of free charges.[34] Several groups have been tackling the challenging task of fabrication both via templating and even 3D printing, as described in Chapter 3.[37,38] A multinational research collaboration led by scientists at Cambridge and Oxford in the United Kingdom has fabricated gyroid structures via electrochemical deposition onto degradable block copolymer films and incorporated them into solar cell architectures and demonstrated improved photovoltaic performance of the cell.[34] The fabricated gyroid structure and illustration of the gyroid cell architecture can also be seen in Figure 8.14.

8.6.2 Molecular Antennas

Within the natural photosynthetic machinery, the molecules responsible for the initial stage, the photoexcitation process within the antenna complexes, are ripe for exploration. Some researchers are looking at natural dyes from fruits, flowers, bacteria, etc., as alternatives.[39] These molecules have the benefit of being biodegradable and more environmentally friendly than synthetic molecules, easy to prepare, and comparatively low cost. However, the natural dye-based DSSCs are limited by short lifetime and relatively low conversion efficiency. One challenge with the predominant existing sensitizer molecules is that they are often ruthenium-centered complexes. Ruthenium is a rare transition metal; it is expensive and dangerous from a health perspective. In addition, the ruthenium complexes, while broadband

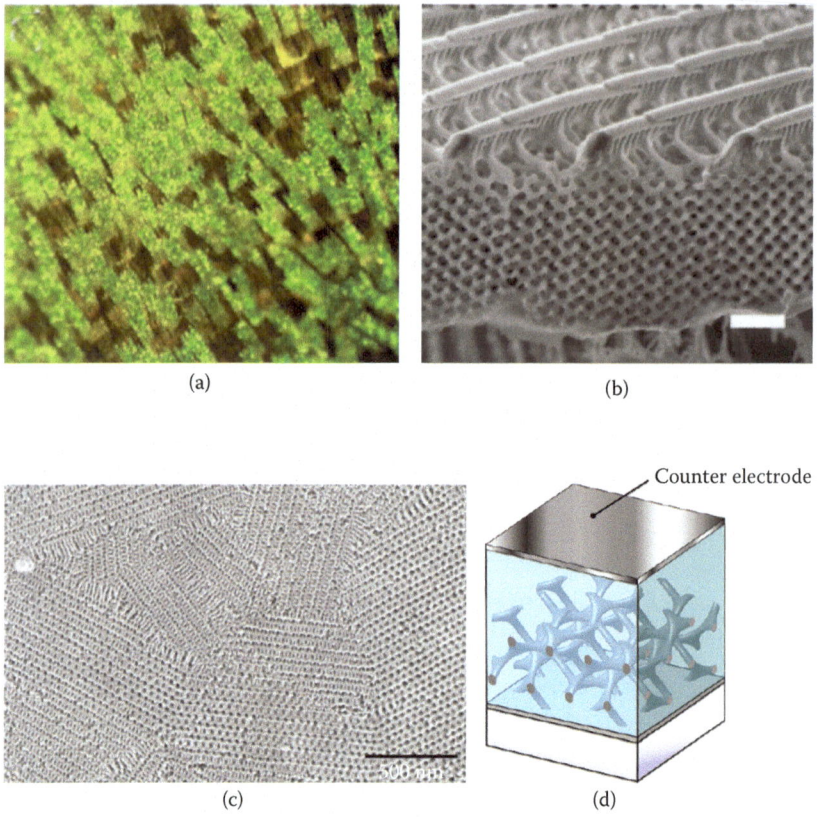

Figure 8.14 The green wing scales of *Callophrys rubi*. (a) In each scale, domains form a mosaic of different colors giving a uniformly green color. (b) Cross-sectional view showing domains of single gyroid structured photonic crystals in different orientations. Note that the cross-ribs and main ribs form disordered connections with the underlying photonic crystals. Scale bar: 1 μm. (c) Bicontinuous bulk heterojunction solar cell architecture. (d) Surface SEM image of replicated anatase TiO_2 gyroid network after oxidative removal of the polymer template. (From Mille, C. et al., *RSC Adv.*, 3(9), 3109–3117, 2013 and Crossland, E.J.W. et al., *Nano Lett.*, 9(8), 2807–2812, 2009.[34,37])

absorbers, have weak extinction coefficients (how strongly they absorb at any given wavelength), particularly in the near IR region.[40] IR is an extremely attractive wavelength band for energy harvesting. A significant portion of the solar light that reaches earth is in the IR. Not only that but the earth absorbs a good portion of that and reemits it in the mid- and long-wave IR. Solar cells that can effectively convert IR light can continue to produce energy at night and on cloudy days. In addition, modern technology creates a lot of waste heat, which is also a target for energy harvesting. Si is transparent to IR light and so cannot access this readily available source. IR absorbers are a particularly attractive target for research and development, as we cannot yet harvest light at those wavelengths with any significant efficiency. Organic dye molecules can be strong absorbers, and their particular wavelength ranges are tunable from ultraviolet (UV) to near IR. However, they are typically narrowband, which means that any particular dye molecule is unlikely to have good absorption over the entire solar spectrum. Many of the research approaches aimed at improving DSSC performance involve improving the band of wavelengths over which the solar cell can harvest photons efficiently. As a single molecule is unlikely to be a strong absorber over the entire spectrum, many approaches involve including multiple dyes in the DSSC. Some create a cocktail of dyes that bind to the TiO_2 and act as cosensitizers. Others create dye multilayers or site-specific adsorption, which is a complex process.

Another strategy is to create an antenna complex, multiple dye molecules bound to a sensitizer dye molecule, modeled after the light-harvesting complexes found in natural photosystems. Proper choice of molecules for this approach must have compatible photonic properties, with complementary absorption spectra. If done successfully, this approach has the potential benefit of increasing the optical density of the cell and broadening the absorption window. The associated antenna molecules can absorb light and then channel it to the sensitizer molecule. One example of these multicomponent approaches is the bichromophore seen in Figure 8.15.[41] Researchers from Northwestern University linked a zinc-porphyrin sensitizer (labeled #2) with a bodipy chromophore (4,4-difluoro-4-bora-3a,4a-diaza-s-indacene, labeled #3). The porphyrin alone has poor absorption of light at wavelengths around 500 nm. The bodipy chromophore can absorb light at around 530 nm, which complements the porphyrin molecules' performance and enables the complex to absorb photons better over a wider range of wavelengths. The quantum efficiency and overall external conversion efficiency were still low but improved over the zinc-porphyrin sensitizer alone.

An advantage of the linked system compared with other approaches (such as the cosensitizers) is that the antenna molecules do not need to be able to inject electrons efficiently, as they are not bound directly to the

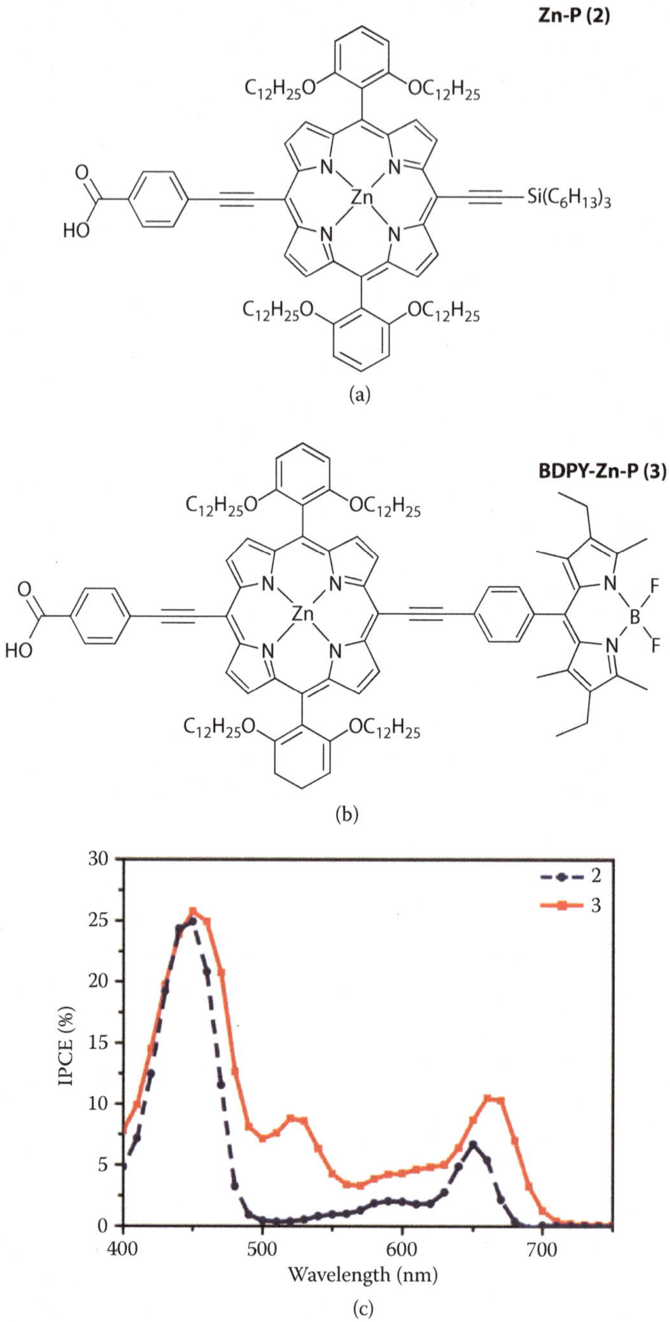

Figure 8.15 (a) The zinc porphyrin derivative (2). (b) The zinc porphyrin-bodipy dyad (3). (c) Incident photon to converted electron versus wavelength comparing the performance of a zinc porphyrin sensitized solar cell to the performance of the zinc porphyrin-bodipy sensitized solar cell under AM1.5 illumination. (From Lee, C.Y. and Hupp, J.T., *Langmuir: ACS J. Surf. Colloids*, 26(5), 3760–3765, 2009.[41])

photoanode.[40] That is, only the molecule that will be directly attached to the photoanode needs to be able to inject electrons well. The other molecules in the antenna complex attach only to the electron-injecting sensitizer molecule. This means that there are more options when it comes to choice of dye molecules. Another advantage is that this approach allows for increased dye loading. The surface area of the photoanode ultimately limits the number of molecules that are able to bind to the surface. Multiple dyes bound to the surface of the photoanode will limit the number of any given molecule that is injecting electrons into the photoanode. In antenna complexes, however, multiple dyes can be combined onto a sensitizer; but as the sensitizer is the only molecule that links to the TiO_2 photoanode, the increased number of dye molecules does not crowd out and restrict the number of sensitizer molecules capable of binding to the surface. Creating the antenna/sensitizer complex also allows for greater control over the intermolecular distances, which are important for efficient energy transfer. Energy transfer efficiency is extremely important to the performance of the solar cell; the molecules need to absorb photons and then move the subsequently energized electrons along. The energy transfer between antenna and sensitizer must also be minimally 80% efficient, according to some authors. These multicomponent arrays can be self-assembled, enabling rapid, lower cost synthesis of highly complex structures.

8.7 Solar Fuels and Artificial Photosynthesis

These same strategies described above for photovoltaic cells are also of interest for solar fuel technologies. Photosynthesis turns sunlight, water, and CO_2 into energy-dense carbohydrates. We take advantage of this process when we synthesize biofuels from corn, switchgrass, or algae. However, cost and land required to grow energy crops, often at the expense of food crops, are two very significant hurdles.[2,42] It is estimated that even if every piece of naturally irrigated, cultivatable land on earth (not currently being used for food) was used to grow energy crops, the output would still only achieve 5–10 TW.[2]

The natural photosynthetic process as a whole is not terribly efficient, reaching at best 2%–4%.[3,7] In these systems, the efficiency challenges lie primarily in the back-end production of biomass, rather than at the front-end photoabsorption process.[3] Coupling solid-state solar harvesters with catalysts to create hybrid structures is one approach that is being explored.[2] Entirely organic approaches, whether synthetic, biological, or genetically engineered, are also being explored. Once the catalysts produce hydrogen, it can be then converted into higher energy density fuels for storage and use. Much of the focus in the field centers on new photoelectrocatalysts that are efficient, robust, made of abundant materials, and low cost.[43] Still, significant work is being done on the photoharvesting piece of the puzzle. There are three primary steps in the formation of photosynthetic hydrogen. First, incoming photons that reach the solar cell are absorbed and the photon energy creates excitons (electron–hole pairs). The electrons and holes are then spatially separated so that they do not recombine and travel to their respective reaction centers. These electrons are then used in the reduction reaction to form hydrogen ions and oxygen. By optimizing the photon-harvesting step so that more electron–hole pairs are formed, photocatalytic efficiency can be improved.[8]

Several approaches that use natural photonic structures to improve conversion efficiency with respect to solar fuels can be found in the literature.[8] One example uses the wing structure found on the male butterfly *Papilio nephelus* Boisduval as a template to grow hierarchically structured photocatalysts. The structures on the black wings of the *P. nephelus* Boisduval butterfly (seen in Figure 8.16) include the inverse V-type ridges and hole arrays described in Chapter 3 and are

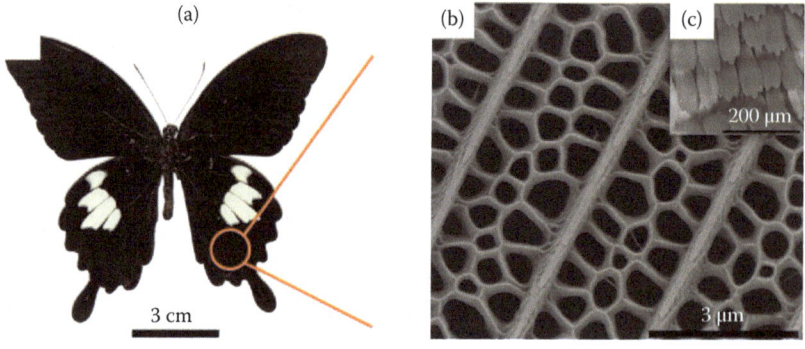

Figure 8.16 (a) Image of the male *Papilio nephelus* Boisduval butterfly. (b) High-magnification field emission scanning electron microscopy (FESEM) image showing the architecture of a black scale; (c) low-magnification FESEM image of orderly arranged wing scales.

(*Continued*)

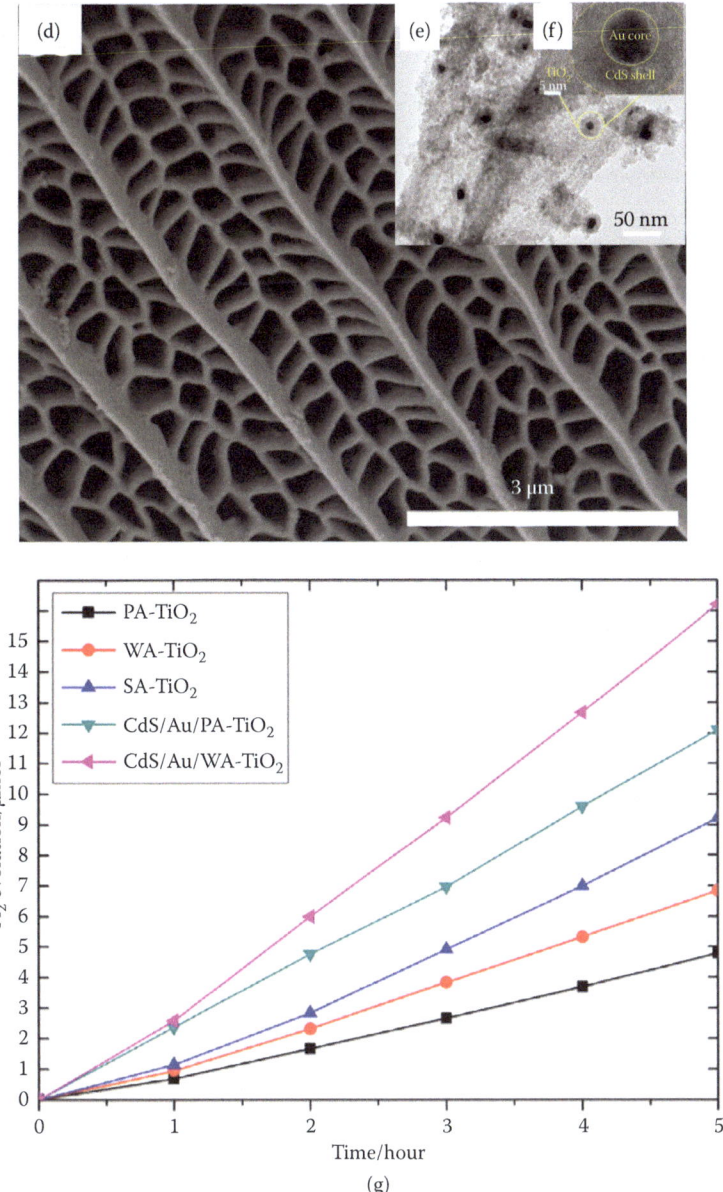

Figure 8.16 (*Continued*) (d) FESEM image of gold-TiO_2 wing architecture replica; (e) transmission electron microscopy image of CdS/Au/TiO_2 wing architecture replica; (f) high-resolution transmission electron microscopy shows CdS/Au with a core-shell structure formed on the TiO_2 wing architecture substrate; (g) H_2 evolution of the samples (50 mg) under ultraviolet and visible light irradiation. WA-TiO_2, TiO_2 wing architecture replica; PA-TiO_2, wing membrane (plate) TiO_2 replica; SA-TiO_2, scale architecture TiO_2 replica; CdS/Au/WA-TiO_2, TiO_2 wing architecture with CdS/Au core-shell nanoparticles; CdS/Au/PA-TiO_2, TiO_2 wing membrane (plate) with CdS/Au core-shell nanoparticles. (From Ding, L. et al., *Int. J. Hydrogen Energy*, 38(20), 8244–8253, 2013.[44])

intriguing photoabsorbing structures. Researchers from Shanghai Jiaotong University used the wings as templates to create nanostructured replicas of the wing architecture, wing membrane (creating a plate structure), and scale architecture out of photocatalytic TiO_2.[44] Versions of the wing architecture and plate architecture replicas were also grown with CdS/Au core-shell nanoparticles deposited onto the TiO_2 to increase photocatalytic performance, inspired by the PSII and PSI architectures of natural photosynthesis. (WA-TiO_2 = TiO_2 wing architecture replica, PA-TiO_2 = wing membrane [plate] TiO_2 replica, SA-TiO_2 = scale architecture TiO_2 replica, CdS/Au/WA-TiO_2 = TiO_2 wing architecture with CdS/Au core-shell nanoparticles, CdS/Au/PA-TiO_2 = TiO_2 wing membrane [plate] with CdS/Au core-shell nanoparticles.) The H_2 evolution of the samples (50 mg) in water, and under UV and visible light, as a function of time shows some insight into performance enhancement of these systems. First, comparing the performance of the three replicas made of titania alone shows that the nanostructured replicas (SA- and WA-TiO_2) outperform the flat (plate) replica. Second, the CdS/Au/TiO_2 replicas outperform the TiO_2 replicas alone by a significant margin, with the nanostructured CdS/Au/WA-TiO_2 having the highest H_2 evolution of the five. In fact, the nanostructured multicatalyst replica improved performance by ~200% compared to the titania replica without structure (PA-TiO_2). Finite difference time domain (FDTD) modeling of the performance of these structures in water was also performed. The data implied that the multicatalyst structure improved H_2 evolution performance compared to titania alone. The data also implied that the nanostructure had a significant effect on the photocatalytic performance, and FDTD analysis indicates that the structure reduced the UV reflectance of the structure, enhancing photoabsorption. Scale-up of this approach, however, could be a challenge as the template size (wing) limits the size of the replica. Unless large-scale fabrication of these and other inspired and optimized structures can be achieved, it is difficult to see this in commercial production.

Another approach utilizing 3D photonic structures used inverse opal structures to trap photons within the crystal and increase the interactions between photons and photocatalysts much like in DSSCs.[8] Inverse opals are attractive because the opal structure can be formed with easy to fabricate and inexpensive materials, such as polymethyl methacrylate (PMMA) spheres. The interstitial spaces can then be filled in with photocatalyst materials and the PMMA dissolved, leaving a nanostructured solid photocatalyst. Noble metal doping of the photocatalyst is being investigated as a way to improve the exciton output and separation and improve performance. Order and disorder in the photonic crystal arrays are also being examined, with research looking at arrays of multiple size spheres. Many of these approaches are showing improvement in catalytic efficiency.

8.8 Hybrid Systems

There have been some approaches to solar fuel that have made it to the popular and science press in the last few years. One of these is the "artificial leaf" technology out of Harvard.[45] The leaf consisted of catalytic molecules deposited on a Si wafer. When the leaf is placed in water in the sunlight, the Si harvests the incoming (incident) photons and provides energy to the catalysts to split the water into oxygen and hydrogen, which can then be used for fuel. One recent advance in this technology development is the use of cobalt-based catalysts, which can break off and reform on the Si surface. This allows the leaf to be, in a sense, self-repairing and enables the device to be used in whatever water is available, even dirty water. However, the efficiency of the entire process is limited by the performance of the solar cell, leaving the artificial leaf operating at around 5% efficiency. Harvard spun this technology off to a company called Sun Catalytix.[46] However, this company has moved away from commercializing this technology at this time, as the cost of the photovoltaic makes the entire photocatalytic system not as competitive as other fuel and energy technologies.

There are other groups also exploring the possibility of self-healing solar cells.[47,48] One of the challenges for solar cells made with organic molecules is lifetime. The organic molecules are often damaged over time and exposure to light and heat. Plants have several mechanisms to reduce damage in their photosynthetic centers, including rejecting a significant portion of incident light.[7] In plants, their photosynthetic compounds are broken down and repaired constantly.[47,48] A team from MIT has demonstrated the ability to trigger assembly and disassembly of a complex system of recombinant proteins, phospholipids, and carbon nanotubes to mimic this repair cycle. The repairing photocell consisted of a photoelectrochemical cell connected to two dializers. The composition of the solution is monitored spectroscopically. The complex consists of lipid bilayer nanodiscs assembled on a single-walled carbon nanotube, as seen in Figure 8.17. Light-converting proteins form the reaction center and attach to the nanodiscs, creating a photoactive complex. The addition of a sodium chlorate surfactant breaks apart the complex, and the damaged components and surfactant are then removed through membrane dialysis. Healthy new complexes then spontaneously reform, regenerating the cell's photocurrent performance as well. Indications are that this repair cycle can be performed repeatedly with no loss of performance. This was the first proof of concept of its kind to mimic the biological repairability of the complex.

Another notable hybrid approach combined biological materials with solid-state materials and showed some interesting properties. One example uses PSI complexes extracted in spinach.[49–51] PSI is a complex of 17 proteins and approximately 180 antenna molecules centered around the reaction center, called P700 (because it is sensitive to 700-nm-wavelength light).[49] PSI is small; it is only 500 kDa or about 10 nm in size. It can be extracted from plants and cyanobacteria and still continue to function, though its lifetime seems limited thus far. PSI extracted from spinach is an efficient converter of sunlight to electrical energy. Extracted PSI has been deposited on a host of

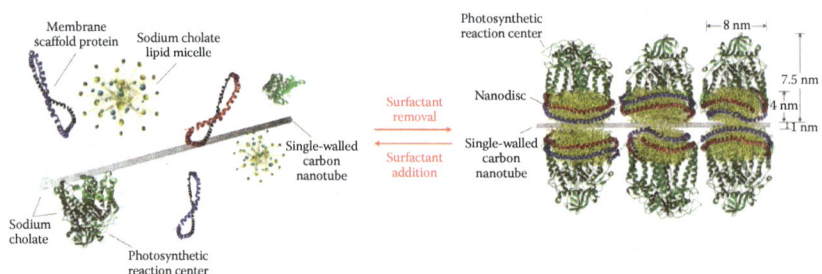

Figure 8.17 Schematic of self-assembled photoelectrochemical complexes. The self-assembly process involves carbon nanotubes and photosynthetic reaction centers and occurs on surfactant (sodium cholate) removal. Membrane dialysis induces spontaneous self-assembly of 1,2-dimyris toyl-*sn*-glycero-3-phosphocholine (DMPC) and membrane scaffold proteins to form nanodiscs, which reconstitute the reaction centers while suspending nanotubes in aqueous solution. The resulting, highly ordered complex is shown on the right-hand side. Addition of sodium cholate completely decomposes the complexes back into the individual components in the initial condition (left side). (From Ham, M.-H. et al., *Nat Chem.*, 2(11), 929–936, 2010.[48])

materials, such as metals, oxides, glass, alumina, and most recently graphene.[49,52] A team from Vanderbilt was able to deposit spinach-extracted PSI onto hole-doped (p-type) Si (p-Si).[51] (Implanting impurities into the Si material adds extra positive charges, creating P-type Si.) Si has the benefit of being ubiquitous in the solar and integrated circuit industries. By controlling the doping type and density, they can engineer the electronic bandgaps so that the valence band of the p-Si matches the HOMO level of the P700 chromophores. The p-Si donates electrons, which are then photoexcited in the P700. The mediator then enables those excited electrons to move to the counter electrode. The doped semiconductor only allows electrons to flow into PSI and will not accept them in the reverse. This functionality mimics some of the behavior seen in the natural antenna systems, which use structural alignment to direct the flow and separation of charge carriers. The PSI proteins on the Si are not aligned, but the doping of the Si enables a similar functionality. The results were pretty impressive. Figure 8.18 shows that for a 1-μm-thick PSI film and methyl viologen mediator (0.2 M concentration), the current density reached ~875 μA/cm². For comparison, Si cells can demonstrate current densities in the mA/cm². In nature, PSI can be extremely long-lived,

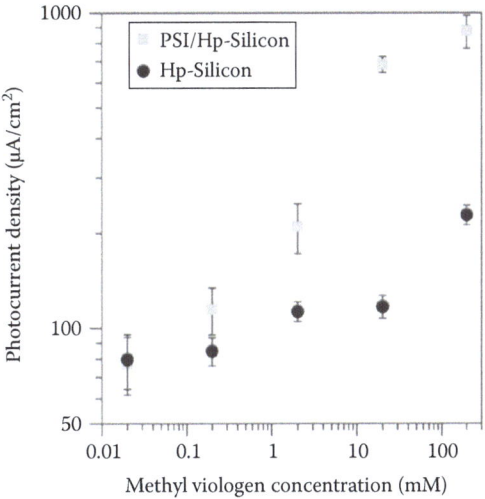

Figure 8.18 Photocurrent density versus mediator concentration. Unmodified heavily p-doped silicon (black) or heavily p-doped silicon modified with a 0.9 ± 0.1 μm film of photosystem I (PSI) (gray) was used as the working electrode. Photocurrent densities ($n = 4$) were determined by taking the difference between the current density in the dark and the current density after 10 seconds of illumination (633 nm high-pass filtered). (From LeBlanc, G. et al., *Adv. Mater.*, 24(44), 5959–5962, 2012.[51])

years in evergreen trees, for example.[50] In the lab, however, lifetime can be a serious issue, with samples only remaining functional on the order of months.

One criticism that spinach PSI has received is that spinach is a food crop. The Vanderbilt team that extracted PSI from spinach also demonstrated biohybrid cells where the PSI has been extracted from the kudzu plant.[52] Because the photosystems in plants like these are more or less similar, what we find in one system should be more or less what we find in another. By studying one as a model system, we can learn about most others. This means that the kudzu-based PSI should be roughly like what we find in the spinach-based PSI, and it is. Kudzu is an invasive vine imported from eastern Asia in 1876. The vine grows rapidly, spreading 500 km²/year and costing hundreds of millions of dollars. There is a reason why it is called the "vine that ate the south"; it currently covers over 810,000 ha of the southern United States. Use of kudzu as an energy crop could potentially turn a noxious weed into our new best friend. The team extracted PSI from kudzu leaves in the same manner as with the spinach. The PSI was obtained by maceration and centrifugation. They then deposited that along with a redox mediator onto a boron-doped Si (p-Si) creating a "wet" cell. Although current densities as high as the spinach biohybrid cell were not seen, the kudzu cell made a promising start, reaching 67 µA/cm².

8.9 Nanoantennas

Organic antenna complexes are modeled after the biological antenna complexes in photosynthetic systems. Another approach inspired by the antenna complexes that feed photoelectrons to reaction centers is to create inorganic antenna architectures. We humans already use antennas quite a bit, for television and radio, in our cell phones, and many other communications applications. Optical antennas are being investigated for a host of interesting applications ranging from imaging to light-emitting structures.[53] Researchers have also been looking at ways to incorporate these inorganic antenna structures into solar cells to increase solar absorption and potentially achieve significant improvements in efficiency. Some researchers have been looking to add nanoparticles as antenna-like additions to traditional Si and thin-film solar cells, as well as the DSSCs we have discussed above. These nanoparticles lengthen the optical path length and enable enhanced absorption with thinner active layers. Another approach is to create a solar cell that does not use the traditional architectures at all. These antenna approaches are exciting as they may ultimately lead to high efficiency, inexpensive, and flexible solar energy collection; however, it requires additional technology development that presents significant challenges to near-term application. Two examples of nanoantennas currently being researched are roll-to-roll processing of "nantennas" from collaborators at the University of Missouri, Idaho National Laboratory (INL), and MicroContinuum, Inc., and the plasmonic nanoantennas from Rice University.

The nanoscale antennas fabricated by the University of Missouri, INL, and MicroContinuum, Inc., are designed to collect light at IR wavelength and convert it into alternating current.[54] Antennas are resonators and their electromagnetic properties (wavelength and bandwidth, for example) are driven by the antenna's material and geometry. The size of the antenna scales with frequency, so it is possible to fabricate structures that can resonate with (and capture) IR wavelengths. However, the antenna structures must be fabricated at very high resolution, something that is still challenging. The incoming photons create a standing wave of electrons within the antenna, generating alternating current at the antenna's resonant frequency.[54] Antennas are designed such that the resonant energy is directed to specific feedpoint for convenient collection and that energy can then be converted to direct current for use by additional conversion circuitry. One of the challenges limiting the applicability of nantennas at this point is actually this conversion circuitry. To harvest IR light and convert it to usable energy, we need integrated rectifying diodes operating at the appropriate frequencies to convert alternating to direct current. However, at IR wavelengths, the antennas operate at terahertz (THz) frequencies. The high-performance electronics operating at these frequencies is a significant challenge.

The INL team has fabricated arrays of IR resonant antenna structures on a flexible, low-cost substrate as a first proof of principle. Potential geometries and materials were explored using e-beam lithography, which is too expensive and time consuming for large-scale application; eventually having a roll-to-roll process is particularly interesting. The nantennas were created using a multistep templating process, where a master stamp is fabricated to create a pattern. The master stamp enabled the team to print 4-in. coupons of metallic antenna arrays that were stitched together (a SEM of an array of nantennas as well as a larger "nantenna sheet" that was stitched together from multiple smaller coupons can be seen in Figure 8.19). The team demonstrated that they were able to design in resonance at desired wavelengths with a >90% difference between

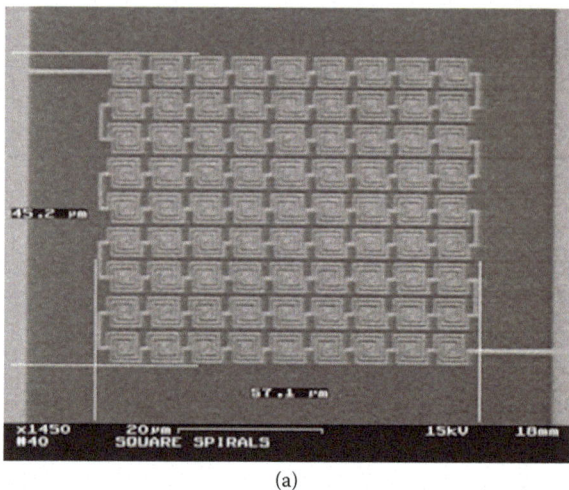

(a)

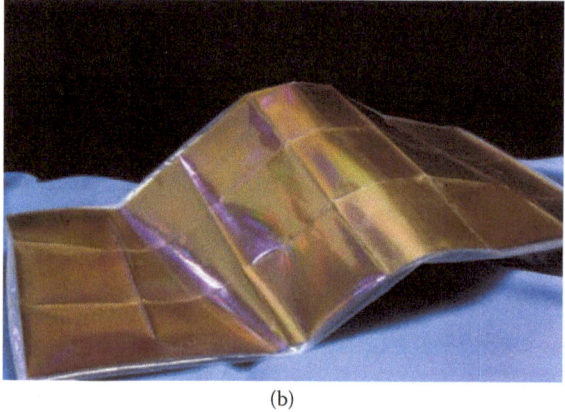

(b)

Figure 8.19 (a) An array of nanoantennas, printed in gold and imaged via SEM. The deposited wire is roughly a thousand atoms thick. (b) Nanoantenna sheet, stitched together from 18 coupons. (From Kotter, D.K. et al., *Solar Nantenna Electromagnetic Collectors*, American Society of Mechanical Engineers, New York, 2009.[54])

in and out of band emissivity. However, our ability to fabricate complete nantenna photovoltaic arrays is limited by several factors: modeling of the antenna elements at these frequencies, understanding the properties of materials that operate at THz frequencies, manufacturing challenges associated with operating at these length scales, and finally the need for ultrafast high-performance diodes. If successful, the combination of very high efficiency energy harvesting over the entire solar band plus the low cost, high-scale manufacturing that the team is developing could have a tremendous impact on the energy industry for solar and waste heat harvesting.

Collaborators at Rice University have demonstrated plasmonic nanoantennas that can also collect light in the IR.[55] Structures were fabricated that combined metallic nanoantenna structures with a semiconductor to create photodiodes that resulted in a measurable photocurrent. Arrays of rectangular gold nanorods, such as those seen in Figure 8.20, were fabricated on a Si substrate. Photons couple to the nanoantennas and excite resonant plasmons, which are collective oscillations in the free electrons of the conductive material. The plasmons decay into energetic electrons. At the interface of the metal and semiconductor is a potential barrier called a Schottky barrier. The hot electrons are energetic enough to get injected over the barrier, resulting in a measurable photocurrent. To demonstrate, arrays of gold nanorods $30 \times 50 \times (110-158)$ nm in height, width, and length, respectively, were fabricated. By varying the length of the nanorods, the resonant wavelengths could be tuned. The gold nanorods were connected to the Si substrate by a titanium (Ti) adhesion layer. To ensure that there was no cross-antenna coupling, the antennas were spaced 250 nm apart and surrounded with an insulating SiO_2 region. A transparent electrode (made from indium tin oxide, ITO) was fabricated on top. The rod geometry has a direct influence on the photocurrent, particularly its wavelength dependence. The team demonstrated this in the short-wave IR, from about 1.2–1.7 µm. Measurements of the photocurrent as a function of wavelength for the different rod lengths can also be seen in Figure 8.20. The Ti adhesion layer plays a large role in antenna performance, as the height of the Ti layer affects the Ti/Si Schottky barrier height. The quantum efficiency, which is affected by the barrier height, antenna

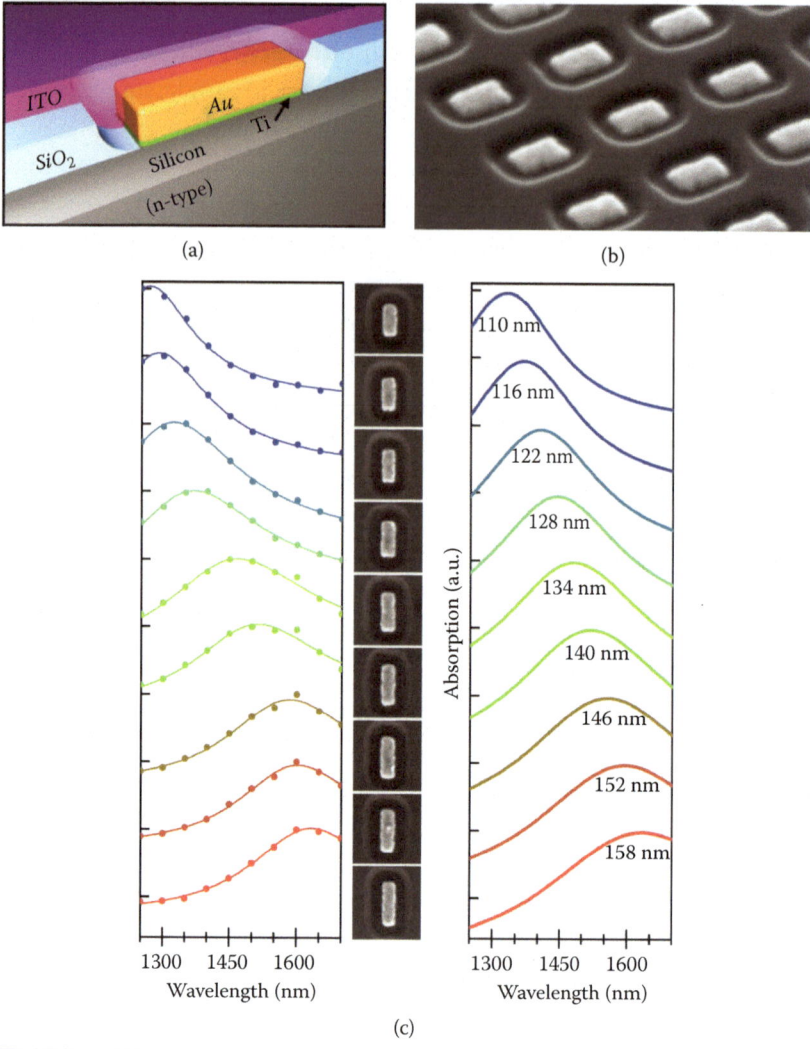

Figure 8.20 (a) Representation of a single Au resonant antenna on an n-type silicon substrate. (b) Scanning electron micrograph of a representative device array prior to indium tin oxide coating, imaged at a 65° tilt angle. (c) Photocurrent spectra extracted from the responsivity spectra (A) (points) fit with Lorentzian curves (solid lines) and corresponding calculated absorption spectra. (From Knight, A. et al., *Nat. Comm.*, 4, 2013.[55])

geometry, and other device parameters, is small in the demonstrated system, only 0.01%. However, there are many parameters to examine for device optimization and QE improvement. This technology is not only attractive for solar cells but also for imaging, photodetection, and a host of other potential applications.

References

1. T. A. Faunce, in *Molecular Solar Fuels* (The Royal Society of Chemistry, 2012), pp. 506–528.
2. N. S. Lewis and D. G. Nocera, *Proc. Natl. Acad. Sci. U.S.A.* **103** (43), 15729–15735 (2006).
3. J. Barber and P. D. Tran, *J. R. Soc. Interface* **10** (81) (2013).
4. J. Gong, J. Liang and K. Sumathy, *Renewable Sustainable Energy Rev.* **16** (8), 5848–5860 (2012).
5. Stephen Luntz, 2014, http://www.iflscience.com/technology/germany-now-produces-half-its-energy-using-solar.
6. Jack Perkowski, 2014, http://www.forbes.com/sites/jackperkowski/2014/06/17/china-leads-in-renewable-investment-again/.
7. R. E. Blankenship, D. M. Tiede, J. Barber, G. W. Brudvig, G. Fleming, M. Ghirardi, M. R. Gunner, W. Junge, D. M. Kramer, A. Melis, T. A. Moore, C. C. Moser, D. G. Nocera, A. J. Nozik, D. R. Ort, W. W. Parson, R. C. Prince, and R. T. Sayre, *Science* **332** (6031), 805–809 (2011).
8. S. Lou, X. Guo, T. Fan, and D. Zhang, *Energy Environ. Sci.* **5** (11), 9195–9216 (2012).
9. "Plagiomnium affine laminazellen" by Kristian Peters—Fabelfroh—photographed by myself. Licensed under Creative Commons Attribution-Share Alike 3.0 via Wikimedia Commons, http://commons.wikimedia.org/wiki/File:Plagiomnium_affine_laminazellen.jpeg-mediaviewer/File:Plagiomnium_affine_laminazellen.jpeg.
10. P. Ball, *Nature* **474** (7351), 272–274 (2011).
11. N. Lambert, Y. N. Chen, Y. C. Cheng, C. M. Li, G. Y. Chen, and F. Nori, *Nat. Phys.* **9** (1), 10–18 (2013).
12. G. S. Engel, T. R. Calhoun, E. L. Read, T. K. Ahn, T. Mancal, Y. C. Cheng, R. E. Blankenship, and G. R. Fleming, *Nature* **446** (7137), 782–786 (2007).
13. M. E. Rumpho, K. N. Pelletreau, A. Moustafa, and D. Bhattacharya, *J. Exp. Biol.* **214** (2), 303–311 (2011).
14. R. Kerney, E. Kim, R. P. Hangarter, A. A. Heiss, C. D. Bishop, and B. K. Hall, *Proc. Natl. Acad. Sc. U.S.A.* **108** (16), 6497–6502 (2011).
15. J. C. Valmalette, A. Dombrovsky, P. Brat, C. Mertz, M. Capovilla, and A. Robichon, *Sci. Rep.* **2** (2012).
16. N. A. Moran and T. Jarvik, *Science* **328** (5978), 624–627 (2010).
17. International Aphid Genomics Consortium, *PLoS Biol.* **8** (2), 1–24 (2010).
18. "Elysia chlorotica (1)" by EOL Learning and Education Group—Elysia chlorotica,Credit: Patrick Krug Cataloging Diversity in the Sacoglossa LifeDeskUploaded by Dominikmatus. Licensed under Creative Commons Attribution 2.0 via Wikimedia Commons, http://commons.wikimedia.org/wiki/File:Elysia_chlorotica_(1).jpg-mediaviewer/File:Elysia_chlorotica_(1).jpg.
19. *Data taken from the National Renewable Energy Laboratory Chart research cell efficiency records*, http://www.nrel.gov/ncpv/images/efficiency_chart.jpg.

20. B. Parida, S. Iniyan, and R. Goic, *Renewable Sustainable Energy Rev.* **15** (3), 1625–1636 (2011).
21. C. M. Hsu, G. F. Burkhard, M. D. McGehee, and Y. Cui, *Nano Res.* **4** (2), 153–158 (2011).
22. K. L. Yu, T. X. Fan, S. Lou, and D. Zhang, *Prog. Mater. Sci.* **58** (6), 825–873 (2013).
23. S. Fischer, C. H. G. Muller, and V. B. Meyer-Rochow, *Zoomorphology* **131** (1), 37–55 (2012).
24. J. W. Leem, J. S. Yu, D. H. Jun, J. Heo, and W. K. Park, *Sol. Energy Mater. Sol. Cells* **127**, 43–49 (2014).
25. Q. B. Zhao, X. M. Guo, T. X. Fan, J. Ding, D. Zhang, and Q. X. Guo, *Soft Matter* **7** (24), 11433–11439 (2011).
26. Q. B. Zhao, T. X. Fan, J. A. Ding, D. Zhang, Q. X. Guo, and M. Kamada, *Carbon* **49** (3), 877–883 (2011).
27. S. A. Boden, A. Asadollahbaik, H. N. Rutt, and D. M. Bagnall, *Scanning* **34** (2), 107–120 (2012).
28. http://www.nrel.gov/ncpv/.
29. H. R. Li, Q. Tang, F. Y. Cai, X. B. Hu, H. H. Lu, Y. Yan, W. Hong, and B. Y. Zhao, *Sol. Energy* **86** (11), 3430–3437 (2012).
30. A. E. Seago, P. Brady, J. P. Vigneron, and T. D. Schultz, *J. R. Soc. Interface* **6 Suppl 2**, S165–184 (2009).
31. J.-Y. Chen, E. Li, and L.-W. Chen, *Prog. Electromagn. Res. M* **17**, 1–11 (2011.).
32. T. Baba, *Nat. Photonics* **2** (8), 465–473 (2008).
33. J. T. Park, J. H. Prosser, S. H. Ahn, S. J. Kim, J. H. Kim, and D. Lee, *Adv. Funct. Mater.* **23** (17), 2193–2200 (2013).
34. E. J. W. Crossland, M. Kamperman, M. Nedelcu, C. Ducati, U. Wiesner, D. M. Smilgies, G. E. S. Toombes, M. A. Hillmyer, S. Ludwigs, U. Steiner, and H. J. Snaith, *Nano Lett.* **9** (8), 2807–2812 (2009).
35. S. Khlebnikov and H. W. Hillhouse, *Phys. Rev. B* **80** (11) (2009).
36. G. E. Schroder-Turk, S. Wickham, H. Averdunk, F. Brink, J. D. F. Gerald, L. Poladian, M. C. J. Large, and S. T. Hyde, *J. Struct. Biol.* **174** (2), 290–295 (2011).
37. C. Mille, E. C. Tyrode, and R. W. Corkery, *RSC Adv.* **3** (9), 3109–3117 (2013).
38. C. Pouya, D. G. Stavenga, and P. Vukusic, *Opt. Express* **19** (12), 11355–11364 (2011).
39. C. Li, F. Wang, and J. C. Yu, *Energy Environ. Sci.* **4** (1), 100–113 (2011).
40. F. Odobel, Y. Pellegrin, and J. Warnan, *Energy Environ. Sci.* (2013).
41. C. Y. Lee and J. T. Hupp, *Langmuir: ACS J. Surf. Colloids* **26** (5), 3760–3765 (2009).
42. K. Bullis, 2013, http://www.technologyreview.com/view/515041/exxon-takes-algae-fuel-back-to-the-drawing-board/.
43. P. D. Tran, L. H. Wong, J. Barber, and J. S. C. Loo, *Energy Environ. Sci.* **5** (3), 5902–5918 (2012).
44. L. Ding, H. Zhou, S. Lou, J. Ding, D. Zhang, H. Zhu, and T. Fan, *Int. J. Hydrogen Energy* **38** (20), 8244–8253 (2013).

45. A. Khan, 2013, http://articles.latimes.com/2013/apr/12/science/la-sci-sn-artificial-leaf-dirty-water-self-healing-20130412.
46. R. van Noorden, 2012, http://www.nature.com/news/artificial-leaf-faces-economic-hurdle -1.10703.
47. D. Chandler, 2010, http://web.mit.edu/newsoffice/2010/self-healing-solar.html.
48. M.-H. Ham, J. H. Choi, A. A. Boghossian, E. S. Jeng, R. A. Graff, D. A. Heller, A. C. Chang, A. Mattis, T. H. Bayburt, Y. V. Grinkova, A. S. Zeiger, K. J. Van Vliet, E. K. Hobbie, S. G. Sligar, C. A. Wraight, and M. S. Strano, *Nat Chem.* **2** (11), 929–936 (2010).
49. D. Gunther, G. LeBlanc, D. Prasai, J. R. Zhang, D. E. Cliffel, K. I. Bolotin, and G. K. Jennings, *Langmuir: ACS J. Surf. Colloids* **29** (13), 4177–4180 (2013).
50. D. Salisbury, 2012, http://news.vanderbilt.edu/2012/09/spinach-power-a-major-boost/?utm_source=vuhomepage&utm_medium = vuhomeslider&utm campaign= 0904-spinach.
51. G. LeBlanc, G. Chen, E. A. Gizzie, G. K. Jennings, and D. E. Cliffel, *Adv. Mater.* **24** (44), 5959–5962 (2012).
52. D. Gunther, G. LeBlanc, D. E. Cliffel, and G. K. Jennings, *Ind. Biotechnol.* **9** (1), 37–41 (2013).
53. P. Bharadwaj, B. Deutsch, and L. Novotny, *Adv. Opt. Photon.* **1** (3), 438–483 (2009).
54. D. K. Kotter, S. D. Novack, W. D. Slafer, P. Pinhero, and Asme, *Solar Nantenna Electromagnetic Collectors* (American Society of Mechanical Engineers, New York, 2009).
55. A. Sobhani, M. W. Knight, Y. M. Wang, B. Zheng, N. S. King, L. V. Brown, Z. Y. Fang, P. Nordlander, and N. J. Halas, *Nat. Comm.* **4** (2013).

9

THE FUTURE OF BIOINSPIRED PHOTONICS

Challenges and Opportunities

9.1 Inspiration from Natural Systems for Conventional and Unconventional Applications

Bioinspired photonics is enabling the start of a technological revolution. Biological photonic systems have evolved to produce, reflect, absorb, and manipulate light, and the discoveries we are making about them are inspiring new scientific approaches, new devices, and entirely new applications. These natural photonic systems have extremely intricate architectures across multiple length scales, from millimeter to nanometer scale, which we describe as hierarchical. The interesting and complex hierarchical structures result in adaptable and fault-tolerant functionality that we cannot yet match in our synthetic designs and fabrication capabilities. These sophisticated optical systems are ancient. While our ability to pinpoint the emergence of other optical effects, such as bioluminescence, for example, is limited due to loss of any soft tissue during fossilization, complex vision systems and structural color seem to have emerged around the time of the Cambrian explosion (around 500 million years ago), and geological evidence of photosynthesis implies a more-than-billion-year history. The study of biological photonics and the inspirational effects on technology have a long and rich story. As scientific study has moved toward intense specialization, the link between biological photonics and synthetic application has remained disjointed. More recently, particularly in the last decade, communication across scientific and technical disciplines has increased significantly, creating a vibrant and motivated multidisciplinary community. The long-developing field is finally coming together. The extremely multidisciplinary nature of the field itself is a major challenge. It is difficult to communicate effectively across disciplines. Our researchers operate at the interface of many subdisciplines and require innovative thinking. Multidisplinary training for future generations is a vital need. Researchers interested in this domain must balance depth and breadth of technical knowledge with the ability to speak the language of multiple disciplines.

Although it should be clear that while there has been tremendous progress in our identification and understanding of the physics, biology, and chemistry involved, as well as the gradual adoption of bioinspired ideas in applications, bioinspired photonics is still a tremendously open field. Throughout this book, we have discussed examples of technical applications impacted by development of bioinspired designs. We have only just begun to understand the tremendous variety of natural biophotonic systems that exist, and there are many things we still do not understand. What we do know has impacted a wide range of potential applications. Structural color provides inspiration for a host of display technologies, art, lighting, and photonic applications where wavelength-specific reflection and color production is important. Dynamic adaptive color systems are important for displays and camouflage. Vision systems inspire new imagers, as well as sensors, and prosthetics. Other inspired applications exist in energy harvesting, sensors, and materials, as well as nanomedicine, neuroscience, and biomedical imaging. Bioinspired photonics may enable a host of new applications as well, such as more environmentally friendly disposable, renewable, and recyclable technologies. There is exciting potential for reconfigurable, flexible, and self-repairing structures and devices, as well as hybrid devices incorporating low-cost functional biomaterials with strategic inorganic high-performance materials. These ideas have the potential to completely change the paradigm for applications creating whole new classes of unconventional devices. Biophotonic structures are often multifunctional, incorporating photonic properties into structures that also enable superhydrophobicity, mechanical robustness, and self-cleaning, for example. Biological photonic structures cause us to rethink our entire design space, particularly when viewed in combination with these other biomaterial structures and effects, such as layered composite biomaterials for mechanical robustness and quantum biology, for example. Bioinspired technologies will have to be either able to compete with conventional designs or engender entirely new applications; market forces are not altruistic and will drive the process.

The field holds great potential for both improving existing devices and enabling entirely new bioinspired photonic systems. Where is the field going from here? There are definitely limits to our current understanding and capabilities. There are still many unanswered questions.

9.2 Fabrication is Still a Challenge

Fabrication is a tremendous challenge. We are not yet able to replicate many of these systems in the laboratory; we cannot match nature's fabrication capability. Fabrication of comparably intricate and complex synthetic structures is tremendously difficult, expensive, and not ready or capable of the scale-up required for commercial application. Developing fabrication techniques as well as new tools to help understand the structures, physics, materials, and function is critical to field. Current fabrication techniques such as top-down lithographic techniques or bottom-up colloidal processes, as well as using actual biostructures as templates, are a solid starting place. Eventually, however, we will need to be able to create these structures at any scale, size, or pattern at low cost and over large areas. Up-and-coming fabrication techniques such as molecular and three-dimensional (3D) printing are still very nascent but may point the way to "fab-less" fab (i.e., the ability to make large amounts of devices without requiring the large fabrication facility [fab] infrastructure). We will want to have extreme local control of inhomogeneity, to control material gradients, and composition even within the nanostructures. We need to be able to build in disorder. Current fabrication techniques are built on the premise that we want perfection. However, biological photonic structures have shown us that disorder can be leveraged to produce desired effects, such as the angle-independent reflection in structural color due to the existence of multidomains. We will want to be able to fabricate these structures out of functional materials, such as optomechanical or electro-optical materials, to create even more functionality. For example, supercrystals have been proposed that incorporate quantum dots into periodic structures.[1] It becomes particularly interesting when you think of the nanoparticle microcolloids discussed in Chapter 2. Finally, biological photonic structures have inspired other approaches to structural color that may be easier to fabricate, such as plasmonic nanostructures seen in Figure 9.1 that

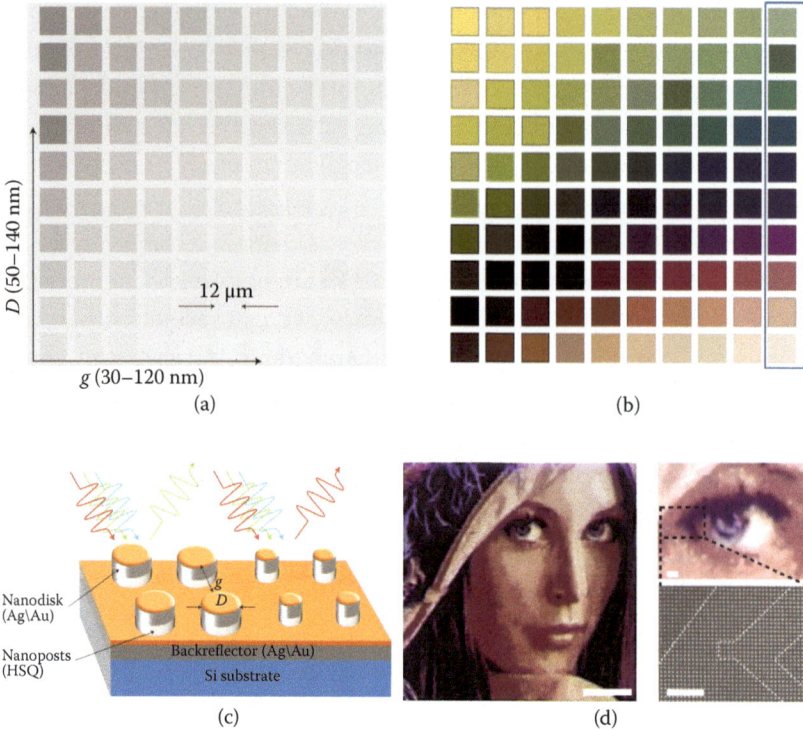

Figure 9.1 Plasmonic structural color. Optical micrographs of arrays of nanostructures with varying diameters D and gaps g. Each 12-μm square in the array consists of a square lattice of nanoposts of periodicity $D + g$. (a) Before metal deposition, the patterns of nanostructures show grayscale variations but do not display any color. (b) Following deposition of thin metal layers of uniform thickness, the full palette of colors is revealed. Nanostructure arrays with similar colors are observed from bottom left to top right, indicating that areas with similar fill factors produce areas with similar colors. Structures with the same periodicity display a wide range of colors (from bottom right to top left). The highlighted column was used to produce the image in (d). (c) Illustration of the interaction of white light with two closely spaced pixels, each consisting of four nanodisks. As a result of the different diameters, D, and separations, g, of the nanodisks within each pixel, different wavelengths of light are preferentially reflected back. (d) Optical micrographs of the Lena image after metal deposition. Right is an optical micrograph of an enlarged region of the image, showing beautiful detail and color rendition on a micrometer-scale level. The accompanying SEM image of the indicated region shows that the nanostructures that make up the image have a similar periodicity of 125 nm but exhibit different colors in the optical image due to a small (30 nm) variation in nanodisk sizes. For clarity, the individual regions of similarly sized disks are separated by the dotted lines. Each pixel consists of a 2 × 2 array of disks with a pitch of 250 nm. Scale bars in (d): 10 μm for the large Lena image and 1 μm for the close-up. (From Kumar et al., *Nature Nanotechnol.*, 7 (9), 557–561, 2012.[3])

have been able to produce a very wide range of colors with high spatial resolution (although only for static images).[2]

We will potentially want to combine both organic and inorganic materials. Heterogeneous integration, which is the very intimate integration of dissimilar materials and devices, is still a great challenge even among inorganic semiconductor devices.[3,4] Very-large-scale fabrication, such as is necessary for silicon-integrated circuits, requires fabrication facilities that cost billions of dollars to build and severely limits the materials and devices that can be integrated intimately at the device level. Currently, whole chips and other components are fabricated in their own facilities and then later assembled together on a common board at the back end of the process. The downside to this is a not-insignificant impact on performance and the inability to choose the best performing device for the job. While there are several large research programs exploring ways to integrated electronic and photonic components at a much more intimate level (at the transistor level, for example), bringing organic materials into the mix causes even more consternation.

In an interesting twist on bioinpsired photonics, researchers from the University of Illinois at Urbana-Champaign and their colleagues were inspired by the overlapping scales found on butterfly wings (and other scaled animals) as a way to introduce both heterogeneous integration and improve scale-up of arrays of varying photonic structures.[5] Butterfly wing scales are photonically active, and the scale structure enables flexibility and functional robustness. Overlapping functional scales are attractive because they offer highly fault-tolerant layouts where loss or failure of one or a few scales has minimal effect on the overall functionality of the system. A tile approach also enables multifunctionality by incorporating scales of different functionality into and within a single system. Tiled systems, when anchored to a substrate, can also provide mechanical flexibility even with full 100% coverage of rigid scales. Images of the final heterogeneous scaled photonic system can be seen in Figure 9.2. The array combines silicon and photonic and plasmonic overlapping scales anchored by posts to the underlying substrate at either the scale centers or edges. Transfer printing was used to create the heterogeneous scale systems. Each scale type was first fabricated separately on a "source" substrate. To prepare for transfer printing and enable release of the scale from its substrate, the scale was undercut etched along the bottom surfaces. A thin layer of photoresist was used around the scale periphery to hold them in place. A soft, elastomeric stamp with pyramidal microtips was used to pick up the scales and transfer them to the new substrate.

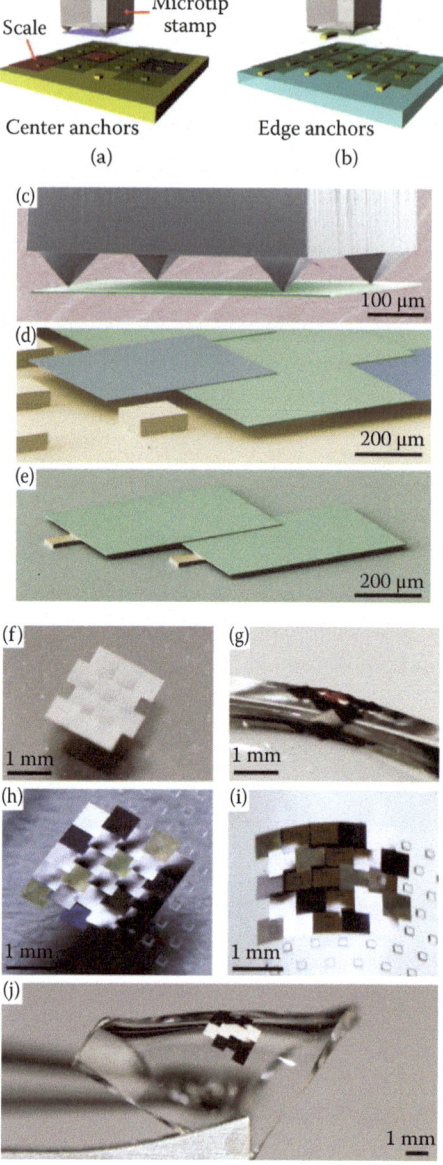

Figure 9.2 (a and b) Illustration of transfer printing procedures for forming assemblies of synthetic scales with anchors located in the centers and edges. (c) Colorized SEM image of a silicon scale suspended on the surface of a microtip stamp; (d) printed photonic, plasmonic, and silicon scales on a Polydimethylsiloxane (PDMS) substrate with center anchors; and (e) printed silicon scales with edge anchors on a silicon substrate. (f–j) Optical images of homogeneous (silicon) (f, g) and heterogeneous (photonic, plasmonic, and silicon) (h, i) collections of scales with center anchors, on PDMS substrates in flat (f, h) and bending (g, i) modes. The appearance of the photonic and plasmonic scales changes with viewing angle. (j) Optical image of homogeneous scale sample bent and twisted using a pair of tweezers. (From Kim et al., *Small*, 8 (6), 901–906, 2012.[2])

The scales were elevated off of the new substrate by posts that act to anchor the scales. Modeling was used to provide design rules. Using this method, large areas of multiple scale types, in a variety of patterns, density of overlap, and area of coverage, can be obtained. The technique also enables these systems to be fabricated on rigid, flexible, or stretchable planar or curved substrates.

9.3 Biological Fabrication

One not-yet fully answered question is how does nature fabricate these complex structures, particularly the intricate architectures of structural color? Biological fabrication would offer many potential benefits. For example, biological fabrication has demonstrated ability to form extremely intricate structures under very mild conditions, particularly compared to semiconductor fabrication requirements. However, very fundamental questions regarding biological growth of photonic systems remain. While we know a good deal now about the photonic mechanisms of structural color production, our understanding of the mechanisms of fabrication of these structures is still very much in the early stages. Only a few key researchers are attempting to elucidate the biophotonics structural development process in birds and insects.[6–10] These structures seem to arise from a combination of biological (cellular and subcellular), chemical, and physical mechanisms, which all contribute to the production of these photonic structures.[6] This makes it very difficult to tease out the details of the biofabrication process, as well as understanding the biological costs and role of genetic control in structural color. Biological control may exist at genetic, enzymatic, intracellular, or intercellular level, or combination of factors. However they are formed, the mechanisms operate with extreme precision and under tight tolerances, as even small changes in the structures (on the order of 20 nm) can result in noticeable color change. Research shows that there is not one single process responsible for the growth of color structures. Study of noniridescent structurally colored feather barbs of blue-and-yellow macaws (*Ara ararauna*) seems to show that structural development is due to intracellular colloidal self-assembly.[6,7] During development, β-keratin protein phase separates from the cytoplasm of the medullary cells of the feather barb rami (the support shaft of the barbules) during keratinization.[7]

In channel-type structures, the photonic architecture arises from a process called spinodal decomposition, where mixed materials rapidly separate to form two distinct but coexisting phases. Spherical nanostructures from feather barbs seem to be formed also via phase separation during protein polymerization, this time by a process called nucleation and growth. There seems to be no biological prepatterning by the cell membrane, endoplasmic reticulum, or cellular intermediate filaments, and we currently know nothing about any potential genetic factors, cellular processes, or cell-to-cell signaling that may influence the process.[6]

In insect systems, cellular membranes seem to play an important role in photonic structure development. The insect epidermal cells secrete the nonliving cuticle, the biopolymer chitin. It is postulated that insect photonic structures may be built from variations in a small number of "standard" developmental pathways within these cells, possibly mediated or controlled by the smooth endoplasmic reticulum.[9] The smooth endoplasmic reticulum interacts with the cellular membrane to form a template. The chitin deposits in the extracellular space and hardens. As the cell dies, only the chitin structure remains. The smooth endoplasmic reticulum systems can switch between structures (e.g., lattice and laminar structures) to induce the formation of many different template geometries. The endoplasmic reticulum seems to be able to form intracellular cubic symmetries as well and is implicated in the formation of the 3D crystal structure (diamond structure) found in the scales of the beetle *Lamprocyphus augustus*.[10] Cubic membrane architectures can be found in the mitochondria and endoplasmic reticulum of many cell types, particularly when diseased or under stress. These stressors cause changes in protein levels, dynamics and interactions, and intracellular ion concentrations, which seem to cause the formation of these highly ordered cubic domains. The cubic membrane structure provides a template for chitin deposition. The researchers at the University of Utah and their colleagues have demonstrated that it is possible to transfer the intracellularly formed large lattice cubic membranes described above into extracellular environments. It had previously been found that amoeba cells could be induced to produce large lattice constant cubic membrane structures by starving the cells. The size of the lattice parameter in these cubic membranes can be changed with the addition or subtraction of other materials. By stripping the amoeba cubic assembly of all proteins and reconstituting them in a liposome, the cubic lattice reformed but with a much smaller lattice constant. When the Utah team attempted the same study but reconstructed the liposomes after stripping only the nonmembrane proteins, the cubic assembly was formed with the large lattice constant and closely resembled the photonic crystal lattices found on the *L. augustus* beetle scales (Figure 9.3).

We do not know very much about how these colors evolved. Artificial selection has been utilized to alter the wavelength of reflection of structural color in butterfly wings.[11] The laboratory model butterfly *Bicyclus anynana* is relatively inexpensive and easy to breed, with brown wings and eye spots (seen in Figure 9.4). A research team led by Yale University used selective breeding to artificially evolve violet scales from *B. anynana*'s normal UV brown scales and compared the mechanism of violet reflection with two other *Bicyclus* butterfly species. These two species, *B. sambulos* and *B. medontias*, have naturally evolved violet/blue scales. The reflectance spectrum of *B. anynana* was measured

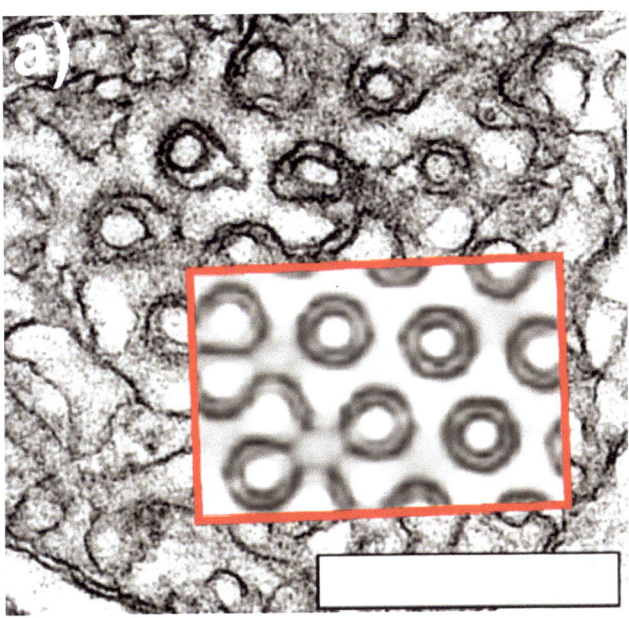

Figure 9.3 Transmission electron microscopy image of a three-dimensional phospholipid-membrane lattice formed after stripping all proteins except membrane proteins. Inset: modeled gyroid crystal faces. Scale bar: 500 nm. (From Bartl et al., *Emerg. Liquid Crystal Technol.*, Vii 8279, 2012.[6])

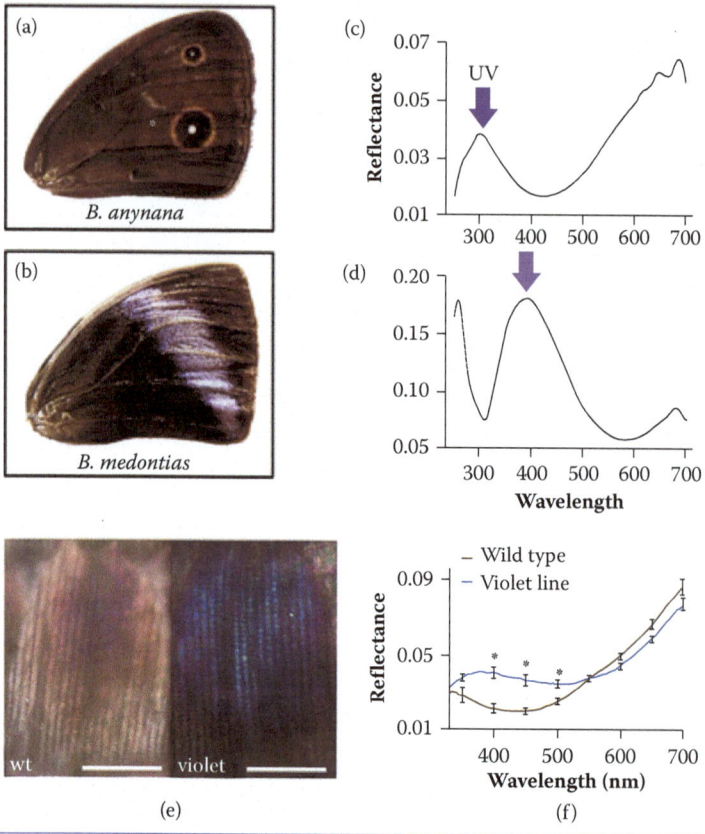

Figure 9.4 (a and b) Dorsal wing images of *Bicyclus. anynana* (a) and *B. medontias* (b) (the region used for artificial selection is marked by white asterisk). (c and d) Graphs of reflectance spectra of the blue/violet wing band showing reflectance peaks in the 400–450 nm range and in the brown-colored homologous region in *B. anynana* with a UV reflectance peak centered at 300 nm (colored arrows). (e and f) Artificial selection of violet structural color in *B. anynana*. (e) Images of wild-type (WT, left) and violet-line (right) ground scales (generation 8) in the selected wing region (scale bars: 20 μm). (f) Representative reflectance spectra of WT and violet-line females. Error bars represent SEM, and asterisks represent statistically significant differences in reflectance. (From Wasik et al., *Proc. Natl. Acad. Sci. U S A*, 111 (33), 12109–12114, 2014.[11])

at the region of dorsal forewing where the other *Bicyclus* species have violet/blue coloring. Each generation butterflies displaying reflectance peak wavelengths near 400 nm were bred with each other. The selection process occurred over eight consecutive generations. (Due to the small size of the study cohort [the low number of adult butterflies], all butterflies in generations 4 and 7 were allowed to reproduce with no artificial selection.) In less than a year (six selected generations), the ultraviolet reflectance peak seen on the *B. anynana* wing shifted to violet. In the first generation of butterflies (the parental population), the average reflectance peak wavelength in the region of interest was 311 nm in males and 341 nm for females. Over the course of the six selection generations (by generation 8), the reflectance wavelength peak had shifted to 362 and 385 nm, respectively, a shift of over 40 nm. The change in reflectance was the result of an increase in thickness of lower lamina of ground scales. This is the same mechanism and similar scale structure that is found on the wings of other *Bicyclus* butterfly species. In unselected *B. anynana*, the cover and ground scales have lower lamina with almost the same thickness (~120 nm). With artificial selection, the lower lamina of both scales became thicker. In the cover scale, the thickness increased from ~126 to ~144 nm. In the ground scale, the thickness increased from ~120 to ~176 nm. Understanding the evolutionary and genetic component to photonic structure development could enable new synthetic biology or directed evolution approaches to device fabrication.[12]

Biological fabrication is accomplished at room temperature under mild aqueous conditions in contrast to modern semiconductor fabrication techniques, which require high temperature, high vacuum conditions, and extremely toxic materials (causing both environmental and health hazard concerns). Understanding how biological photonic structures are formed may enable new methods of complex structure fabrication and new ways of thinking about photonic devices. We hope to be able to harness the control and specificity of biological fabrication to design and build self-assembled nanoscale integrated devices under mild environmental conditions.

The approaches to biofabrication are many and varied. For example, scientists from the University of Washington and their colleagues have demonstrated the ability to computationally design proteins (Figure 9.5), which when mixed self-assemble into specific cage-like nanostructures, and experimentally demonstrated this with multiple architectures.[13] Although we have ample examples of biomineralization to study (such as diatoms and silica), there is a desire to discover how to use biofabrication approaches to template the growth or assembly of other inorganics as well. Bionanocomposite materials would incorporate both biological and inorganic materials.[14] Some are creating devices that use actual biological organisms, such as diatoms, within the synthetic device. Still others are interested in

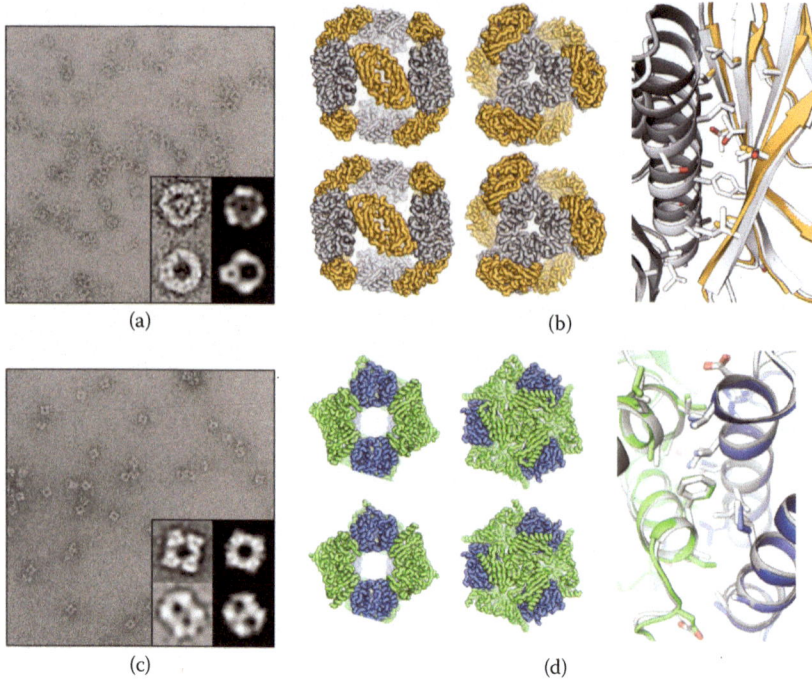

Figure 9.5 Electron micrographs of designed two-component protein nanomaterials. Negative-stain electron micrographs for coexpressed and purified (a) T32-28 and (c) T33-28 are shown to scale (scale bar at top right, 25 nm). For each coexpressed material, two different class averages of the particles (top and bottom) are shown in the insets (left) alongside back projections calculated from the computational design models (right). (b and d) Crystal structures of designed two-component protein nanomaterials. The computational design models (top) and x-ray crystal structures (bottom) are shown at left for (b) T32-28 and (d) T33-28. At right, overlays of the designed interfaces in the design models (white) and crystal structures (gray, orange, green, and blue) are shown. (From King et al., *Nature*, 510 (7503), 103, 2014.[13])

using self-replicating microorganisms (such as viruses and bacteria) to construct inorganic nanoscale objects.[15] "Hybrid materials science" attempts to use biological fabrication techniques to create inorganic materials and devices.[16–18] Approaches like this could serve as a pathway to "green" electronics and photonics by potentially enabling easier fabrication with a much smaller ecological footprint. From an environmental standpoint, extracting and processing semiconductor materials, metals, minerals, etc., to create modern consumer electronic and photonic devices is extremely energy, equipment intensive, and potentially toxic. These materials are also a finite resource. It is becoming increasingly important to find alternate manufacturing techniques for these technology components for improved sustainability (in addition to better recycling of existing material). Fabrication technologies that reduce waste and energy consumption also often have the additional benefit of reduced cost, which makes them attractive for the commercial market. Exploration of other biological and bioinspired systems has the potential to combine with biophotonics systems to create entirely new devices. Molecular electronics, for example, could create integrated biocircuitry that is self-repairing.[19] If all of these were accomplished, one could easily imagine cellular fabrication of photonic devices. Roll-to-roll processing or spray techniques could be used to lay modified cells on a surface, along with any required cofactors, to grow large-scale photonic arrays. Taken to its extreme, and in combination with other organic and bioinspired systems, we can envision vat synthesis of consumer devices. Given the increased pace of technology improvement and new consumer device production, perhaps exploration of fabrication capabilities that enable devices to be quickly and easily recycled, upgraded, and discarded (in an environmentally sound manner) is a wise investment.

9.4 STEM Education and Outreach

Bioinspired science and technology is an excellent subject for science, technology, engineering, and mathematics (STEM) education. In bioinspired photonics (and more broadly all bioinspired technology), there is something for everyone. It offers the opportunity to link multiple technical disciplines, explain interesting phenomena, and describe technological advances that students may care about while discussing animals and plants that students can relate to. The subject is both fascinating and accessible and lends itself to easy adaptation for communication to any audience. It also offers a wide-open vista of potential paths to follow, both on the exploratory biology side and on the technology development aspect. We have barely begun to explore the photonic systems found in nature. Even though we have greatly increased the pace of our expanding understanding and knowledge, we have truly just scratched the surface. There are likely more unknown and undiscovered photonic systems in nature than the number of systems that have been explored. Searching the Internet for colorful spiders alone results in many intriguing, unstudied creatures, such as the mirror spiders of the *Thwaitesia* species that seems to be able to change the size of wide-band (silver metallic) reflecting domains on their abdomens, beautiful purple and gold spiders, and the very small, but multicolored peacock and rainbow spiders (*Maratus volans* and *M. robinsoni*, respectively) the latter of which have clearly visible iridescent reflection.[20]

Many of the examples of biological photonic systems found in this book are from fairly exotic-seeming creatures. However, one does not need to travel far to find examples of interesting, and still unexplored, systems. In fact, you really only need to walk out into your back garden. Perhaps the most conspicuous of the backyard photonics are the examples of structural color that can be found once you know what to look for. Figure 9.6 shows some examples of iridescent and metallic reflective insects found in both Virginia and Michigan. Images of backyard grasshoppers and jumping spiders can be used to discuss eyes. Discussing backyard photonics brings the subject to students in a relatable way. People learning about biological photonics (students and educators alike) seem surprised to discover that it is so pervasive. (During a seminar, I once made comments about exotic creatures vs. photonics in your backyard only to have an audience member who happened to be from Costa Rica speak up and say that the *Morpho* butterflies described in Chapter 3 were from his backyard.)

Another topic that can arise from discussion of bioinspired photonics is the issue of climate change and the risk of extinction of species. It has been declared that we are now likely in the sixth mass extinction of plants and animals.[21,22] We are experiencing the loss of thousands of species a year. As one author said, "We're currently experiencing the worst spate of species die-offs since the loss of the dinosaurs 65 million years ago."[22] Dozens of species go extinct every day. It is estimated that 30%–50% of current extant species will be extinct by the midcentury. The typical "background" rate of extinction is one to five species per year, and the current rate we are experiencing is accelerated 1,000 – 10,000 × over this background rate. From a purely practical standpoint in the context of biophotonics, loss of species means fewer organisms to study, and the potential loss of photonic structures, which might revolutionize technology in the future.

THE FUTURE OF BIOINSPIRED PHOTONICS 375

Figure 9.6 Photographs of garden insects showing interesting biological photonic phenomena. (a) Iridescence and bright white on wing of monarch butterfly. (b) Close up of eyes of a jumping spider. (c) Metallic green, silver, white and yellow in microdomains on the abdomen of a venusta orchard spider. (d) Garden butterfly. Note the hint of purple on the edges of the large white wing spot. (e) Eye of a monarch butterfly (f) Eye and ocelli of a grasshopper (g) Metallics on a fly. (h) Golden metallics on a sweat bee.

9.5 Importance of Multidisciplinary and Basic Research

While much of this book centers on the question of how we can employ bioinspired photonics in applications, the field of bioinspired photonics also deals in very fundamental questions to which we, as scientists, want answers. Basic science is incredibly important to our ability to continue to develop technology and innovation. Investment in basic research is vital to our ability to continue the progress, innovation, and resulting economic development that technology brings. It is the foundation upon which all applied development rests; it is the driver of innovation. Everything in applied science depends on the understanding and exploration accomplished via basic research investment. An objective measure of the value of research (basic or applied) is a challenging thing. The value of science is not always easy to determine. In difficult economic climates, politicians often ask to put a dollar amount on the value of basic research. Lately, we have been asked what the "return on investment" is for our science dollars. Unfortunately, science funding in the United States has been nearly stagnant over the last decade and has begun declining in recent years.[23] Within the U.S. Department of Defense (DoD), basic research dollars have been relatively protected; but in the United States, spending on basic research has been reduced significantly from its high of $40B in 2009. Grant funding is difficult to obtain. Success rates at funding agencies are low. High-risk efforts are not often rewarded. Young researchers are finding it more difficult to find funding. The peer-review system rewards established laboratories, with well-known principal investigators. In funding agencies, research budgets are constantly under attack and regularly require significant justification. Dollars that are not spent fast enough can be taken away, despite the value of the science for which those dollars were planned. "Budget drills" regularly ask about impact for budget cuts of varying severity; questions about 10%, 15%, and even 30% cuts are not unheard of. Given the political climate, and the increasing number of research proposals submitted each year, funding programs are no longer easily able to provide decades-long funding to a researcher. Basic research ideally is the quest for new knowledge and exploration of unknown. But basic research is also sometimes perceived as an "unnecessary luxury."[24] Basic research and applied research are inextricably intertwined. Most research is a hybrid of new knowledge generated and subsequent exploration. Major innovation is rarely possible without the prior generation's efforts. As some have said, the research agenda set by those who participate, so we cannot leave the work to others.[24] The funding landscape has gotten more complicated in the United States, as well, with funding not only by the federal agencies and departments that fund science (National Institutes of Health, National Science Foundation, and DoD) but also by private investors, foundations, and industry (often in partnership with universities). Each entity involved has its own criteria, expectations, and requirements. Recently, there has been an increase in research investment by philanthropists and private enterprise, often sponsoring large research centers such as the $250M Wyss Institute for Biologically Inspired Engineering at Harvard University. Basic research is science for the common good. Budget cuts are especially hurtful to basic research, which does not often have patent or industry dollars to supplement it but is the foundation upon which future technology and economic development depends.

In the United States, some politicians and scientific organizations have attempted to highlight the role that basic research plays in economic benefit and to emphasize that those benefits are often unexpected. Since 2012, the Golden Goose award has been given to those researchers whose seemingly obscure studies have led to major breakthroughs and resulted in significant societal impact.[25] The purpose of the award itself, which was conceived by Congressman Jim Cooper (D-TN), is to demonstrate and to educate members of Congress and the general public about the value and economic benefits of federally funded basic research. The name of the award, the Golden Goose Award, is a reference to the Golden Fleece awards, which existed between 1975 and 1988. This "award" was given to federally funded research grants that were highlighted as examples of government waste. Bioinspired photonics is an excellent example of long-simmering basic research that over time has transitioned into applied research despite having no practical application at the beginning. Shimomura, the discoverer of the bioluminescent protein aequorin and its companion fluorescent protein GFP (green fluorescent protein), labored for 20 years in very fundamental research before his work was absorbed and expanded into the applications we know today.[26] Within the first year's awardees for the Golden Goose Award

(2012) were the Nobel laureates Shimomura, Chalfie, and Tsien for their groundbreaking work on GFP.

Bioinspired photonics is an extremely multidisciplinary field, and this can be a challenge that can sometimes delay the growth of a field. Biologists, chemists, physicists, material scientists, and engineers all play a role in the field of biological and bioinspired photonics. The increased specialization of the traditional scientific disciplines challenges our ability to cross these disciplinary boundaries. Researchers in each of these specializations often work in small, close-knit communities and speak different scientific languages. They may attend different conferences, publish in different journals, and work in different departments. In order for biophotonics to succeed, we need to be able to bridge the gaps between disciplines.[27] It is difficult and time consuming to cross into a new field. One must be willing and able to become fluent in more than one technical language, achieve a necessary depth of understanding, and learn the methods and procedures of that new discipline. Multidisciplinary teams may instead assemble and create a unique common language and framework for discovery and innovation.[28] In the last decade or so, many multidisciplinary conferences and new journals have emerged. The digitalization of journal catalogs has enabled the (comparatively) easy location of interesting journal articles, even in the most esoteric of publications.

Another challenge that exists is in the organizational structure of many research institutions. In many research settings, personnel are stovepiped organizationally, with each division, department, or discipline somewhat isolated from each other with different budgets, processes, and success criteria. Increased collaboration and mixing between disciplines requires people who can bridge not only the science but also the departmental differences. It takes a great deal of extra effort on the researcher's part, and on the part of their management. It often requires gaining the interest and attention of the leadership in both domains. Hiring practices as well as performance goals, promotion, and resource allocation criteria often serve to complicate or inhibit multidisciplinary research.[28] In many cases, peer review for funding and article publication are heavily geared toward single-topic experts, complicating the multidisciplinary researchers' ability to publish and find research dollars. For multidisciplinary research, highly collaborative teams must come together to provide the varying but necessary expertise and capabilities. Many institutions claim to support multidisciplinary science.[29,30] While graduate students are often encouraged to join in cross-disciplinary studies, and many new certifications and degrees have emerged to entice students, a recent study of recent PhD graduates with interdisciplinary dissertations found that multidisciplinary researchers are not well rewarded. There is a cost, a not-insignificant pay gap, to "boundary spanning." This is particularly true for young researchers, assistant professors, and young investigators who are just establishing their research and must face tenure expectations. Because of these challenges, young investigators may avoid cross-disciplinary research. Once academics reach full professor, however, much of the risk associated with multidisciplinary research fades, and evidence shows that this group tends to disproportionately pursue multidisciplinary research.

The potential benefits of multidisciplinary research are large and worth the extra effort required. New insights, and new understanding, often emerge at the intersections between disciplines. New tools are most often transformational when drawn from outside the discipline that developed them.[28] Researchers who are deeply technical in their original field are often able to ask pointed questions that challenge the assumptions of experts in another field. They may be able to achieve success because they had the expertise required to do the hard work and did not know that what they were attempting was "impossible." We need to find ways to facilitate and reward multidisciplinary research organizationally. We need to enhance communication, collaboration, and cross-fertilization of ideas between communities. Science, and the field of bioinspired photonics, needs researchers who are willing to take these risks.

References

1. A. S. Baimuratov, I. D. Rukhlenko, V. K. Turkov, A. V. Baranov, and A. V. Fedorov, *Sci Rep* **3** (2013).
2. S. Kim, Y. Su, A. Mihi, S. Lee, Z. Liu, T. K. Bhandakkar, J. Wu, J. B. Geddes, III, H. T. Johnson, Y. Zhang, J.-K. Park, P. V. Braun, Y. Huang, and J. A. Rogers, *Small* **8** (6), 901–906 (2012).
3. K. Kumar, H. Duan, R. S. Hegde, S. C. W. Koh, J. N. Wei, and J. K. W. Yang, *Nature Nanotechnol.* **7** (9), 557–561 (2012).

4. M. J. Rosker, V. Greanya, T.-H. Chang, and IEEE, in *2008 IEEE Csic Symposium* (2008), pp. 4–7.
5. S. Raman, C. L. Dohrman, T.-H. Chang, and IEEE, *Proc. 2012 IEEE Int. Symp. Radio-Frequency Integration Technol. (Rfit)*, 1–6 (2012).
6. M. H. Bartl, M. R. Dahlby, F. P. Barrows, Z. J. Richens, T. Terooatea, and M. R. Jorgensen, *Emerg. Liquid Crystal Technol.* Vii **8279** (2012).
7. M. G. Meadows, M. W. Butler, N. I. Morehouse, L. A. Taylor, M. B. Toomey, K. J. McGraw, and R. L. Rutowski, *J. Roy. Soc. Interface/Roy. Soc.* **6 Suppl 2**, S107–113 (2009).
8. R. O. Prum, E. R. Dufresne, T. Quinn, and K. Waters, *J. Roy. Soc. Interface /Roy. Soc.* **6 Suppl 2**, S253–265 (2009).
9. H. Ghiradella, *J. Morphol.* **202** (1), 69–88 (1989).
10. H. T. Ghiradella and M. W. Butler, *.Roy. Soc. Interface/Roy. Soc.* **6 Suppl 2**, S243–251 (2009).
11. B. R. Wasik, S. F. Liew, D. A. Lilien, A. J. Dinwiddie, H. Noh, H. Cao, and A. Monteiro, *Proc. Natl. Acad. Sci. U S A* **111** (33), 12109–12114 (2014).
12. G. M. Church, M. B. Elowitz, C. D. Smolke, C. A. Voigt, and R. Weiss, *Nat. Rev. Mol. Cell Biol.* **15** (4), 289–294 (2014).
13. N. P. King, J. B. Bale, W. Sheffler, D. E. McNamara, S. Gonen, T. Gonen, T. O. Yeates, and D. Baker, *Nature* **510** (7503), 103–108 (2014).
14. M. Darder, P. Aranda, and E. Ruiz-Hitzky, *Adv. Mat.* **19** (10), 1309–1319 (2007).
15. X. Xiong, M. E. Lidstrom, and B. A. Parviz, *J. Microelectromechanical Syst.* **16** (2), 429–444 (2007).
16. L. Nicole, C. Laberty-Robert, L. Rozes, and C. Sanchez, *Nanoscale* **6** (12), 6267–6292 (2014).
17. J. M. Galloway, J. P. Bramble, and S. S. Staniland, *Chemistry* **19** (27), 8710–8725 (2013).
18. M. Irimia-Vladu, *Chem. Soc. Rev.* **43** (2), 588–610 (2014).
19. Nature Editorial, Focus issue on Molecular Electronics, *Nat. Nanotechnol.* **8** (6) 377–467, (2013). http://www.nature.com/nnano/focus/molecular-electronics/index.html.
20. C. Brooke, http://www.thefeaturedcreature.com/shockingly-beautiful-purple-and-gold-species-of-jumping-spider-found-in-thailand/, http://www.thefeaturedcreature.com/beauty-beast-all-in-one-spideramazing/, http://sgmacro.blogspot.com/2013/07/transformation-of-mirror-spider.html.
21. E. Kolbert, *The Sixth Extinction: An Unnatural History.* (Henry Holt and Co., New York, 2014).
22. http://www.biologicaldiversity.org/programs/biodiversity/elements_of_biodiversity/extinction_crisis/.
23. AAAS, http://www.aaas.org/page/historical-trends-federal-rd-Overview.
24. International council of scientists, http://www.icsu.org/publications/icsu-position-statements/value-scientific-research/the-value-of-basic-scientific-research-dec-2004.
25. http://www.goldengooseaward.org.
26. M. Zimmer, *Chem. Soc. Rev.* **38** (10), 2823–2832 (2009).

27. Nat Photon **6** (9), 567–567 (2012).
28. J. Trewhella, http://globalhighered.wordpress.com/2009/06/26/multidisciplinary-research-an-essential-driver-for-innovation/.
29. S. Jaschik, https://http://www.insidehighered.com/news/2013/10/31/study-finds-phds-who-write-interdisciplinary-dissertations-earn-less.
30. K. Kniffin and A. Hanks, "Boundary Spanning in Academia: Antecedents and Near-Term Consequences of Academic Entrepreneurialilsm" Cornell Higher Education Research Institute (CHERI), 2013, http://www.ilr.cornell.edu/cheri/upload/cheri_wp158.pdf.

Index

A

A. argenteus scales, 21
Acanthocriemus nigricans, 275
Active biological photonic systems, 121
Acyrthosiphon pisum, 327
Aequorea Victoria, 248, 249
Air Force Research Laboratory (AFRL), 243, 244
ALD, *see* Atomic layer deposition
Alternate display spaces, 122, 123
Ambystoma maculatum, 327
Amorphous carbon butterfly material, 115–116
Amorphous-silicon (α-Si), 283
Anaximander, 13
Ancient structures of biological photonics, 31–33
Ancyluris meliboeus, 71
Animal eye structure, 168–170
Animals
  examples of, 2
  photosynthesis in, 327–328
Anomalocaris, 10–11, 194–195
Anterior median eyes, 172
Antireflective coatings (ARCs), 330
Antireflective structures, 216–219, 331–334
Aperiodic photonic structures, 61–62
Applications of bioinspired photonics, 362
Apposition compound eyes, 177–181
Aptenodytes patagonicus, 48
Ara ararauna, 367
Architeuthis, 172
ARCs, *see* Antireflective coatings
Arctic reindeer, 193
Argyrophorus argenteus, 108, 111
  dorsal wing surfaces of, 109
  structure, 110
Aristotle, 13
Artificial eye prosthetics, 200–201
Artificial photosynthesis, 345–349
Artificial retinas, 201
Aspherical lens, 171
Atmosphere, light traveling in, 271
Atmospheric window, 271
Atomic layer deposition (ALD), 77
Attacus atlas, 204, 206
Aulacoseira, 296

B

Backyard photonics, 374
B. anynana, 370–371
Bats, 274–275
BCC, *see* Body-centered cubic
BCC Bravais lattice symmetry, 95
BCP, *see* Block co-polymer
Beers-Lambert law, 271
Beetles, 275–278
Begonia pavonina, 35
Bicyclus anynana, 369–370
Bicyclus butterfly species, 369–371
Biocompatible optics, 169
Bioinspired dynamic photonic structures, 121
Bioinspired materials, 26, 27
Bioinspired optofluidic lasers, 254
Bioinspired photonic systems
   biological and, 1–6
   evolution, 7–12
   historical perspective and advent of microscopy, 13–18
   multidisciplinary and basic research, importance of, 376–379
   trade, tools of, 18–25
   in twenty-first century and challenge of multidisciplinary science, 25–27
Bioinspired technologies, 362
Bioinspired webs, 110–111
Biolasers, 255
Biological eyes, 169–170
   apposition compound eyes, 177–181
   brittle star, 183–184
   compound eyes, 176–177
   gradient Index Lenses (GRIN) lenses, 175–176
   refracting superposition eye, 182–183
   simple eye model, 170–174
   superposition compound eyes, 181–182
Biological fabrication, 367–373
Biological materials, complexity in, 229–231
Biological organisms, 371
Biological photonic structures, 31
   describing, 36–37
   light and biology in action, 34–36
   light extraction, improving, 62–66
   next generation applications, 31–33
   templated growth and replication, 58–59
Biological photonic systems, 6, 121, 123, 361
Biological sensors, 267
Bioluminescence, 245–247
   applications for, 250–255
   green fluorescent proteins (GFP), 248–250
   luciferase and luciferin, 247–248
Bioluminescent jellyfish, genes, 2
Bioluminescent resonance energy transfer, 249
Biomaterials for photonics, *see* Photonics, biomaterials for
Bionanocomposite materials, 371
Biophotonic bioluminescent proteins, 5
Biophotonic crystal structures, 336–341
Biophotonics, 5
Biosilicate organisms, 236–241
Bitis gabonica rhinoceros, 112–113
Black reflective color, nanostructures in, 111–116
Block co-polymer (BCP), 160
Blood-sucking bug, 275

Blue-and-yellow macaws, noniridescent structurally colored feather barbs of, 367
Blue reflection, 53
B. medontias, 369, 370
Bobtail Squi, 241
Bodipy chromophore, 343
Body-centered cubic (BCC), 88, 95
Bombyx mori, 232, 233
Book of Optics, 13
Bottom–up approaches, 150
Bragg reflectors, 253
  reflection on multilayer, 49
Brainbow, 253
Brittle star, 183–185
Broadband reflection, chirped multilayers, 49
B. sambulos, 369
Buprestidae, 275
Butterflies, 284–285
  generation of, 371
Butterfly wings as sensors, 300–314
Butterfly wing scale, 365

C

Cabbage White butterflies, 107
Cacostatia ossa, 194
Cadmium sulfide (CdS), 58
Callophrys rubi, 92, 100, 341, 342
Cambrian era, 11
Cambrian explosion, 10, 11, 194, 361
Cambrian worm, 12
Cameraria ohridella, 183, 184
Cameras, 197
Camouflage patterns, cephalopods, 128–129
Canadia spinosa, 11
Capillary condensation, 308
Capillary microfluidic device, 106
Cataract, 201
CdS, *see* Cadmium sulfide

Center of Advanced European Studies and Research, 281
Cephalopod chromatophore, 134
Cephalopods, 2, 126–130
Cephonodes hylas, 194
Chameleons, 137
Channa punctatus, 294
Channel-type structures, 368
Charidotella egregia, 23, 133–134
Chelicerae, 42
Chemical vapor deposition (CVD), 286
Chicken retina, 189, 190
Chionoecetes opilio, 226
Chirality, 81
Chirped multilayers, 47–51
Chitin, 225–231, 286, 291
  deposition, 368
  and refractive index, 46–47
Chitin powder paper, 227–228
Chitosan, structures of, 226
Chlamydomonas reinhardtii, 257
Chlorophila obscuripennis, 53
Chloroplast, 321, 322
Cholesteric liquid crystals, 81
Chromatic aberration, 175
Chromatophore-inspired structures, 138–144
Chromatophores, 130, 134–137
Chromophore in GPF, 249
Chrysina aurigans, 32
  broadband metallic effect, 48–49
  metallic broadband reflection, 34
Chrysina chrysagyrea, 3
Chrysina gloriosa, 34
Chrysina limbata
  broadband metallic effects, 48–49
  metallic broadband reflection, 34
Chrysina resplendens, 82–83
Ciliary muscles, 171
Clade, 10
Cluster eye, 210

Coefficients of thermal expansion (CTE), 286
Coeligena iris, 230
Coevolution process, 9, 10
Coherence, 326–327
Coleoid cephalopods, 126
Colloidal fabrication, 151
Colloidal photonic crystals, examples of, 105
Colloidal structures, 151
Color
 changing organisms, 121
 optimization, 50
 structure, 362
Compound eye lens arrays, 201–205
 planar eye structures, 206–209
Compound eyes, 176–177
 imaging systems, 209–215
 strange, 183–184
 variants on, 182–183
Conduction band, 323
Cones, 188
Confocal microscopy, 19
Conventional applications of bioinspired photonics, 361–363
Conventional display technology, 122–123
Convergent evolution process, 9
Corneal eye, 171, 172
Coscinodiscus wailesii, 298, 300
Crab shell, transforming, 227
Crossribs, 92
Crotalinae subfamily, *see* Crotaline snakes
Crotaline snakes, 270, 272
Crotalus o. oreganus, 272
Crypsis, 35
Crystal jellie, 249
Crystalline lens, 175
Crystallographic symmetry, 87–88
CTE, *see* Coefficients of thermal expansion

Cubic membrane architectures, 368
Cubic structures, 87, 89–91
CurvACE compound eye, 215
Cuttlefish, 128, 129
CVD, *see* Chemical vapor deposition
Cyphochilus beetle
 matte reflection, 34
 structures, 110
Cyphochilus spp, 108, 111
Cytoelastic sacculus, 134

D

DCE, *see* Dichloroethylene
Desmodus rotundus, 274
Diamond structures, 92
Diatom frustules, 295
Diatoms, 236–238, 295–300
Dichloroethylene (DCE), 304, 309
Diffraction grating, Queen of the Night tulip, 59
Diptera, 40, 41
Disordered hyperuniformity, 190
Display industry, expanding, 121–124
Display technologies, established, 121–122
Disruptive pattern, cephalopods, 129
DoD, *see* U.S. Department of Defense
Domed tunable-focus microlens array, 205
Doryteuthis opalescens, 148–149, 242
Dragonflies, 2–3
 secondary structures, 191–194
Dry etch, 217
Dye-Sensitized Solar Cells (DSSC), 335–336
 biophotonic crystal structures, 336–341
 molecular antennas, 341–345
 structure of, 339

Dynamic adaptive color systems, 362
Dynamic biological photonics, architectures of, 130–134
Dynamic iridophore control, marine animals capable of, 146, 147
Dynamic photonic systems, 130

E

Elastomeric inverse opal photonic crystal structures, 153
Elastomeric lens array, 212, 213
Electric fields, 156
Electroactive polymers, 138
Electrofluidic device, 140
Electromagnetic radiation, 271
Electromagnetic spectrum, 4
Electronic band structure, 323, 324
Elysia chlorotica, 327, 328
Emerald-patched Cattleheart butterfly, 92
Emerald swallowtai, 51
Endoplasmic reticulum, 368
Energy, power and, 319–320
Entimus imperialis, 92
Epicarp, 81
Escherichia coli, 243, 253
Eukaryotes, 7
Eupholus magnificus, 60
Euplectella, 238
Euprymna scolopes, 241
EUV, *see* Ultra-high-resolution extreme UV
Evolutionay processes, 7–12
Extinction coefficient, 230–231
Eye prosthetics, artificial, 200–201
Eyes
  animal, 168
  biological, 169
Eyeshine, 193

F

Fabrication
  approaches, 24–25
  biological, 367–373
  of organic devices, 223
  process, 217, 334
  techniques, 363–367
  technologies, 373
Face-centered cubic (FCC), 88–90, 104, 155
Facets, 177
Far IR (FIR), 267
FCC, *see* Face-centered cubic
FDTD, *see* Finite difference time domain
FEA, *see* Finite element analysis
FIB, *see* Focused ion beam
Field-induced modulation, 156–162
15 kPa compression, 153
Finite difference time domain (FDTD), 348
Finite element analysis (FEA), 288
Finite element method, 103
FIR, *see* Far IR
Fire-beetle, thermal sensors inspired by, 278–284
Flexible displays, 122–123
Fluorescence microscopy, 19
Fluorescence photodetector array (FPDA), 207, 209
Focused ion beam (FIB), 77, 84, 92, 276
Foldable displays, 122
Fossil evidence of complex optical eye, 194–196
Fossil fuels, 319–320
Fossilization, 361
Fossils, 98–99
Fovea, 172
FPDA, *see* Fluorescence photodetector array

Fresnel reflection loss, 331
Front end optics, *see* Biological eyes
Frustules, 237–238
Full width at half maximum (FWHM), 281
Functional scales, overlapping, 365
FWHM, *see* Full width at half maximum

G

Garden insects, 375
Gas sensors, 292–295, *see also* Vapour sensors
  butterfly wings as sensors, 300–314
  diatoms, 295–300
Genetic drift, 8
GFP, *see* Green fluorescent proteins
Golay cell, 278, 279
Golden Goose award, 377
Golden tortoise beetle, case of, 133–134
Goniometer setup, 22
Gradient index (GRIN) lenses, 175–176, 197–200
Grätzel cell, 335
"Green" electronics, 372
Green fluorescent proteins (GFP), 224, 248–250
Greta Oto, 40
GRIN lenses, *see* Gradient index lenses
Ground scales, butterfly, 51
Growth, protein polymerization, 368
Guanine nanocrystals, 43
Gwangju Institute of Science and Technology, 217
Gyroid structures, 92–99

H

Handedness of polarization, 81
Hapalochlaena lunulata, 146
Helical pitch of liquid crystal, 82
Helicoidal multilayers, 79–82
Helium ion microscopy (HIM), 21, 92
Hemicordulia tau, 179
Hemispherical polydimethylsiloxane, 212
Heterogeneous integration, 364–365
Heterogeneous scaled photonic system, 365
HF, *see* Hydrogen fluoride
HgCdTe, *see* Mercury cadmium telluride
High-resolution TEM (HRTEM), 19–20
HIM, *see* Helium ion microscopy
Hinea brasiliana, 62
Homo, 7
"Honeycomb" structure, 94, 96
HRTEM, *see* High-resolution TEM
Human eye, anatomy of, 171
Human eye lens, hierarchical structure of, 175
Human retina, 186, 187
"Hybrid materials science," 372
Hybrid multilayers, 44
Hybrid systems, 349–352
Hydraulically deformable polymer film, 212
Hydrogen fluoride (HF), 288
Hylastes nigrinus, 212

I

Image-forming process, 255
Image-forming vision, 167, 175
Imaging systems, compound and simple eye, 209–215
Imaging techniques, 19
Incident photon to converted electron (IPCE) efficiency, 326

Infrared (IR) radiation sensors, 267–270
  bats, 274–275
  beetles, 275–278
  butterflies, 284–285
  snakes, 270–274
  thermal expansion and optical sensor structures, 285–292
  thermal sensors inspired by fire-beetle, 278–284
Infrared (IR) wavelength, 4, 126
Inorganic blazed grating, example of, 75
Insect epidermal cells, 368
Insect photonic structures, 368
Insect systems, 368
Inspired synthetic 3D structures, 100–107
Intercalated multilayer structures, 82–85
Intracellular cubic symmetries, 368
Intraocular lens (IOL), 198
Inverse replicas, 100
Inverse V-type, 113
Inverted retina, 187
IOL, *see* Intraocular lens
IPCE efficiency, *see* Incident photon to converted electron efficiency
Iridescence
  biological photonic crystals, 34
  single layer thin film, 38
Iridescent reflective insect, 374, 375
Iridophores, 130–131
  dynamic structural color, 144–146
ITO, *see* Transparent electrode layer

J

Japetella heathi, 132
Jewel Beetle, 275

K

Kaist, *see* Korea Advanced Institute of Science and Technology
Keratin, 46–47
Keratinization, 367
β-Keratin protein phase, 367
Korea Advanced Institute of Science and Technology (Kaist), 206

L

Lamprocyphus augustus, 92, 102, 368, 369
Lamprolenis nitida, 71
Lantern, 62
LEDs, *see* Light-emitting diodes
Lens array fabrication process, 202
Lens-on-lens arrays, 202
Lepidoptera, 108
Lepidopteran scales, 92
Leucophores, 130–131
  dynamic structural color, 146–149
Leucosomes, 131
Light absorbing layers, importance of, 86
Light-emitting diodes (LEDs), 217
  extraction efficiency, improving, 62, 64
  reflection on multilayer, 49
Light extraction, 62–64, 66
Lighting panels, 122
Light-interacting structures, evolutionary time line of, 7
Liquid crystal materials, 162
Lithographic techniques, 64
Loligo plei, 135
Long-wave IR (LWIR), 267
Luciferase, 247–248
Luciferin, 247–248
Lux operon, 247
LWIR, *see* Long-wave IR

Lycaenid butterfly, 92
Lycosa leuckartii, 173

M

Magnetic field-based approaches, 156
Mandrillus sphinx, 60, 61
Mantis shrimps, 4
Maratus volans, 374
Margaritaria nobilis, 44–45
Marine animals eye, 173
Marrella splendens, 11
Mechanical deformation, 153–156
Medical biophotonics, 33
Megaphragma mymaripenne, 179, 180
Megascolia procer javanensis, 38
  wing of, 39–40
Melanin, 86, 98
Melanophila acuminata, 275, 276
Mercury cadmium telluride (HgCdTe), 268
Mesonychoteuthis, 172
Mesoporous Bragg stack (meso-BS layer), 340
Metachrosis, 142
Metallic reflective insect, 374, 375
Microdroplet, 156
Micrographia, 14, 15
Microlens array, 203, 204
Microornamentation, 113
Micropits, 273
Microribs, 92
Microscopy techniques, 19–21
  historical perspective and advent of, 13–18
Microspheres, 156, 157
Microvilli, 188
Mid-wave IR (MWIR), 267
Migration, 8
Mirasol IMOD pixel, 161
Misumena vatia, 124, 125
Modern display technology, 121

Modern electron microscopy techniques, 19
Molecular antennas, 341–345
Molecular electronics, 373
Molecular printing, 363
Morpho butterfly, 2, 294
  index of refraction, 46
  wings, 76–79, 288
Morpho didius, wings of, 78, 79
Morpho genus, 34, 46
Morpho peleides, wings of, 76, 77
Morpho rhetenor, wings of, 77
Morpho sulkowskyi, 288, 300, 304, 309
Moth-eye arrays, 194
Moth-eye structures, 331, 332
Mottle pattern, cephalopods, 128–129
M. robinsoni, 374
Müller cells, 187
Multihyperuniformity, 191
Multilayer blazed gratings, 72
Multilayer structures, 41–45, 47
Musca domestica, 209
Mutation, 8
MWIR, *see* Mid-wave IR

N

NADPH, *see* Nicotinamide adenine dinucleotide phosphate
Nanoantennas, 352–356
Nanofabrication techniques, 24
Nanofabrication technology, 216
Nanophotonics, 33
Nanostructures
  in black, 111–116
  in white, 107–111
Narrow angle reflectance, tilted structures and, 71–75
Natronomonas pharaonas, 258
Natural photosynthetic process, 345
Near IR (NIR), 267, 268

Nectar guide, 10
Nectar robbing, 10
Nemopilema nomurai, 191
NETD, *see* Noise equivalent temperature difference
Nicotinamide adenine dinucleotide phosphate (NADPH), 325
Nipple arrays, 194
NIR, *see* Near IR
Nitric oxide (NO), 296
Nitrogen dioxide (NO_2), 298
Nitzschia closterium, 238
Noise equivalent temperature difference (NETD), 268–269
Nose leaf, 274
Novel structural architectures, 50
Novel techniques, 104
Nucleation, 368

O

Ocelli, 179, 191, 192
Octopus dofleini, leucophores of, 146, 148
Octopus vulgaris, 128, 132, 135–136
Ommatidia, 176, 177
Ommatidial corneal lens, 177
One-dimensional layered structures
 chirped multilayers, 47–51
 multilayers structures, 41–45, 47
 sculpted multilayers, 51–55
 single layer thin films, 38–41
Onitis aygulus, 181
Onychoteuthis banksii, 132
Opal, 86, 104
 structure, 158
 synthetic, 104
Opalux Inc., 153, 158
Ophiocoma wendtii, 183, 185
Opsins, 12, 255–258
Optical antennas, 352
Optical device, 85

Optical effects, 34, 35, 361
Optical eye structures, fossil evidence of, 194–196
Optical micrographs of nanostructures arrays, 364
Optical microscope epi-mode image, 21
Optical microscopy, 19
Optical sensor structures, thermal expansion and, 285–292
Optoelectronic device, 85
Optofluidic biolasers, 254
Optogenetics, 258
Organic antenna complexes, 352
Organic liquid crystal lasers, 81–82
Organic multilayers, photonic structure, 44
Organic optical systems, 223
Ornithoptera goliath, 114, 115, 333
Ossicles, 183
Ostrich's eye, 172
Ovalipes molleri, 48

P

P700, 350
Pachliopta aristolochiae, 113, 284
Pachyrrhincus congestus pavonius, 90, 91
Panamanian tortoise beetle, 133–134
Papilio blumei, 51–53
Papilio nephelus, 346, 347
Papilionid butterfly, 92
Papilio palinurus, 52, 53
Papilio ulysses, 51, 113, 114
 structures, wing scale, 37
Papillae, 128
Papilliochrome II, 72
Paracheirodon innesi, 146, 147
Paradise whiptai, 146, 147
Parallel evolution process, 9
Paridis sesostri, 92, 95, 100, 103

dorsal wings of, 94
optical properties of, 96
Passing cloud display, cephalopod, 129
Passing wave display, cephalopod, 129
PCA, *see* Principal component analysis
PDMS, *see* Polydimethylsiloxane
Peacock feathers, arrays in, 56–58
Peer-review system, 376
Phelsuma, 34, 43, 62
Phelsuma geckos, multilayer structures, 43
Phelsuma lizards, optical effects, 34
Phidippus johnsoni, 42
Photodetectors, 22
Photoelectrochemical complexes, 350
Photoexcited electrons, 330
Photonic band structures, 87–88
Photonic crystal lattices, 369
Photonic crystals, 17
Photonics, biomaterials for, 223–225
    biosilica, 236–241
    chitin, 225–231
    opsins, 255–258
    reflectins, 241–245
    silk, 231–236
Photonic system, 36–37
Photophores, 242
Photoreceptors, 184–188
    sensors in fire-seeking beetles, 12
Photoresist, 365
Photosynthesis, 7, 321–323, 325–327
    in animals, 327–328
    solar fuels and, 345–349
Photosynthetic process, 325
Photosystem I (PSI), 326, 350–352
Photovoltaic cells, 324
Photovoltaics, 329–331
    effect, 323–324

Photuris firefly
    lanterns of, 62
    scale structure, 63, 64
Pierella luna, 34, 71, 72
Pieris brassicae, 284
Pieris rapae, 107–108
Pieris rapae crucivora, 108
Pitch (P), 79
Plagiomnium affine, 321
Planar eye structures, 207
Planar fabrication techniques, 211
Planar polymeric lens, 203, 205
Plasmonic nanostructures, 363–364
Plasmonic structural color, 364
Plusiotis resplendens, 81, 82
PMMA, *see* Poly(methylmethacrylate)
p-n junction, 330
Polarization effects, 81
    in beetles and fruits, 34
Polarization imaging sensors, 216
Pollia condensata, 34, 80, 81
Poly(methylmethacrylate) (PMMA), 233, 235
Polycarbonate film, 202
Polycrystals, 88
Polydimethylsiloxane (PDMS), 77, 155
    lens, 210, 212
Polyommatus, 308
Polyommatus bellargus, 304, 305
Polyommatus coridon, 304, 305, 308
Polyommatus icarus, 304, 305
Portia, 173, 174
Principal component analysis (PCA), 301, 302
Principal eyes, 172–174
Prosopocera lactator, 89
Protein-based crystalline lens, 171
Protein polymerization, 368
Pseudomyagrus waterhousei, 89–91
Pyramidal microtips, elastomeric stamp with, 365

Q

Qualcomm's Mirasol™, 161–162
Quantum biology, 326–327
Quantum coherence, 326
Quantum effects, 327
Quasicrystals, 60–62
Quasi-ordered crystals, 60–62
Quasi two-/three-dimensional
  structures, 71
  helicoidal multilayers, 79–82
  intercalated structures, 82–85
  Morpho butterfly, wings of, 76–79
  tilted structures and narrow angle
    reflectance, 71–75

R

Rachis, 56
Rangifer tarandus, 193
Reactive ion etching (RIE), 286
Red hummingbird, 3
Reflectance spectra, 39
  peacock feather, 56
Reflectins, 241–245
Reflective displays, 123
Refracting superposition eye, 182–183
Refractive index (RI), 230–231
  modulation, 151–153
  of titania, 100
Refractive indices, 46–47
Reindeer tapeta, 193
Rhabdomeres, 187, 188
RI, *see* Refractive index
RIE, *see* Reactive ion etching
Rods, 188
Roll-to-roll processing, 373
Ruthenium, 341

S

Scales, butterfly, 51
Scanning electron microscope
  (SEM), 17, 38, 90, 94, 134
cuticle, damaged portion of, 133
of grating structure and
  subsequent multilayer
  structure, 74
of *Megaphragma mymaripenne*,
  179
microscopy, 19–21
misfit scales cuticle structure, 63
of pigments sac, 136
of synthetic layered structure, 82
wing of male *Megascolia procer
  javanensis*, 39–40
Scatterometry, 22–23
Schottky barrier, 355
Science, technology, engineering,
  and mathematics (STEM),
  373–375
Sculpted multilayers, 51–55
Secondary eyes, 172–174
Secondary structures, dragonflies,
  191–194
Self-replicating microorganisms,
  372
SEM, *see* Scanning electron
  microscope
Semiconductor fabrication
  techniques, 371
Sensilla, 277
Sensors, 265–267
  butterfly wings, 300–314
  gas and vapor sensors, *see* Gas
    sensors; Vapour sensors
  infrared sensor, *see* Infrared (IR)
    radiation sensors
Sepia officinalis, 129–131, 137,
  146
Short-wave IR (SWIR), 267
SiGe, *see* Silicon germanium
Silica, 236, 237, 240
Silica nanostructure fabrication
  processes, 236
Silicon germanium (SiGe), 84
Silicon oxide (SiO), 84

Silk, 231–236
Silk inverse opals (SIOs), 233, 235
Silverfish, 14–15
Simple eye model, 170–175
Single-crystal silicon (Si) solar cell, 329
Single layer thin films, 38–41
Single-wall carbon nanotubes (SWCNTs), 280, 290, 291
SiO, *see* Silicon oxide
SIOs, *see* Silk inverse opals
Si wafer, 79
"Small Silver colour'd Book Worm," *see* Silverfish
Smooth endoplasmic reticulum, 368
Snakes, 270–274
Solar fuels, 345–349
Solar power generation technology, 1
Solar power, harvesting, 320–321
Solenopsis fugax, 212
Sol–gel silica template, 102
Spectacular structural color, examples of, 3
Spectrometry setup, diagram of, 22
Spectrophotometric measurements, 90
Spectroscopy, 22–23
Spherical aberration, 176
Spherical nanostructures from feather barbs, 368
Spinodal decomposition, 368
Sponges, 240
Spray techniques, 373
STEM, *see* Science, technology, engineering, and mathematics
Stroma, 321
Structural color, 149–151
   ancient structures, 31–33
   biological photonic structures, 31, 36–37
   of biophotonics, 362
   field-induced modulation, 156–162
   mechanical deformation, 153–156
   modeling of, 65
   one-dimensional layered structures, *see* One-dimensional layered structures
   production, photonic mechanisms of, 367
   refractive index modulation, 151–153
   sparkly, vibrant, bright, and shiny colors, 34–36
   two-dimensional structures, *see* Two-dimensional structures
Structural color II, 71
Suberites domuncula, 240
Substrate layer, 335
Sun Catalytix, 349
Super black materials, 111
Supercrystals, 363
Superposition compound eyes, 177, 181–182
SWCNTs, *see* Single-wall carbon nanotubes
SWIR, *see* Short-wave IR

T

Taiwanese beetle, 83
Tapetum lucidum, 191
Technology, examples of, 2
TEM, *see* Tunneling electron microscopy
Temperature sensitivity (TS), 280
Thalassiosira rotula, 298
Thermal expansion, 278
   and optical sensor structures, 285–292
Thermal sensors, 275
   inspired by fire-beetle, 278–284
Thermography, 275

Thin-film technologies, 331
Three-dimensional omnidirectional microlasers, 82
Three-dimensional phospholipidmembrane lattice, 369
Three-dimensional (3D) photonic crystals, fabricating, 85
Three-dimensional (3D) printing, 363
Three-dimensional (3D) responsive photonic crystal, 150
Three-dimensional (3D) stereolithography, 103
Three-dimensional (3D) structures, 85–89, *see also* Quasi two-/three-dimensional structures
  cubic structures, 89–91
  diamond structures, 92
  gyroid structures, 92–99
  inspired synthetic 3D structures, 100–107
Thwaitesia, 374
Thylakoids, 321
Tile approach, 365
Tiled systems, 365
Tilted structures, quasi two-/three-dimensional structures, 71–75
Time-of-flight secondary ion mass spectrometry (ToF-SIMS), 309
Titania, 102
  refractive index of, 100
ToF-SIMS, *see* Time-of-flight secondary ion mass spectrometry
Top–down approaches, 150
Top–down lithographic techniques, 363
Topology optimization, 50
Transfer printing, 365, 366
Transgenic animals, 251

Transmission electron microscopy (TEM) of *C. egregria cuticle*, 133
Transparent displays, 122
Transparent electrode layer (ITO), 161
Triatoma infestans, 275
Trichogramma, 179
Trichogramma evanescens, 179, 180
Trigonophorus rothschildi varians, 83–85
Troides magellanus
  butterfly, optical effects, 34
  dorsal view of male, 72, 73
Troides plateni, 113, 284
TS, *see* Temperature sensitivity
Tunneling electron microscopy (TEM), 19–21, 229, 238, 276, 279
  chelicera, 42
  of *Nitzschia closterium* frustules, 237
  of reflectin platelet stack, 243
Twisted multilayers, *see* Helicoidal multilayers
Two-component protein nanomaterials, electron micrographs of, 372
2D grating structures on rose petals, 59, 60
Two-dimensional structures, 55
  arrays in peacock feathers, 56–58
  light extraction, 62–64, 66
  quasi-ordered 2D structures and quasicrystals, 60–62
Type I opsins, 256

U

UC, *see* University of California
Ultra absorptive, 111
Ultra-high-resolution extreme UV (EUV), 72

Ultralow reflectance, 111
Ultrathin synthetic broadband reflector approach, 111
Ultraviolet (UV), 124, 146, 156
Unconventional applications of bioinspired photonics, 361–363
Unconventional display technologies, 124–125
Uniform pattern, cephalopods, 128
University of California (UC), 285
Up-and-coming fabrication techniques, 363
Urosaurus ornatus, 125
U.S. Department of Defense (DoD), 376
UV, *see* Ultraviolet

V

Valence band, 323
Vampire bats, 274
Vapour sensors, 292–295, *see also* Gas sensors
   butterfly wings as sensors, 300–314
   diatoms, 295–300
Very-large-scale fabrication, 365
Very long-wave IR (VLWIR), 267
Vibrio fischer, 247
Visible wavelength of human eye, 4
Vision systems, 362
   antireflective (AR) structures, 216–219
   artificial eye prosthetics, 200–201
   biological eyes, *see* Biological eyes
   compound and simple eye imaging systems, 209–215
   compound eye lens arrays, 201–209
   gradient index (GRIN) lenses, 197–200
   inspiring, 167–170
   photoreceptors, 184–188
   polarization sensors, 216
   secondary structures, 191–194
   spectral sensitivities, 188–191
VLWIR, *see* Very long-wave IR

W

Watermark-ink (W-INK) concept, 152, 153
Waveguiding effect, 240
White reflective color, nanostructures in, 107–111
Wing interference patterns (WIP), 40, 41
W-INK concept, *see* Watermark-ink concept
WIP, *see* Wing interference patterns
Wiwaxia corrugate, 11–12

Z

Zebrafish, 137
Zinc oxide (ZnO), 58
Zinc-porphyrin sensitizer, 343, 344
ZnO, *see* Zinc oxide